Lecture Notes in Computer Science   1717

Edited by G. Goos, J. Hartmanis and J. van Leeuwen

Springer

Berlin
Heidelberg
New York
Barcelona
Hong Kong
London
Milan
Paris
Singapore
Tokyo

Çetin K. Koç   Christof Paar (Eds.)

Cryptographic Hardware and Embedded Systems

First International Workshop, CHES'99
Worcester, MA, USA, August 12-13, 1999
Proceedings

 Springer

Series Editors

Gerhard Goos, Karlsruhe University, Germany
Juris Hartmanis, Cornell University, NY, USA
Jan van Leeuwen, Utrecht University, The Netherlands

Volume Editors

Çetin K. Koç
Oregon State University
Department of Electrical and Computer Engineering
Corvallis, OR 97330, USA
E-mail: koc@ece.orst.edu

Christof Paar
Worcester Polytechnic Institute
Department of Electrical and Computer Engineering
Worcester, MA 01609, USA
E-mail: christof@ece.wpi.edu

Cataloging-in-Publication data applied for

Die Deutsche Bibliothek - CIP-Einheitsaufnahme

Cryptographic hardware and embedded systems : first international workshop ;
proceedings / CHES'99, Worcester, MA, USA, August 12 - 13, 1999. Çetin K. Koç
; Christof Paar (ed.). - Berlin ; Heidelberg ; New York ; Barcelona ; Hong Kong
; London ; Milan ; Paris ; Singapore ; Tokyo : Springer, 1999
(Lecture notes in computer science ; Vol. 1717)
ISBN 3-540-66646-X

CR Subject Classification (1998): E.3, C.2, C.3, B.7.2, G.2.1, D.4.6, K.6.5,
F.2.1, J.1

ISSN 0302-9743
ISBN 3-540-66646-X Springer-Verlag Berlin Heidelberg New York

© Springer-Verlag Berlin Heidelberg 1999
Printed in Germany

Typesetting: Camera-ready by author
SPIN: 10704347   06/3142 – 5 4 3 2 1 0   Printed on acid-free paper

Preface

These are the proceedings of CHES'99, the first workshop on Cryptographic Hardware and Embedded Systems. As it becomes more obvious that strong security will be an important part of the next generation of communication, computer, and electronic consumer devices, we felt that a new type of cryptographic conference was needed. Our goal was to create a forum which discusses innovative solutions for cryptography in practice. Consequently, the focus of the CHES Workshop is on all aspects of cryptographic hardware and embedded system design. Of special interest were contributions that describe new methods for efficient hardware implementations and high-speed software for embedded systems, e.g., smart cards, microprocessors, or DSPs. We hope that the workshop will help to fill the gap between the cryptography research community and the application areas of cryptography.

There were 42 submitted contributions to CHES'99, of which 27 were selected for presentation. All papers were reviewed. In addition, there were three invited speakers.

We hope to continue to make the CHES Workshop a forum of intellectual exchange in creating the secure, reliable, and robust security solutions of tomorrow. We thank everyone whose involvement made the CHES Workshop such a successful event, and in particular we thank Murat Aydos, Dan Bailey, Brendon Chetwynd, Adam Elbirt, Serdar Erdem, Jorge Guajardo, Linda Looft, Pam O'Bryant, Marie Piergallini, Erkay Savaş, and Adam Woodbury.

Corvallis, Oregon
Worcester, Massachusetts
August 1999

Çetin K. Koç
Christof Paar

Acknowledgements

The program chairs express their thanks to the program committee, the companies which provided support to the workshop, and also the referees for their help in getting the best quality papers selected for the workshop.

The program committee members:

- Gordon Agnew, University of Waterloo, Canada
- David Aucsmith, Intel Corporation, USA
- Ernie Brickell, CertCo, USA
- Wayne Burleson, University of Massachusetts at Amherst, USA
- Burt Kaliski, RSA Laboratories, USA
- Jean-Jacques Quisquater, Université Catholique de Louvain, Belgium
- Christoph Ruland, Universität Siegen, Germany
- Victor Shoup, IBM Research, Switzerland
- Michael Wiener, Entrust Technologies, Canada

The companies which provided support to the CHES Workshop:

- Compaq - Atalla Security Products — http://www.atalla.com
- Intel Corporation — http://www.intel.com
- secunet Security Networks AG — http://www.secunet.de
- SITI — Secured Information Technology, Inc.
- Technical Communications Corporation — http://www.tccsecure.com

The referees of the CHES Workshop:

- Tolga Acar <tacar@novell.com>
- Gordon Agnew <g.agnew@coulomb.uwaterloo.ca>
- Dan Bailey <bailey@wpi.wpi.edu>
- Joe Buhler <jpb@reed.edu>
- Wayne Burleson <burleson@galois.ecs.umass.edu>
- Brendon Chetwynd <spunge@ece.wpi.edu>
- Erik DeWin <dewin@esat.kuleuven.ac.be>
- Adam Elbirt <aelbirt@nac.net>
- Dave Farrel <Dave.Farrel@gsg.gte.Com>
- Kris Gaj <kgaj@gmu.edu>
- Guang Gong <ggong@cacr.math,uwaterloo.ca>
- Jim Goodman <jimg@mtl.mit.edu>
- Jorge Guajardo <guajardo@ece.wpi.edu>
- Klaus Huber <huber@tzd.telekom.de>
- Burt Kaliski <burt@rsa.com>
- David King <dave.king@gsc.gte.com>
- Çetin K. Koç <koc@ece.orst.edu>
- Yusuf Leblebici <leblebic@wpi.edu>
- Gerardo Orlando <Gerardo.Orlando@gsc.gte.Com>

- Christof Paar <christof@ece.wpi.edu>
- Sachar Paulus <paulus@secude.com>
- Jean-Jacques Quisquater <jjq@dice.ucl.ac.be>
- Francisco Rodriguez-Henriquez <rodrigfr@ece.orst.edu>
- Hans-Georg Rück <rueck@werner.exp-math.uni-essen.de>
- Christoph Ruland <ruland@nue.et-inf.uni-siegen.de>
- Erkay Savaş <savas@ece.orst.edu>
- Frank Schaefer-Lorinser <lorinser@tzd.telekom.de>
- Tom Schmidt <toms@math.orst.edu>
- Victor Shoup <sho@zurich.ibm.com>
- Alex Tenca <tenca@ece.orst.edu>
- Michael Wiener <michael.wiener@entrust.com>

Table of Contents

Arithmetic Algorithms I

Power Attacks I

Invited Talk

True Random Number Generators

Cryptographic Algorithms on FPGAs

Arithmetic Algorithms II

Power Attacks II

Elliptic Curve Implementations

New Cryptographic Schemes and Modes of Operation

We Need Assurance

Brian D. Snow

National Security Agency, USA

Abstract. Today's commercial cryptographic products have sufficient functionality, plenty of performance, but not enough assurance. Further, in the near term future, I see little chance of improvement in assurance, hence little improvement in true security offered by industry. The malicious environment in which security systems must function absolutely requires the use of strong assurance techniques. Most attacks today result from failures of assurance, not function.

Am I depressed? Yes, I am. The scene I see is products and services sufficiently robust to counter many (but not all) of the "hacker" attacks we hear so much about today, but not adequate against the more serious but real attacks mounted by economic adversaries and nation states. We will be in a truly dangerous stance: we will think we are secure (and act accordingly) when in fact we are not secure.

Assurance techniques (barely) adequate for a benign environment simply will not hold up in a malicious environment.

Despite the real need for additional research in assurance technology, we fail to fully use that which we already have in hand! We need to better use those assurance techniques we have, and continue research and development efforts to improve them and find others.

Recall that assurance are confidence-building activities demonstrating that system functions meet a desired set of properties and only those properties, that the functions are implemented correctly, and that the assurances hold up through manufacturing, delivery, and life-cycle of the system.

Assurance is provided through structured design processes, documentation, and testing, with greater assurance coming through more extensive processes, documentation, and testing. All this leads to increased cost and delayed time-to-market – a severe one-two punch in today's marketplace.

I will briefly discuss assurance features appropriate in each of the following five areas: operating systems, software modules, hardware features, third party testing, and legal constraints.

Each of us should leave today with a stronger commitment to quality research in assurance techniques with strong emphasis on transferring the technology to industry. It is not adequate to have the technique; it must be used. We have our work cut out for us; let's go do it.

Factoring Large Numbers
with the TWINKLE Device
(Extended Abstract)

Adi Shamir

Computer Science Dept.
The Weizmann Institute
Rehovot 76100, Israel
shamir@wisdom.weizmann.ac.il

Abstract. The current record in factoring large RSA keys is the factorization of a 465 bit (140 digit) number achieved in February 1999 by running the Number Field Sieve on hundreds of workstations for several months. This paper describes a novel factoring apparatus which can accelerate known sieve-based factoring algorithms by several orders of magnitude. It is based on a very simple handheld optoelectronic device which can analyse 100,000,000 large integers, and determine in less than 10 milliseconds which ones factor completely over a prime base consisting of the first 200,000 prime numbers. The proposed apparatus can increase the size of factorable numbers by 100 to 200 bits, and in particular can make 512 bit RSA keys (which protect 95% of today's E-commerce on the Internet) very vulnerable.

Keywords: Cryptanalysis, Factoring, Sieving, Quadratic Sieve, Number Field Sieve, optical computing.

1   Introduction

The security of the RSA public key cryptosystem depends on the difficulty of factoring a large number n which is the product of two equal size primes p and q. This problem had been thoroughly investigated (especially over the last 25 years), and the last two breakthroughs were the invention of the Quadratic Sieve (QS) algorithm [P] in the early 1980's and the invention of the Number Field Sieve (NFS) algorithm [LLMP] in the early 1990's. The asymptotic time complexity of the QS algorithm is $O(e^{ln(n)^{1/2}ln(ln(n))^{1/2}})$, and the asymptotic time complexity of the NFS algorithm is $O(e^{1.92\ ln(n)^{1/3}ln(ln(n))^{2/3}})$. For numbers with up to about 350 bits the QS algorithm is faster due to its simplicity, but for larger numbers the NFS algorithm is faster due to its better asymptotic complexity.

The complexity of the NFS algorithm grows fairly slowly with the binary size of n. Denote the complexity of factoring a 465 bit number (which is the current record - see [R]) by X. Then the complexity of factoring numbers which are 100 bits longer is about 40X, the complexity of factoring numbers which are 150 bits longer is about 220X, and the complexity of factoring numbers which are 200 bits longer is about 1100X. Since the technique described in this paper can increase the efficiency of the NFS algorithm by two to three orders of magnitude, we expect it to increase the size of factorable numbers by 100 to 200 bits, or alternatively to make it possible to factor with a budget of one million dollars numbers which previously required hundreds of millions of dollars. The main practical significance of such an improvement is that it can make 512 bit numbers (which are the default setting of most Internet browsers in e-commerce applications, and the maximum size deemed exportable by the US government) easy to crack.

The new factoring technique is based on a novel optoelectronic device called TWINKLE. [1] Designing and constructing the first prototype of this device can cost hundreds of thousands of dollars, but the manufacturing cost of each additional device is about $5,000. It can be combined with any sieve-based factoring algorithm, and in particular it can be used in both the QS and the NFS algoritms. It uses their basic mathematical structure and inherits their asymptotic complexity, but improves the practical efficiency of their sieving stage by a large constant factor. Since this is the most time consuming part of these algorithms, we get a major improvement in their total running time.

For the sake of simplicity, we describe in this extended abstract only the new implementation of the sieving stage in the simplest variant of the QS algorithm. Most of the new ideas apply equally well to improved variants of the QS algorithm and to the general NFS algorithm, but the details are more complicated, and will be described only in the full version of this paper.

---

[1] TWINKLE stands for "The Weizmann INstitute Key Locating Engine".

2   The QS Factoring Algorithm

Given the RSA number $n = pq$, the QS algorithm tries to construct two numbers y and z such that $y \neq \pm z \pmod{n}$ and $y^2 = z^2 \pmod{n}$. Knowledge of such a pair makes it easy to factor n since $\gcd(y - z, n)$ is either p or q. To find such y and z, we generate a large number of values $y_1, y_2, \ldots, y_m$, compute each $y_i^2 \pmod{n}$, and try to factor it into a product of primes p_j from a prime base B consisting of the k smallest primes $p_1 = 2, p_2 = 3, \ldots, p_k$. Numbers $y_i^2 \pmod{n}$ which have such factorizations into $\prod_{j=1}^{k} p_j^{e_j}$ are called *smooth*. If the number of smooth modular squares found in such a way exceeds k, we can use Gauss elimination to find a subset of the vectors $(e_1, e_2, \ldots, e_k)$ of the prime multiplicities which is linearly dependent modulo 2. When the corresponding $y_i^2 \pmod{n}$ and their factorizations are multiplied, we get an equation of the form $\prod_{i=1}^{m} (y_i^2)^{b_i} = \prod_{j=1}^{k} p_j^{c_j} \pmod{n}$ where all the b_i's (which define the subset) are 0's and 1's and all the c_j's (which are the sums of the prime multiplicities) are even numbers. We can now get the desired equation $y^2 = z^2 (\bmod\ n)$ by defining $y = \prod_{i=1}^{m} y_i^{b_i} \pmod{n}$ and $z = \prod_{j=1}^{k} p_j^{c_j/2} \pmod{n}$.

The key to the efficiency of the QS algorithm is the generation of many small modular squares whose smoothness is easy to test. Consider the simplest case in which we use the quadratic polynomial $f(x) = (a + x)^2 \pmod{n}$ where $a = \lfloor \sqrt{(n)} \rfloor$, and choose $y_i = a + i$ for $i = 1, 2, \ldots, m$. Then it is easy to see that for small m the corresponding $y_i^2 = f(i) \pmod{n}$ are half size modular squares which are much more likely to be smooth numbers than random modular squares.

The simplest way of testing the smoothness of the values in such a sequence is to perform trial division of each value in the sequence by each prime in the basis. Since the $f(i)$'s are hundreds of bits long, this is very slow.

The QS algorithm expresses all the generated $f(1), \ldots, f(m)$ in the non modular form $f(i) = (a + i)^2 - n$ (since m is small), and determines which of these values are divisible by p_j from the basis B by solving the quadratic modular equation $(a + i)^2 - n = 0 \pmod{p_j}$. This is easy, since the modulus p_j is quite small. [2]

The quadratic equation mod p_j will have either zero or two solutions d_i' and d_i''. In the first case we can deduce that none of the $f(i)$'s will be divisible by p_j, and in the second case we can deduce that $f(i)$ will be divisible by p_j if and only if i belongs to the union of the two arithmetic progressions $p_j * r + d_j'$ and $p_j * r + d_j''$ for $r \geq 0$.

The smoothness test in the QS algorithm is based on an array A of m counters, where the $i - th$ entry is associated with $f(i)$. The sieving algorithm zeroes all these counters, and then loops over the primes in the basis. For each prime p_j, and for each one of its two arithmetic progressions (if they exist), the algorithm scans the counter array, and adds the constant $log_2(p_j)$ to all the counters $A(i)$

---

[2] We ignore the issue of the divisibility of $f(i)$ by higher powers of p_j, since except for the smallest primes in the basis this is extremely unlikely, and we can explicitly add the powers of the first few primes to the basis without substantially increasing its size.

whose indices i belong to the arithmetic progression (there are about m/p_j such indices). At the end of this loop, the value of $A(i)$ describes the (approximate) binary length of the largest divisor of $f(i)$ which factors completely over the prime base B. The algorithm then scans the array, finds all the entries i for which $A(i)$ is close to the binary length of $f(i)$, tests that these $f(i)$'s are indeed smooth by trial division, and uses them in order to factor n.

Typical large scale factoring attacks with networks of PC's may use $m = 100,000,000$ and $k = 200,000$. The array requires 100 megabytes of RAM, and its counters can be accessed at the standard bus speed of 100 megahertz. [3] Just scanning such a huge array requires about one second. Well optimized implementations of the QS algorithm perform the sieving in 5 to 10 seconds, and find very few smooth numbers. They then choose a different quadratic polynomial $f'(x)$, and repeat the sieving run (on the same machine, or on a different machine working in parallel). This phase stops when a total of $k + 1$ smooth modular squares are collected in all the sieving runs, and a single powerful computer performs the Gauss elimination algorithm and the actual factorization in a small fraction of the time which was devoted to the sieving.

In the next section we describe the new TWINKLE device, which is an ultrafast optical siever. It costs about the same as a powerful PC or a workstation, but can test the smoothness of 100,000,000 modular squares over a prime base of 200,000 primes in less than 0.01 seconds. This is 500 to 1000 times faster than the conventional sieving approach described above.

3 The TWINKLE Device

The TWINKLE device is a simple optoelectronic device which is housed in an opaque blackened cylinder whose diameter is about 6 inches and whose height is about 10 inches. The bottom of the cylinder consists of a large collection of LEDs (light emitting diodes) which twinkle at various frequencies, and the top of the cylinder contains a photodetector which measures the total amount of light emitted at any given moment by all the LEDs. The photodetector alerts a connected PC whenever this total exceeds a certain threshold. Such events are related to the detection of possibly smooth numbers, and their precise timing is the only output of the TWINKLE device. Since these events are extremely rare, the PC can leisurely translate the timing of each reported event to a candidate modular square, verify its smoothness via trial division, and use it in a conventional implementation of the QS or NFS algorithms in order to factor the input n.

The standard PC implementation of the sieving technique assigns modular squares to array elements (using space) and loops over the primes (using time). The TWINKLE device assigns primes to LEDs (using space) and loops over the

---

[3] Note that the faster cache memory is of little use, since the sieving process accesses arithmetic progressions of addresses with huge jumps, which create continuous cache misses.

modular squares (using time), which reverses their roles. This is schematically described in Fig. 1.

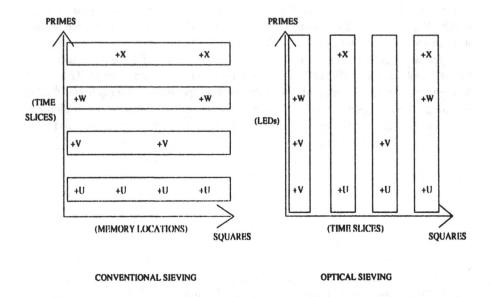

Fig. 1. Conventional vs. optical sieving: the boxed operations are carried out at the same time slice

Each LED is associated with some period p_j and delay d_j, and its only role is to light up for one clock cycle at times described by the arithmetic progression $p_j * r + d_j$ for $r \geq 0$. To mimic the QS sieving procedure, we have to use nonuniform LED intensities. In particular, we want the LED associated with prime p_j to generate light intensity proportional to $log_2(p_j)$ whenever it flashes, so that the total intensity measured by the photodetector at time i will correspond to the binary size of the largest smooth divisor of the $f(i)$ [4] We can achieve this by using an array of LEDs of different sizes or with different resistances. However, a simpler and more elegant solution to the problem is to construct a uniform array of identical LEDs, to assign similar sized primes to neighbouring LEDs, and to cover the LED array with a transparent filter with smoothly changing grey levels. [5] Note that the dynamic range of grey levels we have to use is quite limited, since the ratio of the logs of the largest and the smallest primes in a typical basis does not exceed 24:1.

To increase the sensitivity of the photodetector, we can place it behind a large lense which concentrates all the incoming light on its small surface area. The light

---

[4] Again, we ignore the issue of the divisibility of $f(i)$ by higher powers of the primes.

[5] For example, we can assign primes to LEDs in row major order and use a filter which is dark grey at the top and completely transparent at the bottom, or assign primes to LEDS in spiral order and use a filter which is darkest at its center.

intensity measurement is likely to be influenced by many sources of errors. For example, the grey levels of the filter are only approximations of the logs, and even uniformly designed LEDs may have actual intensities varying by 20% or more. We can improve the accuracy of the TWINKLE device by measuring the actual filtered intensity of each LED in the manufactured array, and assigning the sequence of primes to the various LEDs based on their sorted list of measured intensities. However, the QS and NFS factoring algorithms are very forgiving to such measurement errors, and in PC implementations they use crude integer approximations to the logs in order to speed up the computation. There are two possible types of errors: missed events and false alarms. To minimize the number of missed events we can set a slightly lower reporting threshold, and to eliminate the resultant false alarms we can later test all the reported events on a PC, in order to find the extremely rare real events among the rare false alarms. For typical values of the parameters, the average binary size of the smooth part of candidate values is about one tenth of their size, and only a tiny fraction of all candidate values have ratios exceeding one half. As a result, the desired events stand out very clearly as isolated narrow peaks which are about ten times higher than the background noise.

We claim that optical sieving is much better than conventional counter array sieving for the following reasons:

1. We can perform optical sieving at an extremely fast clock rate. Typical silicon RAM chips in standard PC's operate at about 100 megahertz. LEDs, on the other hand, are manufactured with a much faster Gallium Arsenide (GaAs) technology, and can be clocked at rates exceeding 10 gigahertz without difficulty. Commercially available LEDs and photodetectors are used to send 10 gigabits per second along fiber optic cables, and GaAs devices are widely used at similar clock rates as routers in high speed networks.

2. We can instantaneously add hundreds of thousands of optical contributions, if we do not need perfect accuracy. Building a digital adder with 200,000 inputs which computes their sum in a single clock cycle is completely unrealistic.

3. The optical technique does not need huge arrays of counters. Instead of using one memory cell per sieved value, we use one time slice per sieved value. Even with the declining cost of fast memories, time is cheaper than space.

4. In the optical technique do not have to scan the array at the beginning in order to zero it, and do not have to scan the array again at the end in order to find its high entries - both operations are done at no extra cost during the actual sieving.

In the remaining sections we flesh out the design of each cell and the architecture of the whole device. We based this design on many conversations with experienced GaAs chip designers, and used only commercially available technologies. We may be off by a small factor in some of our size speed and cost estimates, but we believe that the design is realistic, and that someone will try it out in the near future.

4 Cell Design

The LED array is implemented on a single wafer of GaAs. Each cell on this wafer contains one LED plus some circuitry which makes it flash for exactly one clock cycle every exactly p_j clock cycles with an initial delay of exactly d_j clock cycles. The high clock rate and extremely accurate timing requirements rule out analog control methods, and the unavoidable existence of bad cells in the wafer rules out a prewired assignment of primes to cells. Instead, we use identical cells throughout the wafer, and include in each cell two registers, A and B, which are loaded before the beginning of the sieving process with values corresponding to p_j and d_j, respectively. For a typical sieving run over $m = 100,000,000$ values, we need only $log_2(m) \approx 27$ bits in each one of these registers.

The structure of each cell (described in Fig. 2) is very simple. Instead of using counters (with their more complicated designs and additional carry-induced delays), we use register B as a maximal length shift register based on a single XOR of two of its bits. It is driven by the clock, and runs until it enters the special state in which all its bits are "1". When this is recognized by the AND of all the bits of register B, the LED flashes, and register B is reloaded with the contents of register A (which remains unchanged throughout the computation). The initial values loaded into registers A and B are not the binary representations of p_j and d_j, but the (easily computed) states of the shift register which are that many steps before the special state of all "1". That's the whole cell design!

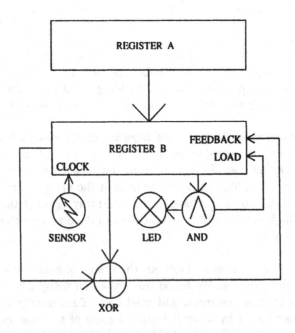

Fig. 2. A single cell in the array

An important issue in such a high speed device is clock synchronization. Each clock cycle lasts only 100 picoseconds, and all the light pulses must be synchronized to within a fraction of this interval in order to correctly sum their contributions. Distributing electrical clock pulses (which travel slowly over long, high capacity wires) at 10 gigahertz to thousands of cells all over the wafer without skewing their arrival times by more than 10-20 picoseconds seems to be a very difficult problem. We solve it by using another optical trick. Since it is easy to construct in GaAs technology a small photodetector in each cell, we use optical rather than electrical clock distribution: a strong LED placed opposite the wafer, which flashes at a fixed rate of 10 gigahertz, and its pulses are almost simultaneously picked up by the photodetectors in all the cells, and used to drive the shift registers in a synchronized way. Since light passes about 3 centimeters in 100 picoseconds, we just have to place the clocking LED and the summing photodetector sufficiently far away from the wafer to guarantee sufficiently similar optical delays to and from all the cells on the flat wafer. To avoid possible confusion between clock and data light pulses, we can use two different wavelengths for the two purposes.

Computing the AND of 27 inputs requires a tree of depth 3 of 3-input AND gates, which may be the slowest cell operation. To speed it up, we can use a systolic design which carries out the tree computation in 3 consecutive clock cycles. This delays the detection of the special state by 3 clock cycles but keeps all the flashing LEDs perfectly synchronized. To compensate for the late reloading of register B, we simply store a modified value of p_j in register A.

An improved cell design is based on the observation that about half the primes do not yield arithmetic progressions, whereas each prime in the other half yields two arithmetic progressions with the same period p_j. In standard PC implementations this has little effect, since we still have to scan on average one arithmetic progression per prime in the basis. However, in the TWINKLE design the two cells assigned to the same p_j can share the same A register (which never changes) to reload their separate B shift registers. In addition, the two cells can share the same LED and flash it with the OR of the two AND gates, since the two arithmetic progressions are always disjoint. We call such a combination a double cell, and use it to reduce the average number of registers per prime in the basis from 2 to 1.5. Since these registers occupy most of the area of the cell, this observation can increase the number of primes we can handle with a single wafer by almost 33%.

5 Wafer Design

We would like to justify our claim that a single wafer can handle a prime base of 200,000 primes (which is the actual size used in recent PC-based factorizations). A standard 6 inch wafer has a total usable area of about $16 * 10^9$ square microns. Commercially available LED arrays (such as the arrays sold by Oki Semiconductors to manufacturers of laser printers - see http://www.oki.co.jp/OKI/home/English/New/OKI-News/1998/z9819e.html for further details) have a linear density of 1200 LEDs per inch. At this density, each LED occupies a $20\mu \times 20\mu$ square with an area of $400\mu^2$, and we can fit about 40,000,000 LEDs on a single wafer. However, most of area of each double cell will be devoted to the three 27 bit registers. Crude conservative estimates indicate that we can very comfortably fit each one of these 81 bits into an area of $1,600\mu^2$ using commercially available GaAs technology. We can thus fit the whole double cell into an area of less than $160,000\mu^2$, and pack 100,000 double cells into a single wafer. Such a wafer will be able to sieve numbers over a prime base of 200,000 primes.

A simple reality check is based on the computation of the total amount of memory on the wafer. The 100,000 double cells contain $81 \times 100,000$ bits, or about one megabyte of memory. The other gates (XOR, AND) and diodes (LEDs, photodetectors) occupy a small additional area. This is a very modest goal for wafer scale designs.

The cost of manufacturing silicon wafers in a commercial FAB is about $1,500 per wafer, and the cost of manufacturing the more expensive GaAs wafers is about $5,000 per wafer (excluding design costs and assuming a reasonably large order of wafers). This is comparable to the cost of a strong workstation, but provides a sieving efficiency which is several hundred times higher.

The TWINKLE device does not have a yield problem, which plagues many other wafer-scale designs: During the sieving process each cell works completely independently, without receiving any inputs or sending any outputs to neighbouring cells. Even if 20% of the cells are found to be defective in postproduction inspection, we can use the remaining 80% of the cells. If necessary, we can place two or more wafers at the same distance opposite the same summing detector, in order to compensate for defective cells or to sieve over larger prime bases.

After determining the number of cells, we can consider the issue (which was ignored so far) of loading registers A and B in each cell with some precomputed data from a connected storage device. Silicon memory cannot operate at 10 gigahertz, and thus we have to slow down the clocking LEDs facing the GaAs wafer during the loading phase. The A registers which contain the primes assigned to each LED can be loaded only once after each powerup, but the B registers which contain the initial delays have to be loaded for each sieving run. The total size of the 200,000 B registers is about 675 kilobytes. Such a small amount of data can be kept in a standard type of silicon memory, and transfered to the wafer in 0.002 seconds on a 27 bit bus operating at 100 megahertz. This is one fifth the time required to carry out the actual sieving at the full 10 gigahertz clock rate, and thus it does not create a new speed bottleneck.

The proposed wafer design has just 31 external connections: Two for power, two for control, and 27 for the input bus. The four modes of operation induced by the two control wires consist of a test mode (in which the various LEDs are sequentially flashed to test their functionality and measure their light intensity), LOAD-A mode (in which the various A registers are sequentially loaded from the bus), LOAD-B mode (in which the various B registers are sequentially loaded from the bus), and sieving mode (in which all the shift registers are simultaneously clocked at 10 gigahertz). We can briefly freeze the optical clocking during mode changes in order to enable the slow electric control signals to propagate to all the cells on the wafer before we start operating in the new mode.

Another important factor in the wafer design is its total power consumption. Strong LEDs consume considerable amounts of power, and if a large number of LEDs on the wafer flash simultaneously, the excessive power consumption can skew the intensity of the flashes. However, each tested number can be divisible by at most several hundred primes from the basis, and thus we have a small upper bound on the total power which can be consumed by all the LEDs at any given moment in the sieving process.

6 The Geometry of the TWINKLE Device

The TWINKLE device is housed in an opaque cylinder with the wafer at the bottom and the summing photodetector and clocking LED at the top. Its diameter is determined by the size of the wafer, which is about 6 inches. Its height is determined by the uniformity requirements of the length of the various optical paths.

To determine this height, we recall that light travels about 3 centimeters in a single clock cycle which lasts 100 picoseconds. To make sure that all the received light pulses are synchronized to within 15% of this duration, we want the length of the optical paths from the clocking LED to any point in the wafer and from there to the summing photodetector to vary by at most 0.5 centimeter. The simplest arrangement places both elements at the center of the top face of the cylinder, but this penalizes twice LEDs located at the rim compared to LEDs located at the center, and requires a cylinder whose length is about 110 centimeters. A better arrangement uses several clocking LEDs placed symmetrically around the rim of the top face, and a single photodetector at the center of this face. A simple geometric calculation shows that the required uniformity will be attained in a cylinder which is just 25 centimeters (10 inches) long.

7 Concluding Remarks

The idea of using physical devices in number theoretic computations is not new. D. H. Lehmer managed to factor (relatively small) numbers and solve other diophantine equations by pedalling on a device based on toothed wheels and bicycle chains of various lengths (a replica of this ingenious contraption from the 1920's is located at the Boston Computer Museum). His device even included a photodetector to alert the rider when the solution was found, but its mode of operation was of course completely different from our implementation of the quadratic sieve.

The TWINKLE device proposed in this paper demonstrates the incredible speed and almost unbounded parallelism which is offered by today's optoelectronic techniques. We believe that they will find many additional applications in cryptography and cryptanalysis.

Acknowledgements: I would like to thank Moty Heiblum and Vladimir Umanski for many useful discussions of GaAs technology.

References

[LLMP] A. K. Lenstra, H. W. Lenstra, M. S. Manasse, and J. M. Pollard, *The number field sieve*, Vol. 1554 of Lecture Notes in Mathematics, 11-42, Springer Verlag, 1993.

[P] C. Pomerance, *The quadratic sieve factoring algorithm*, Proceedings of EURO-CRYPT 84 (LNCS 209), 169-182, 1985.

[R] Hermann J. J. te Riele, email announcement, February 4 1999, available at http://jya.com/rsa140.htm.

DES Cracking on the Transmogrifier 2a

Ivan Hamer and Paul Chow

Department of Electrical and Computer Engineering
University of Toronto
Toronto, Ontario, Canada M5S 3G4
ivan.hamer@utoronto.ca
pc@eecg.toronto.edu

Abstract. The Cryptographic Challenges sponsored by RSA Laboratories have given some members of the computing community an opportunity to participate in some of the intrigue involved with solving secret messages. This paper describes an effort to build DES-cracking hardware on a field-programmable system called the Transmogrifier 2a. A fully implemented system will be able to search the entire key space in 1040 days at a rate of 800 million keys/second.

1   Introduction

The RSA Cryptographic Challenges sponsored by RSA Laboratories [1] have provided some interesting opportunities for those in the computing area to become involved in the mystery and intrigue of discovering secret messages. One of the challenges was to break a straightforward version of the Data Encryption Standard, more commonly known as DES [2]. The brute-force approach is to search the entire key space consisting of 2^{56} or about 7.2×10^{16} keys.

This paper describes a project to implement a DES cracking system in a general-purpose programmable hardware system called the Transmogrifier 2a (TM-2a) [3, 4]. The TM-2a is a unique system of field-programmable gate arrays being developed at the University of Toronto that is intended for doing prototyping of hardware. A brief description of the TM-2a will be given in Section 2.

In the remainder of this section, a brief overview of DES will be given and a review of other attempts at cracking DES will be given. Section 3 will describe our implementation of DES on the TM-2a. We will conclude and give some future work in Section 4.

1.1   Overview of DES

The simplest form of the DES algorithm takes a 56-bit encryption key and uses it to encode a 64-bit block of plain text data into a 64-bit block of output cipher text. Between an initial and final permutation, there are 16 essentially identical stages. In the first stage, one half of the data as well as the key goes through a function, **F**, and the result is exclusive-ored with the other half. For each successive stage, the same thing happens with the halves reversed. Figure 1 shows the data flow. Function **F** is shown in Fig. 2. The data and the key first go through the expander and reducer that do simple selection and/or replication of the input bits to generate two 48-bit words. These two words are then exclusive-ored to form a single 48-bit word, which then goes through a table lookup called the *S-box substitution*. The *S-box* substitution is shown in Fig. 3. It consists of eight 6-bit input, 4-bit output lookup tables. The lookup tables are predetermined functions that, along with the permutations, does most of the coding of the data. The same algorithm is used for decoding. Hence, if we run the output through the circuit again, we should get the same as what we started with.

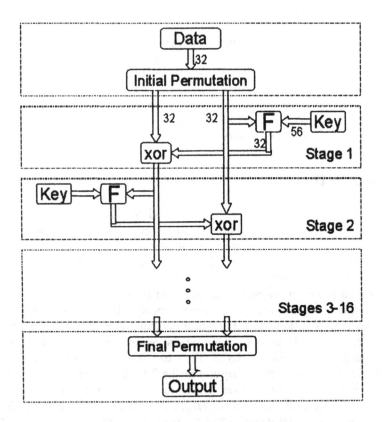

Fig. 1. The basic DES pipeline.

1.2 Other Attempts

The DES standard has long been criticized as being susceptible to an exhaustive key search and there has been much discussion and many recent attempts to show that it is weak.

One of the earliest analyses of a practical machine for doing this was done by Wiener [5] in 1993. In his appendix, there is a very detailed gate-level design

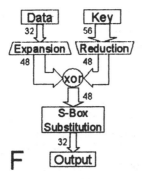

Fig. 2. The F function.

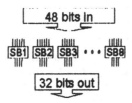

Fig. 3. The S box substitution.

of a chip that could be implemented in a CMOS technology. He estimates that the chip can test keys at a rate of 50 million keys per second. This chip can be used as the basis of a machine that can reduce the search time down to hours or minutes depending on the available budget. A review of numerous other designs was also given by Wiener.

Recently, the evolution of the world-wide web has made it possible to network together thousands of computers, ranging from low-cost personal computers to high-end workstations, all working on portions of the key space [6, 7]. This was how the first RSA DES Challenge was solved in about 4 months [6].

A real hardware system, called *Deep Crack*, was constructed by the Electronic Frontier Foundation (EFF) for under $250,000 and it was able to win the second RSA DES Challenge in 56 hours [8, 9].

A world-wide web group, hosted by Distributed.Net [7], and EFF combined their technologies to solve the final DES Challenge in a record 22 hours and 15 minutes [10].

The use of FPGAs as a means of building hardware to crack cryptosystems has been suggested by many in the past and we only cite a few here [11, 12]. FPGAs are an obvious technology because of the relatively low cost. Although our system of FPGAs will not come close to meeting the speeds of the EFF *Deep Crack* or Distributed.Net systems, we present it here as another data point showing what can be done with some programmable hardware, which puts it somewhere between an application-specific hardware approach, and a large network of general-purpose computers.

We first describe our hardware system.

2 The Transmogrifier 2a

The Transmogrifier 2a (TM-2a) [3, 4] is a second-generation field-programmable system that is constructed with field-programmable gate arrays (FPGAs). The TM-2a is a flexible rapid-prototyping system that offers high capacity and high clock rates. It is intended to be flexible enough to implement a wide variety of systems. A simple way to visualize the TM-2a is to think of building a large FPGA using existing FPGAs and field-programmable interconnect chips (FPICs).

Figure 4 shows the resources available on one TM-2a board. There are two Altera 10K100 [13] logic devices and four I-Cube IQ320 [14] FPICs. Attached to each FPGA is up to 4MB of RAM. The FPICs provide a programmable routing network that can be used to connect the FPGAs on the board to each other and to FPGAs on other boards. Each board also has a low-skew, programmable clock generator and a *housekeeping* FPGA that is used to monitor the system and communicate over a bus to the host system. When the host is on a network, then the TM-2a can be programmed and run remotely.

There can be up to 16 boards in a system. Assuming that each FPGA can hold a circuit of about 60K gates, the size of a 16-board system is about 2-million programmable gates.

The TM-2a is being used at the University of Toronto to prototype a number of hardware concepts such as 3-dimensional procedural texture mapping, head tracking, and image processing. When the RSA DES Challenge was announced, the TM-2a seemed like an obvious system for building a DES cracker.

3 DES on the TM-2a

In this section we describe the implementation of our DES cracking system on the TM-2a hardware. We first give a small primer on the Altera 10K series logic devices architecture and the design methodology that we use. Some of the history behind the development of the hardware is given before we describe the final implementation. We end with a summary of our results.

3.1 The Altera 10K Logic Device

The main building block of the Altera 10K logic device is called a *Logic Element* or LE. Each LE has a number of resources of which the important ones for us were the 4-input, 1-output look-up table (4-LUT), the cascade chain, and the programmable register. The LEs are grouped in blocks of eight called LABs with local routing within the LABs. The LABs are arranged in the chip as a matrix with another routing structure connecting the LABs. A 10K100 has 52 columns and 12 rows for a total of 4992 LEs.

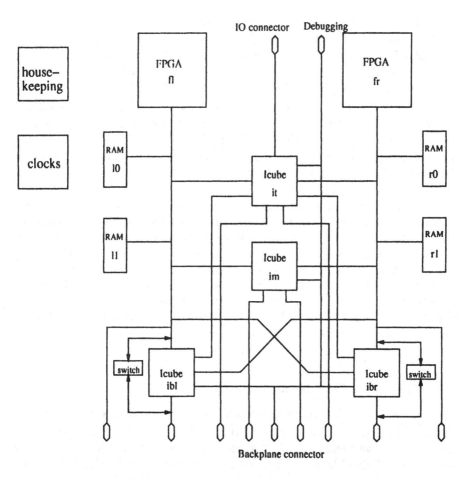

Fig. 4. Resources available on one of the boards in the TM-2a.

There are several ways to describe circuits that will be programmed in the device. These include schematics and various hardware description languages (HDLs). We chose to use AHDL, which is Altera's proprietary HDL, instead of a language such as VHDL. With AHDL, it is easier to control the logic mapping and therefore get more efficient and faster designs than with a more generic language. The actual synthesis and place and route is done using Altera's design system called Max+Plus II.

3.2 Early DES Work on the TM-2

Based on the work of Wiener [5] we understood that the goal was to build a pipeline capable of having a throughput of one key crack per cycle. Our first attempt [15] was based on an earlier version of the hardware, called the TM-2. The TM-2 was built at a time when the largest available FPGA was the Altera 10K50, which has roughly half the capacity of the 10K100 used in the TM-2a. Our TM-2 system has two boards, and four 10K50 FPGAs. On this system it was only possible to build half of the DES pipeline in a single 10K50. Therefore, we could only build two complete pipelines on the original TM-2 system. At that time the TM-2 only ran at 6.25 MHz, which was the limiting factor. This meant that we could crack keys at the rate of 12.5 million keys per second taking about 183 years to search the space.

Further analysis [16] of the work by Bernier showed that there were two areas that would limit the performance of the circuit. One was in the *S-box* circuitry and the other was in the interface circuitry that was used to communicate with the host. The interface could be easily decoupled from the rest of the circuit while the *S-box* needed more thought. A more serious problem we discovered was that the 10K100 did not really have double the logic of the 10K50 despite what the part numbers might suggest! The reason has to do with how the FPGA manufacturer counts its gates. This meant that we could not simply combine our two 10K50 circuits to form a single DES pipeline in one 10K100. More analysis and optimization of the circuit area was required.

3.3 The TM-2a DES Implementation

The goal of the TM-2a implementation was to implement a complete DES pipeline in a single 10K100, make it run as fast as possible, and then replicate it so that we could have 32 pipelines running in parallel. By doing this, we would not be limited by the interconnect network and crossing chip boundaries. It would also be much easier to replicate the pipelines across the system.

The top-level organization in a single chip is shown in Fig. 5. The *key maker*, which is some sort of counter, is used to produce keys. The plain text is coded with each key and then compared with the given cipher text. The circuit stops when a match is found.

There are enough resources to build all 16 stages as a large combinational circuit, but clearly this would be very slow. The next obvious step is to pipeline the logic by separating each stage with pipeline registers as shown in Fig. 6.

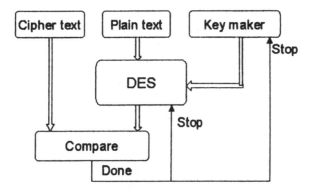

Fig. 5. Top-level structure of a chip.

The problem with this design is that there are not enough resources. As the computation proceeds down the pipeline, it is necessary to also keep the key for that stage in a register meaning that 16 keys will have to be stored. This uses almost 20% of the available LEs in the 10K100. We needed to find a *key maker* that would remember the 16 most recent key values without using so many registers. The next step would be to try and make the S-box logic go faster.

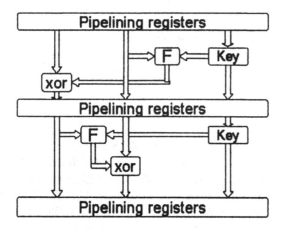

Fig. 6. Simple DES pipeline.

The Key Maker Our solution to the resource problem was to use a Linear-Feedback Shift Register (LFSR). By choosing the feedback taps correctly it is possible to generate each key exactly once. To remember previous keys, it is only necessary to extend the shift register by 15 registers as shown in Fig. 7. As each key is generated, the older ones can be found by sampling the values at a shifted offset from the new key. A possible disadvantage of this is routing the extended bits to the rest of the pipeline, but this ended up not being a factor.

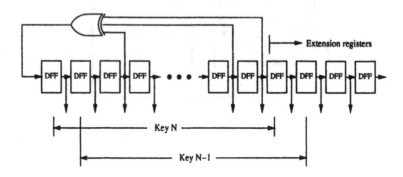

Fig. 7. LFSR with extension to save old keys.

Other advantages result from using an LFSR. An LFSR is much faster and simpler than binary counters, although in our case the key generation was not the critical path. Also, it is straightforward to serially preload the counter without additional logic and extra pins.

A slight disadvantage of the LFSR is because it does not count in a linear sequence. This means that we have to be a bit more careful when dividing the key space across the chips. The simple solution is to fix the key space for each chip by pre-setting the high order five bits when we are using 32 chips. We then build an LFSR that is only 51 bits long instead of 56 bits long.

Pipelining Possibilities Based on our previous work we knew that the *S-box* was the important critical path. Figure 8 shows one stage of the basic 16-stage pipeline and more details of how one bit of the *S-box* is constructed.

An *S-box* is a 6-input, 4-output lookup table, which can be thought of as four 6-input, 1-output lookup tables (6-LUT). The Altera device only has 4-LUTs so we had to find an efficient way to build the 6-LUTs. The straightforward solution is to have four 4-LUTs and a 4:1 multiplexer. The 4:1 multiplexer can be implemented as two levels of 2:1 multiplexers, which means that three levels of 4-LUTs are needed. A solution that uses only two levels and one fewer 4-LUT is shown in Fig. 8. This takes advantage of the AND gate that is available as part of the cascade chain in the LEs. An extra inversion is necessary at the output of the modified S-box but this can be absorbed transparently in the next level of logic.

The modified *S-box* structure can be easily pipelined, almost for free because the output of each 4-LUT can be latched at no extra cost. Only two additional registers are needed to pipeline the 2-bit bus that is connected to the inputs of

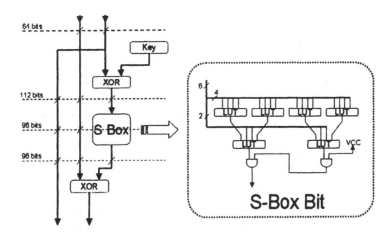

Fig. 8. S-Box details and pipelining options.

the second level of LUTs. However, the true cost of an additional pipeline stage must consider the context of the *S-box* in the full pipeline.

The simple pipeline puts a register between each of the 16 stages. If we wish to add an additional pipeline stage, then there are three possibilities as shown in Fig. 8. The dashed line at the top shows the existing 64-bit pipeline register. The labels on the dashed lines show the number bits that have to be registered if pipelining were done at that level. We do not have enough resources to add registers at all of the levels. The most economic place is at the level that goes through the *S-box* because many of the register bits come for free as mentioned above. However, it is still necessary to register the 64 other bits at that level that do not go through the *S-box*. Unfortunately, adding this extra pipeline stage exceeded the resources available to us so we are left with the original simple pipeline.

The Complete System The full system will consist of 32 complete DES pipelines, each running in one of the 10K100s on the TM-2a. Software running on a host machine will communicate with the hardware to monitor the status of each chip. In addition there is a separate daemon program that monitors the status of the TM-2a. Since the TM-2a is available to everyone on our network, it is essentially a common resource. Users make calls to the monitor to gain access to the machine and to load their circuits. The actual utilization of the TM-2a for other projects is low so we have modified the monitor to determine when the TM-2a is idle. During the idle periods, the DES cracker can be loaded and run. When the TM-2a is needed, then the current state, which is just the current key in the LFSR, is saved so that it can be restored the next time the circuit is loaded.

3.4 Results and Status

Our final design uses 4300 of the 4992 available LEs, which is about 86% of the resources. Adding an extra level of pipelining in the *S-box* was just 4% larger than what could fit in our FPGAs. It would have easily worked if we had 10K130 devices.

The maximum clock speed reported by Max+Plus II is 25MHz. Since we are able to process one key per clock cycle, this gives us 25M keys/second per chip. By using all 32 chips available on TM-2a, a total throughput of 800M keys/second is achieved. To search through the whole key space of 2^{56} keys, it would take 90.1 million seconds, or 25 thousand hours, which is about 1040 days. While this is clearly not fast enough for practical use, it represents a tremendous speed increase compared to what conventional computers can do within the same volume of space. If we could have improved the pipeline with one extra stage in the *S-box*, the speed would have been over 40 MHz giving around 650days to search the key space.

Since much of the structure of the circuit is reasonably regular and the data flows in one direction, we would have liked the option of hand-placing the logic to reduce routing delays. Evidence from other work using other devices shows that amazing speeds can sometimes be obtained, such as a 250 MHz cross-correlator [17]. Hand-placement was not an option with our devices. We do not know how much difference this would have made, given the hierarchical routing structure of the Altera 10K devices but it would have been nice to try. We feel that with a different FPGA architecture, we could have more easily optimized the design for speed.

The TM-2a is estimated to cost about US$3300 per board and about US$60,000 for the 16-board system using prices from the fall of 1998[1]. If the desire is to always be using the current state-of-the-art FPGA then the above numbers are probably a good estimate for a starting point.

However, this is much more than would be needed for a dedicated system of 32 chips. A single-board system with 32 chips using similar technology to ours is estimated to be less than half the cost of a 16-board TM-2a system. The TM-2a is also using technology that is about 2 years old. When we revised the TM-2 design to use the 10K100s, we could have used larger and faster parts but this would have caused too much change to our design, given our desire to make the revision quickly. We would have had to redo our routing network because there would have been more pins, and the faster parts run at lower voltages, meaning our board design would have had to change too much.

It is clear that as the density and speed of FPGAs continues to improve, it will become easier and easier to build a small fast machine to crack DES.

We have successfully run the system on a two-board (four-FPGA) version of the TM-2a. At this point in time, summer of 1999, our 16-board system is being

---

[1] Our numbers are very approximate because we have always been fortunate that Altera was willing to donate the devices that we needed so we have been sheltered from a lot of the true costs.

commissioned. Our DES cracking circuit is the first application to run on it that uses all of the boards.

4 Conclusions and Future Work

In this paper, we have described the implementation of a DES cracking system on a general-purpose field-programmable hardware system. The goal was to demonstrate the capabilities of field-programmable hardware and, in particular, the capabilities of our particular TM-2a field-programmable system. Although the system cannot compete with those that actually were able to solve the DES Challenges, our implementation does show how close technology is to being able to build machines capable of cracking DES without the aid of special-purpose custom hardware or the organizational requirements of coordinating a large number of computers on a network. This technology is available to everyone.

A 16-board TM-2a system can achieve a throughput of 800 million keys per second, which is still about 300 times slower than the last DES Challenge winner that was a combination of the EFF *Deep Crack* custom hardware and Distributed.Net's roughly 100,000 computers. They were testing 245 billion keys per second when the key was found [9]. When compared to just the *Deep Crack* hardware, which can test over 88 billion keys per second, the TM-2a is about 110 times slower. Based on our estimate of about US$30K for a dedicated 32-chip system, spending the same amount as EFF did would give us 8 times more performance, so that the FPGA system would only be about 14 times slower. By using a tool that allows more manual placement and routing and a similar generation of technology to *Deep Crack*, it is possible we could find another factor of 2 to 3 in performance. The difference between programmable and custom hardware then becomes even smaller.

With very few modifications, our DES cracker can be used as an ordinary high speed DES encoder/decoder.

For our own research into FPGA architectures and systems, the DES cracker circuit has given us a large benchmarking circuit. In future we plan to investigate more sophisticated ciphers such as RC5 [18].

Finally, it is clear that DES cracking hardware is quickly becoming within reach of many institutions because of the rapid improvement in FPGA technology.

5 Acknowledgements

Thanks go to Marcus van Ierssel and Dave Galloway for keeping the machines running. The Transmogrifier project benefits from the support of Micronet, a National Centre of Excellence in Canada, ATI Technologies, Altera Corporation, Cypress Semiconductor, and the Natural Sciences and Engineering Research Council of Canada.

References

1. RSA Laboratories. http://www.rsa.com/rsalabs/.
2. Data Encryption Standard. National Bureau of Standards (U.S.), Federal Information Processing Standards Publication 46, National Technical Information Service, Springfield, VA, 1977.
3. David M. Lewis, David R. Galloway, Marcus van Ierssel, Jonathan Rose, and Paul Chow. The Transmogrifier-2: A 1 Million Gate Rapid Prototyping System. *IEEE Transactions on VLSI Systems*, 6(2):188–198, June 1998.
4. http://www.eecg.toronto.edu/EECG/RESEARCH/FPGA.html.
5. Michael J. Wiener. Efficient DES Key Search. In W. Stallings, editor, *Practical Cryptography for Data Internetworks*, pages 31–97. IEEE Computer Society Press, 1996. First presented at the Rump session of Crypto '93 and also available by searching the WWW.
6. http://www.frii.com/~rcv/deschall.htm.
7. http://www.distributed.net.
8. Electronic Frontier Foundation, editor. *Cracking DES: Secrets of Encryption Research, Wiretap Politics & Chip Design*. O'Reilly & Associates, Inc., 101 Morris Street, Sebastopol, CA 95472, 1998.
9. http://www.eff.org/descracker.html.
10. http://www.rsa.com/rsalabs/des3/index.html.
11. Electronic Frontier Foundation, editor. *Cracking DES: Secrets of Encryption Research, Wiretap Politics & Chip Design*, chapter 11. O'Reilly & Associates, Inc., 101 Morris Street, Sebastopol, CA 95472, 1998.
12. Tom Kean and Ann Duncan. DES Key Breaking, Encryption and Decryption on the XC6216. In *IEEE Symposium on FPGAs for Custom Computing Machines*, pages 310–311, 1998.
13. http://www.altera.com.
14. http://www.icube.com.
15. Carolynn Bernier. DES Cracking on the TM-2. Undergraduate summer project report, 1997.
16. Kathleen Lam. Implementation and Optimization of a DES Cracking Circuit on the Transmogrifier-2 and the Transmogrifier-2a. B.A.Sc. thesis, Division of Engineering Science, Faculty of Applied Science and Engineering, University of Toronto, supervised by Professor Paul Chow, 1998.
17. Brian von Herzen. Signal Processing at 250 MHz using High-Performance FPGAs. In *International Symposium on Field Programmable Gate Arrays*, pages 62–68. ACM/SIGDA, 1997.
18. Bruce Schneier. *Applied Cryptography*. John Wiley and Sons, New York, 2nd edition, 1996.

Modelling the Crypto-Processor from Design to Synthesis

W. P. Choi and L. M. Cheng

Department of Electronic Engineering, City University of Hong Kong,
Tat Chee Avenue, Kowloon,
Hong Kong S.A.R., P.R.C.
wpchoi@ee.cityu.edu.hk, eelcheng@cityu.edu.hk

Abstract. In this paper, the modelling of a Crypto-processor in a FPGA chip based on the Rapid Prototyping of Application Specific Signal Processors (RASSP) design concept is described. By using this concept, the modelling is carried out in a structural manner from the design capture in VHDL code to design synthesis in FPGA prototype. Through this process, the turnaround time of the design cycle is reduced by above 50% compare to normal design cycle. This paper also emphasises on the crypto-processor architecture for space and speed trade-off; design methodology for design insertion and modification; and design automation from virtual prototyping to real hardware. In which above 60% of spatial and 75% of timing reduction is reported in this paper.

1   Introduction

The design flow and the techniques of modelling a crypto-processor [1] in FPGA chip based on the RASSP [2,3,4,5] are described in this paper. The modelling is made use of the VHDL platform. This platform has provided the perfect simulation and synthesis media for rapid prototyping. As well as, it also facilitated the design methodology of RASSP which promoting the design upgrades and re-uses. The modelled crypto-processor is designed for use in embedded digital systems which requiring area/speed/power trade-off, as crypto-processor is now commonly used in nowadays' digital devices, such as in Electronic Fund Transfer (EFT) systems and electronics wallet using smart cards.

This paper highlighted the procedures of modelling the crypto-processor from design to synthesis as in the following sections. In section 1.1 & 1.2, the background of the RASSP and the modelled crypto-processors are introduced. In section 2, the design process based on the VHDL virtual prototyping is described from the design specification, executable specification to detailed design. In section 3, the detailed design methodology of the crypto-processors is demonstrated. In section 4, the observations and results of this study are reported. In section 5, conclusions are made.

1.1 Scope of RASSP

Rapid Prototyping of Application Specific Signal Processors (RASSP) [2,3,4,5] is a modern methodology of designing embedded digital system nowadays. It supports the design of processor through a structural framework. The framework of RASSP mainly emphasises on the top-down design, design re-use and model-year design concepts [11]. Implementing these design concepts will result in shorter time-to-market and first-time silicon success fabrication.

In this study, those concepts are demonstrated by using the VHDL modelling via multi-level of abstraction, with all component objects defined in a standard open interface and technology independent specification. Based on this, it is not only provides the architecture reuse library components, but also supports the rapid insertion of a new element into an existing design for upgrades or modifications.

1.2 Cryptography and Crypto-Processors

Nowadays, cryptography is commonly used in commercial and banking sectors as Electronic Commerce created these urgent needs in Electronic Fund Transfer (EFT) application. In this paper, main focus is put on the modelling of symmetric crypto-processor which encrypt fixed-length of data block. The Data Encryption Standard (DES) [6] is often used as a basic building block in the existing cryptosystem, that difference applications are used in different ways. On the contrary, attacks on DES using linear cryptanalysis and differential cryptanalysis, as well as exhaustive search are also well known. Therefore, in order to strengthen the security level of the existing cryptosystem, various kinds of modification and upgrade of the DES algorithm are proposed which using DES components as a building block. Hence, modelling the DES algorithm in a RASSP design framework helps the rapid prototyping of a new design. This is benefited from the reuse of design information and functional block library from previous design, for instance, the Randomised-DES [7,8,9] proposed by T. Kaneko, K. Koyama and R. Terada and the Extended-DES [10] proposed by H.S. Oh and S.J. Han. These are DES-based cryptosystem which used DES components as a building block.

Randomised-DES (RDES) [7,8,9] is a cryptosystem with an n-round DES in which a probabilistic swapping, $SW(Rn, Sn)$, is added onto the right half output of each round as shown in Fig. 1. It has been claimed that the n-round RDES is stronger than the n-round DES against differential cryptanalysis.

Extended-DES [10] is a cryptosystem utilising the iteration F-function of the DES to extend the property of the algorithm in form of a matrix. It defines the input plaintext as 96-bits and the key size as 128-bits, as well as the order of the S-box is randomly arranged. The 128-bits key is divided into two independent key, K_1 and K_2, and used the same key scheduling algorithm of DES for generating the subkeys. The encryption and decryption formulas of EDES are shown in Table 1. With this extended configuration, it is verified to be less vulnerable to attack by differential cryptanalysis.

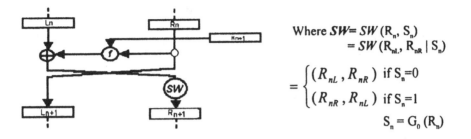

Where $SW = SW(R_n, S_n)$
$= SW(R_{nL}, R_{nR} \mid S_n)$

$$= \begin{cases} (R_{nL}, R_{nR}) & \text{if } S_n = 0 \\ (R_{nR}, R_{nL}) & \text{if } S_n = 1 \end{cases}$$

$S_n = G_0(R_n)$

Fig. 1. The Randomised-DES (RDES)

Table 1. Encryption and Decryption Formulas of EDES

Encryption	Decryption
$A_i = B_{i-1}$	$A_{i-1} = C_i \, \text{Xor} \, f(A_{i-1}, K_{2,i})$
$B_i = C_{i-1} \, \text{Xor} \, f(B_{i-1}, K_{2,i})$	$B_{i-1} = A_i$
$C_i = A_{i-1} \, \text{Xor} \, f(B_{i-1}, K_{1,i})$	$C_{i-1} = B_i \, \text{Xor} \, f(A_i, K_{1,i})$

2   Design Process by VHDL

VHSIC Hardware Description Language (VHDL) provides a media of vendor, platform and technology-independent design method of describing, simulating, and documenting complex digital system. It helps the rapid prototyping application-specific simulatable and sysnthesisable VHDL models of various signal-processing functions. The support of multi-level of abstraction, as well as working at a higher-levels of abstraction, facilitates the design transfers from the system level algorithm to structural implementation. Through out the modelling process in VHDL, it supports a cost-effective means for rapid exploration of area, speed, and power requirements of the processor. It also facilitates the functional trade-offs of algorithm and architectural design alternatives at the very early stages in the design process. The design process of VHDL can be divided into three parts as shown in Fig. 2: they are design specification, executable specification and detailed design.

2.1   Design Specification

Design specification captures customer requirements and converts these system-level needs into processing requirements (functional and performance) by VHDL description. Functional and performance analyses are performed to properly decompose the system level description. The system process has no notion of either hardware functionality or processor implementation. It also specifies an appropriate set of parameters specifying the performance and implementation goals for the processor (size, weight, power, cost, etc.). The traditional approach is to utilise text-based files in a specific format to support extraction of key parameters by the

28

appropriate tools. Nowadays, VHDL is regarded as the unifying design representation language and tool integration approach for describing the design specification. Eventually, the design specification is translated into simulatable functions, which refers to an executable specification.

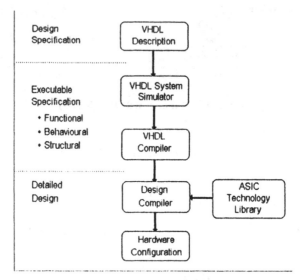

Fig. 2. The Design process by VHDL

2.2   Executable Specification

An executable specification [12] is a behavioural description of a component or system module without describing a specific implementation. The description reflects the particular function and timing of the intended design as looking on the component's interface level. During this process, the system level processing requirements are allocated to functional modules and each module is then verifying its specified functionality against the system requirements. The module is then integrated with other components of the system and to test whether an implementation of the entire system is consistent with the specified behaviour in the design specification. Finally, a virtual prototype is resulted in a detailed behavioural description of the processor hardware.

In this stage, an extensive simulation of all components is carried out in any form of the above models which can be described as functionally, behaviourally or structurally. Simulation is carried out by using the VHDL system simulator and VHDL compiler. It is intended to verify all of the codes during this portion of the processor design. After this process, all modules are fully tested and resulted in a detailed behavioural description of the processor hardware. Thus, the result of the executable specification is the virtual prototype describing the custom modules down to individual components at the behavioural level with emphasis on interface behaviour rather than internal chip structure.

2.3 Detailed Design

With the above processes, the design is modelled and verified through a set of extensive functional and performance simulations using integrated simulators in VHDL platform. At the completion of those simulations, the design is in the form of a fully verified virtual prototype of the system and the timing is also verified to ensure proper performance against the design specification. For the design to be realised in a physical hardware, in this stage, the executable specification of the processor is transformed into detailed designs in Register Transfer Level (RTL) and/or logic level which specifying the actual implementation technology. This process resulted in a detailed technology-dependent hardware layout and artwork, netlist, and test vectors of the entire processor. Making uses of that information, the processor can be put into real hardware for integration, as well as used for silicon fabrication. It is accomplished by using the VHDL design compile and the specific ASIC technology library to generate the vendor-specific hardware configuration details.

3. The Crypto-Processor Model

The crypto-processors are synthesised using the Synopsys VHDL integrated simulator and implemented in a Xilinx FPGA chip. The main task of the synthesis tool is to transfer the design into a virtual prototype with simulation and debugging of system functionality. The implementation tool is to realise the design in real hardware and used for design verification. In this section, the detailed modelling of the baseline algorithm, DES, is demonstrated. The reuse concept is also exercised in the RDES and Extended-DES models. Finally, some observations and results are shown.

3.1 Top-Down Modelling of the DES

To rapidly prototype the DES in VHDL, the procedures described in section 2 is deployed. First of all, the design specification is defined, i.e. the mathematical representations of the algorithm. Then, the algorithm is partitioned into functional modules for synthesis, in which, the algorithm is simulated in form of functional, behavioural and structural models. The modules are refined into smaller component which is implementable in FPGA architecture. Finally, the virtual prototype is transformed to detailed design of FPGA configuration netlist.

Design specification
The design specification of the DES is the standardised algorithm defined in International Standard document [6]. As stated in the document, the DES algorithm is making use of a series of permutation, substitution and exclusive-or operations to scramble the data depending on a binary key. The core of the algorithm computation includes the Initial Permutation (IP), the Expansion Box (E-box), the Substitution Box (S-box), the Permutation Box (P-box), the Inverse Initial Permutation (IP^{l}) and the Exclusive-OR (XOR) operations. By combining the E-box, S-box and P-box with the associated XOR operations, it forms the iteration function (F-box) which is the core

computation unit of the DES. In addition, the Key Schedule (*KS*) associated with the algorithm provides the 48-bits subkeys used in each round of iteration. The KS includes the Permutation Choice-1 and -2 (*PC-1, PC-2*), and a series of shift operations.

According to the DES specifications, all computation of the above units follows a set of operation tables defined in the standard [6]. To capture the design for implementation, each operation table specified in the standard is coded as a functional entity in VHDL description. Eventually, a DES VHDL package, which translates the textual specifications into synthesisable VHDL code, is modelled. The package is then used for program coding, design validation and system integration.

Executable Specification
In this stage, the functionality of the algorithm is validated by simulating and testing the algorithm in VHDL simulator, ultimately completed with a fully verified virtual prototype of the algorithm. To achieve this, the process is conducted through a combination of functionality partitioning and synthesis at all levels of abstraction. The partitioning of the model into smaller modules also facilitated the reuse concept.

The DES algorithm is partitioned into four top-level functional modules, including the *F-box*, the *IP*, *IP'* and the *KS*. In those functional modules, their interfaces between sub-modules, as well as the resource requirements (performance/area) for each component module are specified. Probably, the functional module is further decomposed into lower behavioural level model, so as to form a layered-architecture. This layer approach made the module more manageable, understandable, reusable, and maintainable. This helps to facilitate the design reuse concept and to build an encapsulated library. For instance, the functional module, *F-box*, with the defined interface is shown in Fig. 3, and the partitioned behaviour module of the *F-box* is shown in Fig. 4, with each box represents a behavioural component model in VHDL description entity.

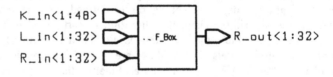

Fig. 3. Interface of F-box

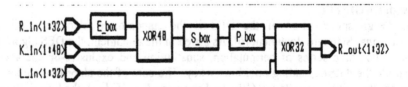

Fig. 4. Partitioned behavioural modules of F-box

The behavioural VHDL description of each module is then designed and tested individually using the pre-defined VHDL DES-algorithm package as the component library. At this level, further model partitioning is also conducted for each sub-module, until a sufficient detail is resolved for physical construction, performance and area requirements in the specific technology platform.

In this case, the S-box is further partitioned in smaller sub-modules with respect to satisfy the area and speed requirements. With this partitioning, the large function is implemented in a minimum of area and the delay in signal path, and thus achieves performance increase both in spatial and speed requirements. Finally, the entire algorithm is integrated by a structural model which interconnects all verified modules together.

In this stage, the functionality of all modules is verified and the timing analysis is computed. This ends up with a technology-independent and fully functional verified virtual prototype of the algorithm. For it is synthesisable in real hardware, say a FPGA chip, it needs to forward to the VHDL design compiler with the specific FPGA technology library to generate the detailed FPGA configuration netlist.

Detailed Design
Detailed design will match the design into a physical reality. The virtual prototype synthesised in the pervious stage is ready for realisation in FPGA. In this stage, the entire design is converted into the FPGA netlist by the VHDL design compiler and the associated FPGA technology library. As a result, the Xilinx Netlist Format (xnf) file is generated. By making use of this netlist in the automated development environment, logic mapping, placement, and routing are done automatically and finally a FPGA configuration file is generated. Then the configuration file is stored inside an EPROM for programming the FPGA in the real hardware prototype.

Finally, the DES algorithm is transformed into the FPGA configuration file. The spatial requirement of the algorithm in pipeline mode occupied 2,176 Configuration Logic Block (CLB) with a signal path delay of 164.96 ns. Detailed synthesis results of all modules are shown in Section 4.

3.2 Design Re-use for the Randomised-DES and Extended-DES

Randomised-DES [7,8,9]
RDES is the case of design insertion in the existing DES algorithm. It is an extension of the DES by inserting a special modular, *SWAP*, in the algorithm as illustrated in Fig. 1. By the modular-design concept applied in the DES design, the VHDL DES-algorithm package library, as well as the verified functional, behavioural and structural models, are reused as the components for constructing the RDES.

In the virtual prototyping stage, only the functionality of the *SWAP* module needs to verify as it is stated in the specification. The insertion of the *SWAP* module only affected the internal structural of the *F-box* which is retained with the same interface structure. Thus, all structural models remained in its defined interface, as well as their original interconnection between modules as in the DES model. The modified structural of the *F-box* model is shown in Fig. 5.

As a result, all of the structural models designed in the pervious DES model is reused. For the entire RDES algorithm design, modification is made only in the structural model of the *F-box*. By this design reuse of the DES algorithm component library, the RDES algorithm is rapidly prototyped within one-man week. This is achieved by the top-down, model-year structural design methodology applied in the designing of the DES algorithm.

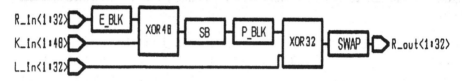

Fig. 5. Modified Structural Model of F-box for RDES.

Extended-DES [10]
EDES is the case of modifying the existing DES algorithm. EDES is just an extension of the DES by increasing its data block length to 96-bits and the key size to 128-bits, in addition with a special arrangement in the order of *S-box*. In such, the processing block size is remained in 32-bits using the same DES iteration *F-box* as its core, and the same key-scheduling algorithm that is used in DES. In this case, only the top-level structure model is needed to re-design and the *S-box* sub-module is needed to re-structure. In the *S-box*, since the functionality of all the smaller components in the lower-level module is verified in the DES-algorithm library. It only needs to re-program the structural model according to the EDES design specification and then reinsert it back to the encapsulated library. The *F-box* module and all other sub-modules within the *F-Box* are not affected at all.

Therefore, the modelling of the EDES required in this case is just to modify the structural level models. All functional models of the processing units are using the standard DES modules extracted from the encapsulated library built during the previous design. As a result, the virtual prototype of EDES re-defined the interconnection between modules in a structural model and this re-design is prototyped by one-man week.

4 Observations & Results

4.1 Space and Speed Requirements

To prototype a design into a physical hardware, such as in FPGA chip, the space and speed constraints in physical device are not negligible. Since all electronic technologies deliver finite spatial resources for building functions and wiring resources for communications which are especially tight with FPGA. By using the top-down design concept to partition design functionality into small modules has facilitated the design optimisation against those constraints.

During the modelling of DES algorithm, the following results are observed. By transforming the functional model directly to detailed design, the resultant requirement in space and delay are higher. In the partitioned behaviour model which module is in form of a small component, the resultant requirement is much lower. The results of the DES modules are tabled in Table 2.

Table 2. Synthesis results of the DES module[1]

Module	Un-partitioned Functional Model		Partitioned Behavioural Model	
	CLB	Timing(ns)	CLB	Timing (ns)
IP	0	0	0	0
IP-1	0	0	0	0
E-box	N/A	N/A	0	0
P-box	N/A	N/A	0	0
S-box	230	33.92	96	6.31
XOR32	N/A	N/A	16	3.26
XOR48	N/A	N/A	24	3.26
F-box	323	42.38	136	10.31
KS	0	0	0	0
SWAP	N/A	N/A	16	3.62

From the table, it is found that the spatial requirement of the partitioned s-box is reduced by 60% and the delay is reduced by over 80%. (CLB's propagation delay is reduced from 7 stages to 2 stages). While in the partitioned F-box, a 60% reduction is achieved. CLB's propagation delay is reduced from 9 stages to 4 stages, above 75% reduction is accomplished. With this result, the algorithm is more feasible for implementation in FPGA chip with benefits in both spatial and timing requirements. Those benefits are also encountered in the case of RDES and EDES implementations.

---

[1] *Timing is measuring under the Xilinx Xfpga_4025e-3 library parameters: path_full, delay_max, max_paths, and WCCOM operation conditions*

4.2   Design Insertion and Modification

Through designing the DES algorithm in model-year architecture, by defining the module with open interface and partitioning functionality into small components, it can facilitate the rapid design insertion and modification. Like the cases of modelling the RDES and EDES, the turnaround time to prototyping those algorithms are reduced rapidly. In modelling of RDES and EDES, above 70% and 40% of the development time is eliminated respectively. This is achieved by the result of using the encapsulated library, as most of the functionality verification is exempted. Thus, the design reuse concept of the encapsulated component library has shown its advantage and significance in this aspect.

4.3   Design Automation

Beside the design methodology, the platforms for simulation, debugging, synthesis, logic placement, routing, test vectors generation and hardware implementation the design are also important. Any one of those elements cannot be omitted in the process of processor design and prototyping. Therefore, a standardised integrated development environment is essential for designer, so as to speedup design process and reduce design transfer/translation cumbersome. In this study, the use of Synopsys VHDL integrated platform has helped a lot in the design automation aspect from the design capture to the synthesis in hardware.

4.4   Hardware Prototype

The designs are realised in a 25,000 logic-gates FPGA chip for testing and integration. Those algorithms are synthesised in both recursive and pipeline mode of operations. The hardware prototype is as shown in Fig. 6.

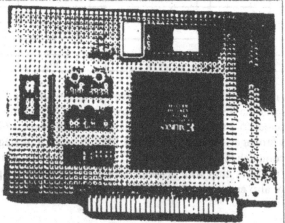

Fig. 6 Prototype of the Hardware

5. Conclusions

Through out the modelling of the crypto-processors in this study, the design concept of rapid prototyping the application-specific signal processors is practised. By the described approach, it not only verifies design functionality early in the design process, but also provides the key to rapid prototyping and upgrading of signal processors, as the same time reduces the development time and costs significantly.

Deployment of model-year design concept in rapid prototype has provides the use of previous models as a baseline for further developments. As in the cases of modelling the RDES and EDES algorithm, which using DES as a baseline, allowed the modification of the functional models in the virtual prototyping stages and allowed partitioning and re-targeting design during the synthesis activities. In this case, above 50% of development time is reduced in modelling the RDES and EDES algorithms.

On the other hand, in the VHDL development environment, design can automatically converts a VHDL description to a gate-level implementation in a given technology; and can automatically transform a synthesis design to a smaller or faster circuit through partitioning. In this experience, above 60% of spatial and 75% of timing reduction is achieved. In addition, capturing the design in VHDL technology-independent functional models for the virtual prototype also enhances reuse of functional primitives and generates the design in different technology. Simultaneously, it also provides a technology-independent documentation for a design and its functionality.

To conclude, the modelling is carried out in a structural manner from the design capture in VHDL code to design synthesis in FPGA prototype. Through those prototyping procedures, the turnaround time of the design cycle is reduced; and through the modular design concept, the feasibility of design upgrade and modification is enabled.

6. References

1 Rita C. Summers, "Secure Computing: Threats and Safeguards", McGraw-Hill, Chapter 5, 1997.
2 Mark A. Richards, Anthony J. Gadient and Geoffrey A. Frank, "Rapid Prototyping of Application Specific Signal Processors", Kluwer Academic Publishers, Boston, February 1997.
3 Mark A. Richards, "The Rapid Prototyping of Application Specific signal Processors (RASSP) Program: Overview and Accomplishments", Proceedings of the 1st Annual RASSP Conference, pp.1-10, August 1994.
4 Jeffrey S, Pridmore and W. Bernard Schaming, "RASSP Methodology Overview", Proceedings of the 1st Annual RASSP Conference, pp.71-85, 1994.
5 Carl Hein, Paul Kalutkiewicz, Todd Carpenter and Vijay Madisetti, "RASSP VHDL Modelling Terminology and Taxonomy – Revision 2.3", RASSP Taxonomy Working Group (RTWG), June 23, 1997.
6 "Data Encryption Standard", Federal Information Processing Standard (FIPS) 46, Nat. Bur. Stand., Jan. 1977.

7   K. Koyama and R. Terada, "How to Strengthen DES-like Cryptosystems against Differential Cryptanalysis", IEICE Trans. Fundamentals, Vol. E76-A, no. 1, Jan. 1993.
8   T. Kaneko, et. al., "Dynamic Swapping Schemes and Differential Cryptanalysis", IEICE Trans. Fundamentals, Vol. E77-A, no. 8, Aug. 1994.
9   Y. Nakao, et. al., "The Security of an RDES Cryptosystem against Linear Cryptanalysis", IEICE Trans. Fundamentals, Vol. E79-A, no. 1, Jan. 1996.
10   Haeng-Soo Oh and Seung-Jo Han, "Design of the Extended-DES Cryptography", Proceeding of 1995 IEEE International Symposium on Information Theory, pp.353, 1995
11   Jeff Pridmore, Greg Buchanan, Gerry Caracciolo and Janet Wedgwood, "Model-Year Architectures for Rapid Prototyping", Journal of VLSI Single Processor 15, 83-96 (1997).
12   Allan H. Anderson and Gray A. Shaw, "Execute Requirements and Specifications", Journal of VLSI Single Processor 15, 49-61 (1997).

A DES ASIC Suitable for Network Encryption at 10 Gbps and Beyond

D. Craig Wilcox[1], Lyndon G. Pierson[1], Perry J. Robertson[1], Edward L. Witzke[1] and
Karl Gass[2]

[1]Sandia National Laboratories
P.O. Box 5800
Albuquerque, New Mexico 87185
{dcwilco, lgpiers, pjrober, elwitzk}@sandia.gov
[2]Utah State University
Albuquerque, NM 87111
kgass@sandia.gov

Abstract. The Sandia National Laboratories (SNL) Data Encryption Standard (DES) Application Specific Integrated Circuit (ASIC) is the fastest known implementation of the DES algorithm as defined in the Federal Information Processing Standards (FIPS) Publication 46-2. DES is used for protecting data by cryptographic means. The SNL DES ASIC, over 10 times faster than other currently available DES chips, is a high-speed, fully pipelined implementation offering encryption, decryption, unique key input, or algorithm bypassing on each clock cycle. Operating beyond 105 MHz on 64 bit words, this device is capable of data throughputs greater than 6.7 Billion bits per second (tester limited). Simulations predict proper operation up to 9.28 Billion bits per second. In low frequency, low data rate applications, the ASIC consumes less that one milliwatt of power. The device has features for passing control signals synchronized to throughput data. Three SNL DES ASICs may be easily cascaded to provide the much greater security of triple-key, triple-DES.

1 Introduction

Since 1977, the United States has had a Data Encryption Standard (DES). DES is a block cipher that operates on 64 bit blocks of data and uses a 56 bit key [2]. It is a Feistel-type cipher. Feistel Ciphers [5][7] operate on left and right halves of a block of bits, in multiple rounds. The block halves are exchanged (left for right) from their usual order after the last round. An important property of Feistel Ciphers is that the function f, employed by a Feistel Cipher to operate on a left or right half-block of data, need not be invertible to allow inversion of the Feistel Cipher. In DES, the function f can itself be considered a product cipher (or substitution-permutation cipher), since that function performs both substitutions (to introduce confusion) and permutations (to introduce diffusion).

Another important property of Feistel Ciphers is that due to their structure, decryption is performed using the same multiple round process as encryption, but using the subkeys (one required per round) in reverse order. By eliminating the need for two different algorithms (one for encryption and one for decryption), hardware implementation is simplified and real estate on a chip is conserved.

A survey of the available integrated circuit implementations of the DES showed only devices with throughputs below 0.5 Gbps, far below the encryption rates required to scale Asynchronous Transfer Mode (ATM) encryption beyond 10 Gbps (SONET OC-192c). The existing implementations appeared to implement the sixteen rounds of the DES algorithm by iterating data through the hardware of a single round 16 times, resulting in low throughputs and the inability to change key variables quickly. To achieve the high throughput and key agility required for high speed ATM cell encryption, a fully pipelined implementation of all 16 rounds of DES with the key variables pipelined along with the data stages was designed and studied.

2 Proof of Principle

In order to study pipelined (and non-pipelined) implementations of the DES, an Excel spreadsheet implementation of the DES key schedule and algorithm was developed. This spreadsheet implementation of DES enabled the designers to familiarize themselves with the algorithm, and to examine multiple options for hardware implementation of the key schedule. After verification of the proper operation of the spreadsheet, the well-tested descriptions of the permutations and "S-boxes" were "cut and pasted" into the hardware description language, minimizing the opportunity for transcription errors.

First, the permutations and S-boxes of a single round were implemented in ALTERA's AHDL; compiled and simulated for ALTERA's 7000, 8000, and 10k families of devices. The simulations indicated that these operations could be computed more swiftly in the 7000 series devices, but only a small portion of the required functionality could be fit into the smaller gate count devices. We found that only four of the sixteen rounds of the DES algorithm would fit into an ALTERA 10K100 device, necessitating four large Programmable Logic Devices (PLDs) to fully pipeline all sixteen rounds. The key was pipelined through each stage along with the data in order to provide full key agility (the ability to change keys on each and every clocked word transfer, if desired). This pipelining of key as well as data also increased greatly the number of I/O pins required to transfer the data between the devices comprising the pipeline.

The simulations of the circuitry in ALTERA 10K100-3 speed grade devices showed that the time required to compute and latch a single round was 50 ns. Once the synchronous pipeline was filled, 64 bits of output every 50 ns would yield approximately 1.3 Gbps throughput with a latency of 800 ns.

For certain "feedback" modes of operation, such as Cipher Block Chaining (CBC) [3], the output of the pipeline must be combined with the next input. This requires the pipeline to "run dry", with only one 64-bit data word traversing all the pipeline stages before the next word can be input to the pipeline. Therefore, the CBC mode throughput for the synchronous pipeline is 64 bits each 800 Ns, or 0.08 Gbps (one sixteenth of the full pipeline throughput). For this reason, an "asynchronous" version of the pipeline (with no latches between stages) was analyzed, in order to maximize the CBC mode throughput by minimizing the pipeline latency. Analysis of this "asynchronous" pipeline showed that the total latency could be reduced to 650 Ns, improving the potential CBC mode throughput to 64 bits per 650 Ns, or 0.098 Gbps. Clearly, CBC and other similar feedback modes of operation are difficult to scale to high speed operation.

For non-feedback modes of operation such as Electronic Codebook (ECB) [3] or Counter Mode [1], the full pipeline can be utilized. In order to achieve 10 Gbps throughput (SONET OC-192c), however, eight such 1.3Gbps pipelines would have to be operated in parallel. ATM cell order must be maintained, and cells of different Virtual Circuits (requiring different key variables) may be interleaved in the cell stream. The processing of more than one ATM cell in parallel therefore introduces great complexity into an encryptor. Since an ATM cell payload is 384 bits, evenly divisible into six 64 bit words, six is the practical limit for parallel operation of 64-bit encryption pipelines for ATM cell encryption. In order to achieve 10 Gbps encryption, the throughput of each of the six pipelines must be at least 1.7 Gbps, which is greater than the 1.3 Gbps predicted by the ALTERA 10K100-3 simulations.

The pipelined DES design was then implemented as a CMOS Application Specific Integrated Circuit in order to achieve the increased throughput required to implement ATM cell encryption at rates greater than 10 Gbps

3 SNL DES ASIC

The Sandia National Laboratories (SNL) DES Application Specific Integrated Circuit (ASIC) is the fastest known implementation of the DES algorithm as defined in FIPS Pub 46-2 [2]. The SNL DES ASIC, over 10 times faster than other currently available DES chips, is a high-speed, fully pipelined implementation providing encryption, decryption, unique key input, or algorithm bypassing on each clock cycle. In other words, for each clock cycle, data presented to the ASIC may be encrypted or decrypted using the key data presented to the ASIC at that cycle or the data may pass through the ASIC with no modification. Operating beyond 105 MHz on 64 bit words, this device is capable of data throughputs greater than 6.7 Gbps, while simulations show the chip capable of operating at up to 9.28 Gbps. In low frequency applications the device consumes less that one milliwatt of power. The device also has features for passing control signals synchronized to the data.

The SNL DES ASIC was fabricated with static 0.6 micron CMOS technology. Its die size is 11.1 millimeters square, and contains 319 total pins (251 signals and 68 power/ground pins). All outputs are tristate CMOS drivers to facilitate common busses driven by several devices. This device accommodates the full input of plain text, 64 bits, and a complete DES key of 56 bits. Additionally, 120 synchronous output signals provide 64 bits of cipher text and the 56 bit key.

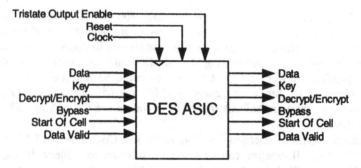

Fig. 1. DES ASIC Block Diagram

Three input only signals control electrical functions for logic clocking (CLK), logic reset (RST), and the tristate output enables (OE). The CLK signal provides synchronous operation and pipeline latching on the rising edge. Both RST and OE are asynchronous, active high signals.

Two synchronous signals, decrypt/encrypt (DEN) and bypass (BYP), determine the DES cryptographic functionality. On the rising edge of each CLK, the logic value presented to the DEN input selects whether input data will be decrypted (logic 1) or encrypted (logic 0). In a similar manner, BYP selects algorithm bypassing (logic 1) or not (logic 0) for each clock cycle. Both of these signals pipeline through the ASIC and exit the device synchronous with the key and data.

Two more signals, start-of-cell (SOC) and data valid (VAL) enter and exit the device synchronous with data and key information. These are merely data bits that may provide any user-defined information to travel with input text and key. These signals are typically used to indicate the start of an ATM cell and which words in the pipeline contain valid data.

ASICs from two wafer lots were shown to operate beyond the maximum frequency (105 MHz) of Sandia's IC Test systems. For 64-bit words, this equates to 6.7 Gb/s. This operational frequency was tested over a voltage range of 4.5 to 5.5 Volts and a temperature range of −55 to 125 degrees C.

3.1 Design

After implementing the DES algorithm in the set of four PLDs, the design was translated into VHDL and synthesized into the Compass library of standard cells. The device (figure 2) was fabricated in Sandia's MDL (Microelectronics Development Laboratory). Two wafer lots were successfully fabricated.

This implementation is a fully pipelined design. It takes eighteen clock cycles to completely process data through the pipeline causing the appropriately decrypted, encrypted, or bypassed data to appear on the ASIC outputs. Additionally, all key and control input signals pass through the pipeline and exit the ASIC synchronized to the ciphertext outputs.

The SNL DES ASIC is the only known fully pipelined implementation of all 16 rounds of the DES algorithm. Pipelining increased the device throughput by dividing the algorithm into equally sized blocks and latching information at the block boundaries. This gives signals just enough time to process through each block between clock cycles, thereby maximizing the operational frequency.

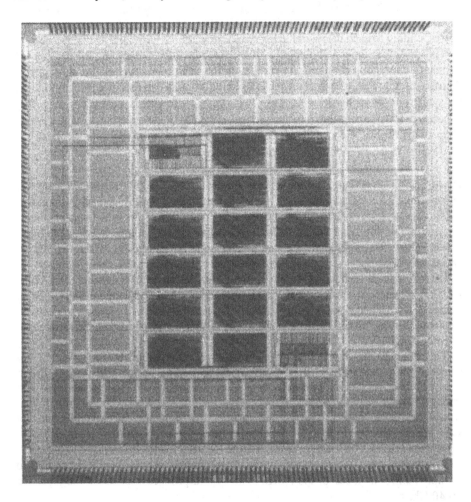

Fig. 2. DES ASIC Die (11.1 x 11.1 mm)

Pipelining the algorithm allows a high degree of key and function agility for this device. Here, agility means that the SNL DES ASIC processes inputs differently on each clock cycle. As an example, the device may encrypt data with one key on one clock cycle, decrypt new input data with a different key on the very next clock cycle, bypass the algorithm (pass the data unencrypted) on the following clock, then encrypt data with yet another independent key on the fourth clock cycle. The control signals used to select these various modes of operation are presented at the output, passing through the device synchronized to the input data and the input key information. All inputs and outputs (control, key, and data) enter and exit the part synchronously. The authors know of no other single-chip implementation with all these features.

Features. This DES ASIC is unique in its ability to encrypt or decrypt with a new key, or pass information unprocessed, on each and every clock cycle. This enables the separate encryption of many virtual channels within a high speed communication system. In addition, it enables cryptosystems to be built with fewer components and lower cost. Also, as stated previously, no other encryption chip outputs the key corresponding to each data word in every clock cycle.

This per-cycle input and output of all variables facilitates cascading the devices for increased encryption strength, and paralleling the devices for even higher throughput. Another unique feature of this device is its ability to pass two user-defined control bits in synchronism with the data being encrypted, decrypted, or bypassed. This capability is indispensable for the design of ATM data encryptors, which must identify the Start of Cell boundaries and for systems that must flag data as "valid" or "not valid" in the encryption/decryption pipeline.

Design Enhancements. Since the initial ASICs were fabricated, several enhancements have been identified. These enhancements would increase throughput, aid in cascading devices, and ease the use at the board level. Projected enhancements include use of improved design tools, improved synthesis options, low-voltage high-speed I/O buffers, improved pin-outs, greater parallelism, and processing in higher-speed technologies.

Several design techniques could improve the design of the existing SNL DES ASIC. For example, recent synthesis developments would allow the DES ASIC to be redesigned with additional pipeline stages. A greater number of pipeline stages with improved timing, would increase the operational frequency and boost the throughput beyond 10 Gb/s.

To enhance the high-frequency operation at the circuit-board level, higher performance input-output buffers would reduce switching noise. The present design uses CMOS level (0 – 5 V) interfaces. Future designs would incorporate these low voltage, low power I/O buffers. Bringing out the clock phased with the output data, would facilitate higher speeds and greater performance by enabling source synchronous clocking. Also, optionally inverting the encrypt/decrypt output would better facilitate encrypt-decrypt-encrypt triple DES (described below).

A redesign into Gallium Arsenide (GaAs) technology should yield a factor of 3 to 4 improvement in speed of the ASIC. This would produce expected throughputs of 30-40 Gbps.

To achieve higher total throughput, multiple SNL DES ASICs can operate in parallel, with each ASIC processing a 64 bit block of the data stream. Figure 3 contains an example of multiple devices, performing DES operations on two blocks of data in parallel.

Because the data outputs of the SNL DES ASIC are tri-stated, there are several ways the ASICs can be used in parallel. The data outputs from both ASICs can be connected to a single output bus in a time-multiplexed fashion and the two DES ASICs can be operated using opposite clock phases to double the data throughput to greater than 13 Gbps. If both ASICs are driven off the same clock edge, the two 64 bit wide data outputs can also be combined into a single 128 bit wide output to achieve the 13 Gbps throughput.

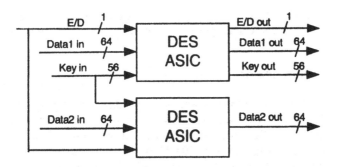

Fig. 3. Parallel DES ASIC Implementation

For encryption of Asynchronous Transfer Mode (ATM) communication sessions where six SNL DES ASICs could operate in parallel on 64 bit blocks to encrypt a 384 bit payload, 40 Gbs (OC-768) rates could be achieved. The authors would expect six parallel DES ASICs made using a GaAs process to support encryption at 160 Gbps and beyond.

Power Consumption. Being a fully static CMOS device, the power usage is proportional to operating frequency. At 105 MHz, the SNL DES ASIC consumes 6.5 Watts of power. While designed to dissipate the heat generated in high-bandwidth applications, the SNL DES ASIC can be operated at much lower data clock rates, consuming very little power, thus enabling many low speed, extremely low power applications.

Table 1. Power Consumption of SNL DES ASIC

MHz Vdd	0.01	1	105
3.0	510 μW	54 mW	*
3.3	*	66 mW	*
5.0	*	165 mW	6.5 W

* untested

In the 1200 to 640,000 bits per second range, the DES ASIC consumes only microwatts of power. The SNL DES ASIC operating at 10 KHz (640,000 bits per second, or ten 64 Kbps voice channels) only consumes 510 microwatts. Iterating around the ASIC to triple encrypt a single 64 Kbps voice channel, the SNL DES ASIC would need to operate at about 3 KHz, requiring well less than half of a milliwatt of power. Triple encrypting lower data rate channels (1200-28800 bps) requires even less power, enabling operation from a small battery or solar panel.

3.2 Package Development

The SNL DES ASIC has been packaged into three different packages including a 360 pin PGA, a 503-pin PGA and a 352 pin BGA. The original 360-pin package was used in initial testing of the DES ASIC performance. It was in this package that the DES chip was shown to operate at over 105 MHz. Sandia had earlier developed a 1.1 million gate PLD board that used 11 Altera 10K100 devices. This board was used in the development of the DES ASIC pipeline design, housing the four 10K100 devices. It was determined that the SNL DES ASIC could be used with the original PLD11 board, being substituted for a single 10K100 device, if a 503-pin equivalent package were available. Sandia designed an FR4 board onto which the DES ASIC was wire bonded and 503 pins could be inserted. The chip-on-board package had to be designed to dissipate up to 5 watts produced by the DES ASIC. This is accomplished by attaching the DES die directly onto a gold plated copper insert that is attached to the FR4 board using a tin-lead solder preform. Pictures of a representative cross section and this package are shown in figures 4 and 5.

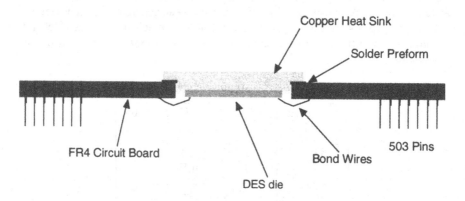

Fig. 4. Cross Section of the 503-Pin Package

The design of this package enables a heat sink and integrated fan to be attached to the back of the copper insert to enable the package to dissipate over 6 watts. The FR4 printed wiring board uses 3 mil copper traces and spaces with 5 mil vias. This design also allowed the board to be used to connect the existing bus signal assignments from the PLD11 board to the appropriate key and text signals on the SNL DES ASIC. Two versions of the package were designed and fabricated. Each has a different wiring schematic designed to fit into a different socket on the PLD11 board. SNL DES ASICs in the 503 pin package were demonstrated at the Super Computing 98 Conference in Orlando, November 1998.

The SNL DES ASIC has also been packaged in a 35 x 35 mm, 352-pin ball grid array (BGA) package. This is an open tool commercial package available from

Fig. 5. Picture of the 503-Pin FR4 Board Package

Abpac Inc. (of Phoenix , Arizona, U.S.A.). The package was chosen not only for its capability to dissipate over 5 watts and smaller size, but also its low cost. Abpac's automated manufacturing capability enabled a reduction of over 20 times in packaging costs. This package is being used in the design of a triple key, triple DES encryption module.

3.3 Applications

Although mainly used as an encryption engine for single DES or triple DES cryptosystems, the SNL DES ASIC has other uses such as a data randomizer. Some encryption algorithms need to hide or obscure relationships between bits or bytes of data prior to encryption. Using the SNL DES ASIC on the front end as a randomizer introduces no significant delay to the host cryptosystem. In a similar vein, this device can be used as a pseudo random number generator as part of a larger cryptosystem.

In a counter mode or filter generator cryptosystem (shown in figure 6), a linear recurring sequence (LRS) generator produces a sequence, which is fed to a non-linear function. The purpose of the non-linear function (in this case, an SNL DES ASIC) is to mask the linearity properties of the LRS. The output of the DES ASIC is then combined with the data, through an Exclusive-OR operation.

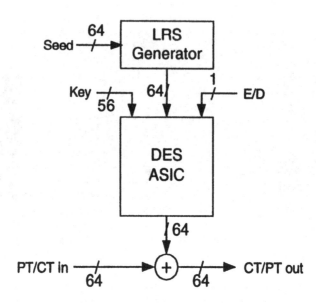

Fig. 6. Encryption/Decryption for Counter Mode

This device can be used to keep encryption/decryption keys synchronized with the data in cascaded triple DES implementations (described below). At the highest data rates, programmable logic devices (PLDs) may not be able to keep pace with the SNL DES ASIC for passing keys with the data. Consequently, SNL DES ASICs operating in bypass mode may be used to pass keys to subsequent encryption chips in step with the data.

Triple DES. Triple DES employs the Data Encryption Standard algorithm in a way sometimes referred to as encrypt-decrypt-encrypt (E-D-E) mode. E-D-E mode using two keys, was proposed by W. Tuchman and summarized by Schneier in [7]. The incoming plaintext is encrypted with the first key, decrypted with the second key, and then encrypted again with the first key. On the other end, the received ciphertext is decrypted with the first key, encrypted with the second key, and again decrypted with the first key to produce plaintext. If the two keys are set alike, it has the effect of single encryption with one key, thereby preserving backward compatibility. While advances have been made in cracking single DES cryptosystems [4], data protected by the SNL DES ASICs using two key, triple DES have a good degree of cryptographic robustness.

Two key, triple DES schemes (with 56 bit keys) can be cryptanalyzed using a *chosen plaintext* attack with about 2^{56} operations and 2^{56} words of memory [6]. (In terms of work, this is on par with two key, double DES, which is susceptible to a *known plaintext* attack with 2^{56} operations and 2^{56} words of memory [6].) Although in theory this is a weakness, Merkle and Hellman [6] state that in practice it is very difficult to mount a chosen plaintext attack against a DES cryptosystem. This makes two key, triple DES significantly stronger than two key, double DES, because an attack would now require 2^{112} operations (and no memory).

Triple DES can also be performed with three independent keys, using a separate key for each of the encryption and decryption operations. Triple DES with three independent keys gives a slightly higher level of protection, but is susceptible to a meet-in-the-middle attack requiring about 2^{112} operations and 2^{56} words of memory [7]. Again, with regards to compatibility, keys one and three could be set to the same value, to interoperate with Tuchman's two key, triple DES, or all three keys could be set alike to interoperate with single DES.

The SNL DES ASIC supports triple DES in several unique ways. For highest throughput speeds, multiple SNL DES ASICS can be cascaded to implement the encrypt-decrypt-encrypt mode. In situations where top performance is not needed, but board real estate, cost, or power consumption are constrained, a single SNL DES ASIC may perform triple DES in an iterative manner.

Because keys and control information march in lock step with the data, a string of three SNL DES ASICs can be cascaded. This would be accomplished by connecting the output data, key out, and control information output pins of one ASIC to the input data, key in, and control information input pins on the next ASIC. To perform E-D-E triple DES, an inverter must be placed on the path of the encrypt/decrypt signal between ASICs. This way, the middle ASIC will always perform the opposite operation (encrypt or decrypt) from the first and last ASICs in the string. PLDs, or SNL DES ASICs set to bypass mode, can be used to provide the proper (18 clock tick) delay, so that the keys for the second and third encryption/decryption operations will arrive in synchronization with the appropriate data. An example of this, using two keys is shown in figure 7.

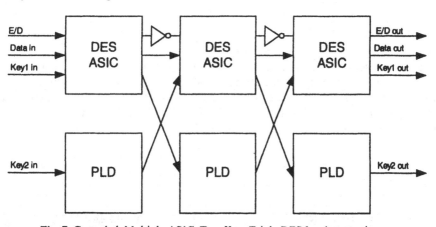

Fig. 7. Cascaded, Multiple ASIC, Two Key, Triple DES Implementation

By applying appropriate glue logic, the SNL DES ASIC can be used to perform E-D-E triple DES in an iterative manner by looping the data, key, and control information around the ASIC, processing the data three times. The glue logic will need to contain a two bit wide, 18 stage delay to count (in synchronization with the data, key, and control information) the number of times a given block of data has been processed. Logic will also be needed to invert the encrypt/decrypt bit between passes through the SNL DES ASIC. An example of this is shown in figure 8.

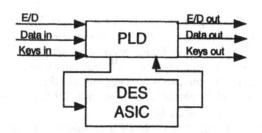

Fig. 8. Iterative, Single ASIC, Triple DES Implementation

4 Summary

For this project, the authors explored how a representative "heavyweight," unclassified encryption algorithm could be optimized and pipelined. This project has yielded a device that could be used in building encryption research prototypes. The project was successful, producing not only a research vehicle, but the fastest known ASIC implementation of DES.

The SNL DES ASIC can support two- or three-key triple DES using a multiple cascaded ASIC configuration at rates of 6.7 Gbps and beyond. It can also support very low power triple DES, iteratively, in a single ASIC configuration.

Acknowledgements

The work described in this paper was performed by Sandia National Laboratories (a multiprogram laboratory operated by Sandia Corporation, a Lockheed Martin Company) under contract number DE-AC04-94AL85000 to the United States Department of Energy. The authors would like to thank Hans Rodriques de Miranda, formerly of RE/SPEC, for his contribution to this work.

Bibliography

1. The ATM Forum Technical Committee, ATM Security Specification Version 1.0, Straw Ballot, STR-SECURITY-01.00, The ATM Forum, Mountain View, CA, December 1997.
2. Data Encryption Standard (FIPS PUB 46-2), Federal Information Processing Standards Publication 46-2, National Bureau of Standards, Washington, D.C., December 30, 1993.
3. DES Modes of Operation (FIPS PUB 81), Federal Information Processing Standards Publication 81, National Bureau of Standards, Washington, D.C., December 2, 1980.
4. http://www.eff.org/descracker, January 1999.
5. Menezes, Alfred J., et al., Handbook of Applied Cryptography, CRC Press, Boca Raton, FL, 1997.
6. Merkle, Ralph C., and Martin E. Hellman, "On the Security of Multiple Encryption," Communications of the ACM, Vol. 24, No. 7, p. 465-467, July 1981.
7. Schneier, Bruce, Applied Cryptography, 2nd edition, John Wiley & Sons, New York, 1996.

Hardware Design and Performance Estimation of the 128-bit Block Cipher CRYPTON

Eunjong Hong, Jai-Hoon Chung, and Chae Hoon Lim

Future Systems, Inc.
372-2 Yangjae-Dong, Seocho-Ku, Seoul, Korea 137-130
E-mail: {ejhong, jhoon, chlim}@future.co.kr

Abstract. CRYPTON is a 128-bit block encryption algorithm proposed as a candidate for the Advanced Encryption Standard (AES), and is expected to be especially efficient in hardware implementation. In this paper, hardware designs of CRYPTON, and their performance estimation results are presented. Straightforward hardware designs are improved by exploiting hardware-friendly features of CRYPTON. Hardware architectures are described in VHDL and simulated. Circuits are synthesized using 0.35 μm gate array library, and timing and gate counts are measured. Data encryption rate of 1.6 Gbit/s could be achieved with moderate area of 30,000 gates and up to 2.6 Gbit/s with less than 100,000 gates.

1. Introduction

The explosive growth in computer systems and their interconnections via networks has changed the way we live. From politics to business, our lives depend on the information stored and communicated using these systems. This in turn has led to a heightened awareness of the need of the information security. To enforce information security by protecting data and resources from disclosure, a secure and efficient encryption algorithm is needed. Since the widely used encryption algorithms, DES or Triple DES, are no more considered secure enough or efficient for future applications, a new block encryption algorithm with a strength equal to or better than that of Triple DES and significantly improved efficiency is needed.

Recently very high bandwidth networking technologies such as ATM and Gigabit Ethernet are rapidly deployed [1]. Network applications such as virtual private network [2] need high-speed executions of encryption algorithms to match high-speed networks. In order to utilize the high-speed networks at link speed, encryption speed as fast as 1 Giga bits per second is required. According to our experiment, the execution speed of Triple DES on Intel Pentium-II, 333MHz is only 19.6Mbps, and the highest performance of Triple DES from the commercially available encryption hardware is about 200 Mbps [3, 4].

CRYPTON [5, 6] is a 128-bit block encryption algorithm proposed as a candidate for the Advanced Encryption Standard (AES) [7]. In the evaluation performed by NIST, its software implementation on Pentium-Pro, 200MHz showed about 40Mbps, the best encryption and decryption speeds among the AES candidates' [8]. Hardware implementations of CRYPTON are expected to be more efficient than software implementations because it was designed from the beginning with hardware implementations in mind. The encryption and decryption use the identical circuitry, and there needs no large logic for S-boxes. Moreover it does not use addition or multiplication but only uses exclusive-OR operations, and the exclusive-OR operations can be executed in parallel. CRYPTON is considered as the most hardware-friendly AES candidate on several researches [9-11].

In this paper, hardware designs of CRYPTON Version 1.0 [6], which were optimized by exploiting the hardware-friendly features of CRYPTON, are presented. To maximize operation parallelism, key generation and data encryption operations are executed simultaneously. Some round loops can be unrolled for speedup without increasing the quantity of logic. S-boxes used for both key scheduling and data encryption are shared to minimize the area. Hardware architectures are described in VHDL and simulated using Synopsys Compiler and Simulator. Circuits are synthesized using 0.35 μm gate array library, and timing and gate counts are measured.

This paper is organized as follows. In Section 2, CRYPTON algorithm is briefly introduced. In Section 3, our design considerations of CRYPTON hardware are described. In Section 4 and 5, the detailed hardware designs of CRYPTON and estimation results are presented respectively. Concluding remarks are made in Section 6.

2. CRYPTON Algorithm

CRYPTON is a SPN (substitution-permutation network)-type cipher based on the structure of SQUARE [12]. CRYPTON represents each 128-bit data block 4 × 4 byte array and processes it using a sequence of round transformations. Each round transformation consists of four parallelizable steps: byte-wise substitutions, column-wise bit permutation, column-to-row transposition, and then key addition. The encryption process involves 12 repetitions of the same round transformation. The decryption process can be made identical to the encryption process with a different key schedule. The high level structure of CRYPTON is shown in Figure 1. For details of CRYPTON algorithm, please refer to [6].

3. Design Considerations of CRYPTON Hardware

3.1 Parallel Execution of Key Generation and Data Encryption

Some block ciphers pre-compute round keys and store them. Then the stored keys are used repeatedly while encrypting. This style needs extra cycles for key setup, and large storage if all 12 round keys of 128-bit length are to be stored. However, CRYPTON can generate keys simultaneously with encryption. The time it takes to generate a round key of CRYPTON is insignificant compared to the time it takes for a round transformation, and this makes it possible to generate round keys with the encryption proceeding. The parallel operation of hardware CRYPTON is illustrated in Figure 2.

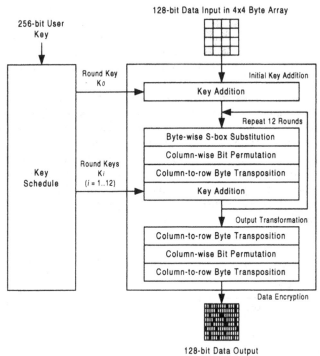

Fig. 1. The Structure of CRYPTON.

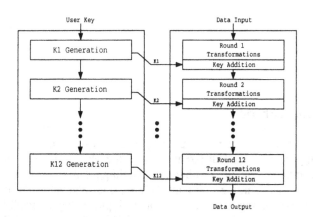

Fig. 2. Parallel Execution of Key Generation and Data Encryption.

3.2 Loop Unrolling

Because CRYPTON consists of 12 repetitions of round transformation, iteration is an inevitable choice for small area designs. Rather than building every round transformation separately, a data flow is made to pass the same hardware block repeatedly. We can build the whole encryption with only a small, basic building block by exploiting iteration. The block diagram for 12-cycle iteration is shown in Figure 3(a).

Although iteration results in small area designs, it accompanies additional path delay taken from multiplexer and register. To reduce the number of pass through multiplexer and register, we can unroll the loop so that one cycle contains double or many times of the round transformation logic. In Figure 3(b), two round transformations are performed in a cycle, but the number of components included in the design is one multiplexer less than that of Figure 3(a). We can find that the design in Figure 3(b) has better speed and area than the design in Figure 3(a). CRYPTON has a good tradeoff between speed and area when 2, 3, 4, 6, or 12 rounds are concatenated in a loop.

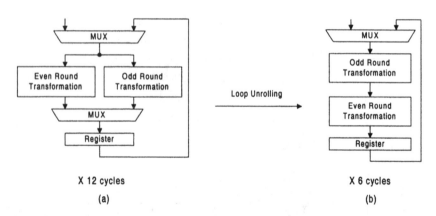

Fig. 3. Loop Unrolling.

3.3 Common Logic Sharing

CRYPTON uses only a few simple operations as its basic components. Because the components appear in several different parts of the algorithm, we can make those parts share a common logic block rather than building many separate, redundant logic blocks. Common logic sharing will contribute to area reduction only if the reduced amount of logic by sharing is larger than the multiplexer logic added. Otherwise, it will only result in increased control burden and longer path delay. CRYPTON has byte substitution operation both in key scheduling module and encryption module. Because byte substitution charges a significantly larger area than 128 2:1 multiplexers do, there is a benefit of adopting common logic sharing as shown in Figure 4.

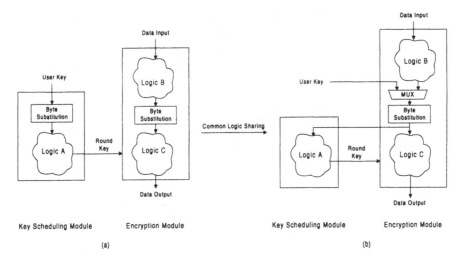

Fig. 4. Common Logic Sharing

4. Hardware Design of CRYPTON

In this section, we propose two hardware designs of CRYPTON.

4.1 Two-Round Model

Two-round model performs two successive transformations for even and odd rounds within a clock cycle. Two-round model is efficient in area-speed tradeoff as we saw in Section 3.2, and it has smaller area than designs with 3, 4, 6, or 12 rounds concatenated.

Two-round model consists of three modules: encryption module, key scheduling module, and control module as shown in Figure 5. The following scheme describes how two-round model works.

1. Load 128-bit input data at "DATA_IN" port, and 256-bit user key at "USER_KEY" port.
2. If decryption is to be performed, apply logic high on "DECR" port until the encryption is finished.
3. Start encryption by applying logic high pulse for one clock cycle at the "START" port.
4. Check the output port "DONE" to see if the encryption is completed.
5. If "DONE" port outputs logic high, read the encryption result through output port "DATA_OUT".
6. "Cycle0" is a signal indicating the key expansion cycle, and "Cycle1" tells the first cycle among 7 is going on. Both signals are used as control inputs for multiplexers.

The 12-round encryption process needs 13 round keys, from the round 0 key to the round 12 key. Generating 13 round keys will take 7 clock cycles because two-round model computes two keys within a clock cycle. Thus, the result is available 7 clock cycles later after logic high on "START" port is latched at the rising edge of clock.

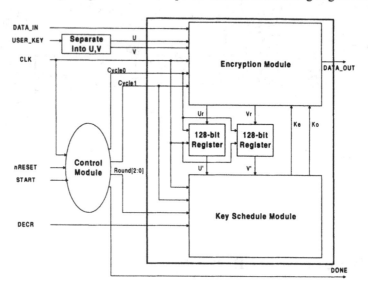

Fig. 5. Top View of Two-Round Model

The encryption module of two-round model is shown in Figure 6(a). In encryption module, two blocks for even round transformation and odd round transformation are cascaded serially, but the sequence of blocks in Figure 6(a) is somewhat different from that of Figure 1. Because round 0 has uniqueness that only key addition is performed in it, it will require extra 128 two-input exclusive-OR gates if we did not take the sequence shown in Figure 6(a).

The most unwieldy part in CRYPTON might be the S-box byte substitution. The byte substitution logic has 16 256-entry tables. As Figure 6(a) shows, encryption module must have two byte substitution blocks because it is distinct for odd rounds and even rounds. In addition, key scheduling module has two byte substitution blocks for user key expansion, which is used only once when a new user key is set. Seeing that most of the other operations takes comparatively small area, incorporating separate byte substitutions for key expansion looks very mismatched for its rare usage. This lack of balance can be corrected by sharing byte substitution blocks between the expansion block of key scheduling module and round transformation block of encryption module. This scheme effectively reduces the total area but needs one more clock cycle only for key expansion when new expanded keys are to be computed. Figure 7(b) shows the new block diagram for encryption module with sharing of byte substitution blocks. In Figure 7(b), outputs "Ur" and "Vr" are fed to the inputs of two 128-bit registers and stored in them. Those stored values in the registers are used repeatedly for generation of round keys unless a new user key is set. The key scheduling area also has change in expansion block. Key scheduling area takes U' and $V'[6]$ – the outputs of expanded key registers – as inputs, and performs operations later than byte substitution blocks.

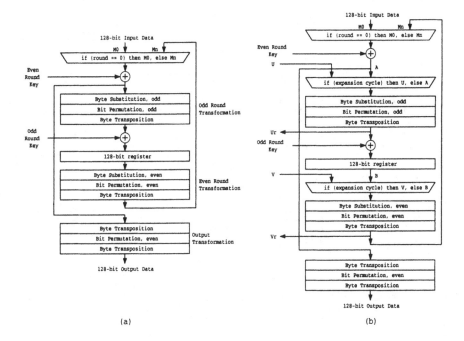

(a)   (b)

Fig. 6. Encryption Model of Two-Round Model

The block diagram for key scheduling module is depicted in Figure 7. Key scheduling module conveys two successive round keys to the encryption module at every clock cycle. One of our design principles is parallel execution mentioned in Section 3.1, and we do not adopt any round-key pre-computation style for speed enhancement.

There is design concern in the final stage of round-key computation. A round key is exclusive-OR sum of a round-constant, masking constants, and expanded keys. There are only 4 masking constants that can be easily built into combination logic, and expanded keys are unknown before the round, thus we have no design choices about them. But there are as many as 13 round-constants of 32-bit length, and one should decide if he will build wired logic out of pre-computed round-constants, or make the design compute the constants from the specified equation at each clock cycle. Computation of constants in fully parallelized design like Figure 8 needs at least 4 32-bit adders and two 32-bit registers. On the other hand using wired logic for the 13 constants introduces a little longer propagation delay in key computation. Because two-round model aims at small area, implementation with wired logic is chosen.

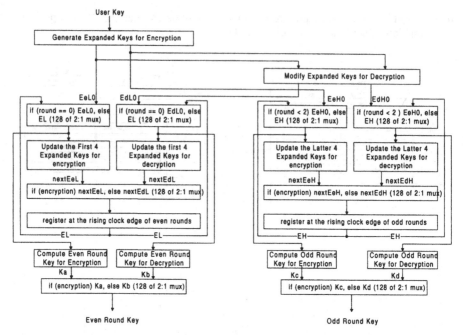

Fig. 7. Key Scheduling Module of Two-Round Model

4.2 Full-Round Model

To make a fast design, pipelining or loop unrolling can be generally applied. In pipelining, the algorithm is partitioned into several stages, and this enables several data blocks to be encrypted simultaneously. This is possible in ECB mode because each result of block encryption is independent from others. However, modes except ECB require the previous result of encryption to be available to complete the present encryption, which makes pipelining useless. Since most of the recent application of block ciphers use chaining or feedback mode, speed enhancement through pipelining is not considered here. Instead, we make full-round model with 12 rounds fully unrolled. This full-round model computes 12 rounds of transformation without looping, and it is the fastest but the largest design among those exploiting loop-unrolling. The loop unrolling of 4 or 6 will have just intermediate values of area and speed between those of two-round model and full-round model. Block diagrams for full-round model are straightforward and shown in Figure 8, Figure 9, and Figure 10

In key scheduling module shown in Figure 10, round-key output Kn for the n-th round will be Ken if encryption is performed, and Kdn if decryption is performed. To achieve high speed, common logic sharing in expansion block is not adopted here. Full-round model tells us area cost of the nearly pure algorithm because it does not need any registers or control unit. However, the 12 multiplexers in key scheduling area could not be removed.

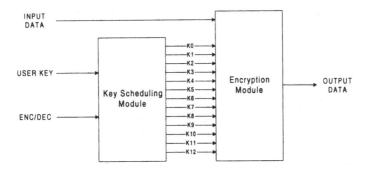

Fig. 8. Top View of Full-Round Model

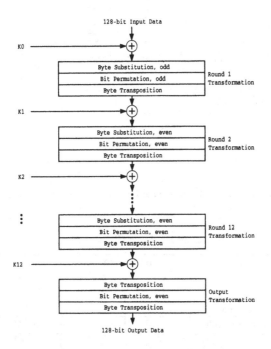

Fig. 9. Encryption Module of Full-Round Model

5. Estimation Results

Our estimations are all based on Synopsys DesignCompiler and Hyundai 0.35 µm gate array library. Table 1 shows area and speed of two-round model and full-round model each with two speed-area tradeoffs. In this paper, all estimations are for typical cases. "Gate" means a 2-input NAND gate, equivalent to 4 transistors.

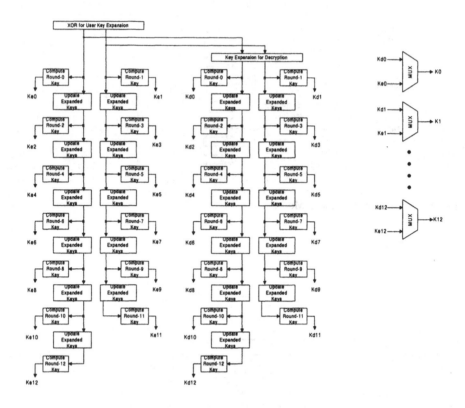

Fig. 10. Key Scheduling Module of Full-Round Model

Table 1. Speed and Area Estimations of Hardware CRYPTON.

Estimated items		Two-round model		Full-round model	
		Area critical	Speed critical	Area critical	Speed critical
Gate count (Cell Area)	total gates	18,322	28,179	46,259	93,929
	encryption module	8,267	17,958	33,598	74,857
	key scheduling [1]module	7,930	8,078	12,661	19,072
Minimum clock period (T_{clk})		18.97 ns	10.23 ns	74.03 ns	44.30 ns
Key setup time		on the fly	on the fly	10.13 ns	7.91 ns
Time to switch keys		18.97 ns	10.23 ns	10.13 ns	6.13 ns
Time to encrypt one block		132.79 ns ($7 \times T_{clk}$)	71.61 ns ($7 \times T_{clk}$)	74.03 ns	44.30 ns
Throughput		898Mbps[1]	1.66Gbps[2]	1.61Gbps	2.69Gbps

---

[1] 1 Mbps = 1,024 × 1,024 bps = 1,048,576 bps

[2] 1 Gbps = 1,024 × 1,024 × 1,024 bps = 1,073,741,824 bps

The two results for two-round model have identical logic design, but are different only in optimization strategies. The one indicated as "Area Critical" was optimized to reduce as much area as possible. On the other hand, the result of "Speed Critical" was obtained by minimizing the clock period. As shown in "Time to switch keys", one clock is spent on expanding the user key when the user key is switched to a new value. Since the design uses one clock source for its whole area, the time to switch keys will be at least "minimum clock period", but the actual propagation delay in key expansion is less than the minimum clock period. Round keys are generated on the fly, and if the user keys remain same, 7 clock cycles are needed for both encryption and decryption.

The same optimization strategies were applied to full-round model. However, in speed critical optimization, the path from data input to the final output was optimized rather than minimizing the clock period as in two-round model.

Although results of full-round model were entered into the same format of table with two-round model, the numbers should not be compared directly because the two designs have different architectures. The following three explanations qualify the meaning of each item for full-round model.

- *Time to switch keys*: time from the insertion of a new user key to the generation of the round 0 key (this is because time taken to compute the next round key is much shorter than time to compute one round transformation. By the time γ operation of 1st round has been performed, all round keys from 1st round to 12th round are available).
- *Key setup time*: time needed to compute the whole 12 round keys.
- *Time to encrypt one block*: The longest path delay from data input to the final data output (it is the time taken to encrypt one block when all round keys are set up and ready to be used).

In two-round model, "Total Area" is larger than the sum of "encryption module" and "key scheduling module" because area of buffer for expanded keys and control unit is missing. Full-round model does not have any extra logic except encryption module and key scheduling module, and thus the sum is exactly matched.

Two main issues in logic design are speed and area. Because speed and area are objects of tradeoff in most cases, we can get the best results on one criterion by optimizing for it while ignoring the other. On the other hand, the result is the worst one for the ignored criterion. We can find the approximate upper and lower bounds of speed and area of CRYPTON by once optimizing for area and then for speed.

The main trade-off between space and time takes place in optimization of S-box. We could obtain as high speed as 2.6Gbps by growing the area of S-box, but the total number of gates was over 90,000. Resorting to speed-area tradeoff, two-round model faster than full-round model was possible contrary to our initial scheme of using loop unrolling to make a fast design. In our estimation, two-round model was found more moderate and practical both in area and in speed than full-round model. Although optimization in the synthesis tool was very useful to achieve a goal in speed or area, a better result was possible by modifying the design itself.

6. Conclusions

In this paper, we designed and proposed two hardware architectures of CRYPTON exploiting the inherent hardware-friendly features of CRYPTON. The architectures were described in VHDL and circuits were synthesized using 0.35 µm gate array with several speed-area tradeoffs. 0.9 Gbps with the smallest area of 18,000 gates and the fastest speed of 2.6 Gbps with less than 100,000 gates could be achieved. The speed of 2.6 Gbps is faster than the commercially available fastest Triple-DES chip with an order of magnitude. This is enough speed to support the Gigabit networks. Since CRYPTON has good scalability in gate count, a designer can select a proper speed-area tradeoff from the large set choices.

References

1. L. Geppert and W. Sweet, "Technology 1999 Analysis & Forecast – Communications," *IEEE Spectrum*, January 1999, pp.29-34.
2. D. Kosiur, Building and Managing Virtual Private Networks, Wiley, 1998.
3. Analog Devices, ADSP-2141L SafeNet DSP Preliminary Technical Data Sheet, REV.PrB, June 1999.
4. VLSI Technology, VMS115 IPSec Coprocessor Data Sheet, Rev2.0, January 1999.
5. C.H. Lim, "CRYPTON: A New 128-bit Block Cipher," *Proceedings of the First Advanced Encryption Standard Candidate Conference*, Ventura, California, National Institute of Standards and Technology (NIST), August 1998.
6. C.H. Lim, "A Revised Version of CRYPTON – CRYPTON Version 1.0," *Proceedings of the 1999 Fast Software Encryption Workshop*, March 1999.
7. NIST, Request for Candidate Algorithm Nominations for the Advanced Encryption Standard (AES), Federal Register, Vol.62, No.177, September 12, 1997.
8. M. Smid and E. Roback, "Developing the Advanced Encryption Standard," *Proceedings of the 1999 RSA Conference*, January 1999.
9. E. Biham, "A Note on Comparing the AES Candidates," *Proceedings of the Second Advanced Encryption Standard Candidate Conference*, Rome, Italy, National Institute of Standards and Technology (NIST), March 1999.
10. C.S.K. Clapp, "Instruction-level Parallelism in AES Candidates," *Proceedings of the Second Advanced Encryption Standard Candidate Conference*, Rome, Italy, National Institute of Standards and Technology (NIST), March 1999.
11. B. Schneier, et. al., "Performance Comparison of the AES Submissions," *Proceedings of the Second Advanced Encryption Standard Candidate Conference*, Rome, Italy, National Institute of Standards and Technology (NIST), March 1999.
12. J. Daemen, L. Knudsen, and V. Rijmen, "The Block Cipher Square," *Proceedings of the 1997 Fast Software Encryption Workshop*, Lecture Notes in Computer Science (LNCS) 1267, Springer-Verlag, 1997, pp.149-171.

Fast Implementation of Public-Key Cryptography on a DSP TMS320C6201

Kouichi ITOH[1], Masahiko TAKENAKA[1], Naoya TORII[1], Syouji TEMMA[2], and Yasushi KURIHARA[2]

[1] FUJITSU LABORATORIES LTD. 64 Nishiwaki, Ohkubo-cho, Akashi 674-8555 Japan.
{kito, takenaka, torii}@flab.fujitsu.co.jp
[2] FUJITSU LTD. 4-1-1 Kami-kodanaka, Nakahara-ku, Kawasaki 211-8588 Japan.
{temma, kurihara}@cl.mfd.cs.fujitsu.co.jp

Abstract. We propose new fast implementation method of public-key cryptography suitable for DSP. We improved modular multiplication and elliptic doubling to increase speed. For modular multiplication, we devised a new implementation method of Montgomery multiplication, which is suitable for pipeline processing. For elliptic doubling, we devised an improved computation for the number of multiplications and additions.

We implemented RSA, DSA and ECDSA on the latest DSP (TMS320C6201, Texas Instruments), and achieved a performance of 11.7 msec for 1024-bit RSA signing, 14.5 msec for 1024-bit DSA verification and 3.97 msec for 160-bit ECDSA verification.

1 Introduction

Public-key cryptography is an important encryption technique. It can be applied to many practical uses such as electronic commerce systems and WWW systems for enabling digital signatures and key agreement. The server systems for them are required to process a vast number of public key operations.

Additionally, for communicating with various kinds of clients, the server systems are required to provide various public-key cryptography functions, such as RSA [15] , Diffie-Hellman key agreement [5], DSA [16] and elliptic curve cryptography (ECC) [9][12]. These functions are under standardization in IEEE P1363 [17].

In this paper, we describe a fast implementation method using DSP as a cryptographic engine for server systems. In public-key cryptography, modular multiplications are the most time-consuming operations. A DSP can compute these operations efficiently with a fast hardware multiplier. Furthermore, a DSP can be used as the hardware engine for various algorithms since it is programmable.

In the past, fast public key cryptographic implementations on DSPs have been reported [1][2][6]. They concentrated on the implementation of RSA using the latest DSP at the time. We implemented RSA, DSA and ECDSA over prime fields based on the IEEE P1363 draft, and propose new implementation methods suitable for DSP. Our methods concern modular multiplication and elliptic doubling.

For modular multiplication, we devised a fast implementation method for Montgomery multiplication [14]. Our method is suitable for pipeline processing.

For elliptic doubling, we devised a new method which reduces the number of multiplications and additions in comparison with that specified in the IEEE P1363 draft. In general, the running time of addition is considered negligible compared with that of multiplication. But in fact, the running time of addition is not negligible on a processor such as a DSP, which has a fast hardware multiplier.

There are some reports concerning the fast implementation of ECC [3][4][13]. They used the special elliptic curve domain parameters (EC domain parameters) for speeding up. On the other hand, our implementation can use any EC domain parameters for the server systems. The server systems require high performance and communicating with client systems that use various types of EC domain parameters.

We implemented public-key cryptography functions with our method on the latest DSP TMS320C6201 (Texas Instruments). This DSP can operate eight function units in parallel and has a performance of 1600 MIPS at 200 MHz. The performance achieved in our implementation was 11.7 msec for 1024-bit RSA signing, 14.5 msec for 1024-bit DSA verification and 3.97 msec for 160-bit ECDSA verification.

We describe our improvement method for Montgomery multiplication in section 2, our elliptic doubling method in section 3 and the performance in section 4.

2 Fast implementation method of Montgomery multiplication

2.1 Montgomery multiplication

Basic algorithm. Set $N > 1$. Select a radix R co-prime to N such that $R > N$ and such that computations modulo R are inexpensive to process. Let N' be integers satisfying $0 < N' < R$ and $N' = -N^{-1} \pmod{R}$. For all integers A and B satisfying $0 \le AB < RN$, we can compute $REDC(A, B) = ABR^{-1} \pmod{N}$ with Algorithm 1.

Algorithm 1. *Montgomery multiplication algorithm REDC.*
input : A, B, R, N.
output : $Y = ABR^{-1} \pmod{N}$.
101 $N' := -N^{-1} \pmod{R}$
102 $T := AB$
103 $M := (T \pmod{R})N' \pmod{R}$
104 $T := T + MN$
105 $T := T/R$
106 *if $T \ge N$ then return $T - N$ else return T*

If R is a power of 2, line 105 can be computed fast with shift operations.

Modular multiplication with Montgomery method. Since $REDC(A, B) = ABR^{-1}$ (mod N), it can not compute modular multiplication directly. But on reviewing $REDC(AR, BR) = ABR$ (mod N), it can be seen that REDC can compute modular multiplication by converting A (mod N) to AR (mod N). After this conversion, a series of modular multiplications can be computed fast with REDC. For example, we show an m-ary exponentiation [7] with REDC in Algorithm 2, where e is a k-bit exponent and e_i is an m-bit integer which satisfies $e = \sum (2^m)^i e_i$.

Algorithm 2. *m-ary exponentiation method with REDC.*
input : A, e, N, R
output : $Y = A^e$ (mod N)

201 $A' := A \times R$ (mod N)
202 $T[0] := 1 \times R$ (mod N)
203 *for $i := 1$ to $2^m - 1$*
204    $T[i] = REDC(T[i-1], A')$
205 *next i*
206 $Y := 1 \times R$ (mod N)
207 *for $i := \lceil k/m \rceil - 1$ down to 0*
208   *for $j := 1$ to m*
209     $Y := REDC(Y, Y)$
210   *next j*
211    $Y := REDC(Y, T[e_i])$
212 *next i*
213 $Y := Y \times R^{-1}$ (mod N)
214 *return Y*

REDC routine with single-precision. To implement REDC on general processors, multi-precision computation must be divided into iterations of single-precision computation. In [10], many types of REDC routines are constructed with single-precision computation. Algorithm 3 shows a Finely Integrated Operand Scanning (FIOS) type of REDC routine in [10].

We will use the following notations. Capital variables such as A or B, mean a multi-precision integer. Small letter variables such as a_i, b_j or $tmp1$ mean a single-precision integer of w-bit length.

A multi-precision integer, for example A, is expressed as the series of single-precision variables $(a_{g-1}, a_{g-2}, \ldots, a_0)$. The expression such as (a, b) means the concatenation of single-precision variables a and b. We also use the expression such as (A, b), which means the concatenation of a multi-precision variable A and a single-precision variable b.

In Algorithm 3, the block-shift is executed by reading from y_i and writing to y_{i-1}. Note that the w-bit variables $tmp3$ and c_1 have 1-bit value.

Algorithm 3. *REDC routine with single-precision computation. (FIOS [10].)*
input: $A = (a_{g-1}, a_{g-2}, \ldots, a_0)$, $B = (b_{g-1}, b_{g-2}, \ldots, b_0)$, $N' = (n'_{g-1}, n'_{g-2}, \ldots n'_0)$, $R = (2^w)^g$.
output: $Y = (y_g, y_{g-1}, \ldots, y_0) = ABR^{-1}$ (mod N).

301 $Y := 0$
302 *for* $j := 0$ *to* $g - 1$
303 $(tmp2, tmp1) := y_0 + a_0 \times b_j$
304 $m := tmp1 \times n_0' \pmod{2^w}$
305 $(tmp4, tmp1) := tmp1 + m \times n_0$
306 $(c_1, c_0) := tmp2 + tmp4$
307 *for* $i := 1$ *to* $g - 1$
308 $(tmp3, tmp2, tmp1) := y_i + (c_1, c_0) + a_i \times b_j$ *single-precision multiplication*
309 $(tmp4, y_{i-1}) := tmp1 + m \times n_i$ *single-precision reduction*
310 $(c_1, c_0) := tmp4 + (tmp3, tmp2)$ *carry computation*
311 *next* i
312 $(c_1, c_0) := (c_1, c_0) + y_g$
313 $y_{g-1} := c_0$
314 $y_g := c_1$
315 *next* j
316 *if* $Y \geq N$ *then* $Y := Y - N$
317 *return* Y

2.2   Proposed method

To speed up Algorithm 3 on a DSP, let us consider improving the core loop in lines 308-310 suitable for pipelining. For the improvement, we considered the following problems:

(1) *Single-precision multiplication* in line 308 cannot execute until *single-precision reduction* in line 309 and *carry computation* in line 310 finish.
(2) The contents of the computation are different among *single-precision multiplication*, *single-precision reduction* and *carry computation*.
(3) The result of *carry computation*, (c_1, c_0) in line 310, has $(w + 1)$-bit length value so that it must be processed as a multi-precision variable.

We reviewed the computation to solve these problems. Figure 1 shows the construction of the core loop. On reviewing the carry processing in Fig.1, carry of the *single-precision multiplication* and carry of the *single-precision reduction* are added to $C = (c_1, c_0)$, and C is input to the carry of *single-precision multiplication* in the next loop. To review this processing, we combine the computation in the core loop as follows:

$$(C, y_{i-1}) := y_i + C + a_i \times b_j + m \times n_i$$

From this equation, we can divide the carry C into the carry c_1 for the $a_i \times b_j$ and the carry c_2 for the $m \times n_i$ as follows:

$(c_1, tmp1) := y_i + c_1 + a_i \times b_j$ *single-precision multiplication*
$(c_2, y_{i-1}) := tmp1 + c_2 + m \times n_i$ *single-precision reduction*

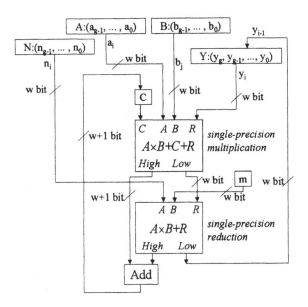

Fig. 1. Construction of core loop in Algorithm 3.

From these equations, we can see that problems (1), (2) and (3) are solved as follows:

Problem (1) is solved because both carry c_1 and c_2 feed back to themselves, which enables *single-precision multiplication* to start computing without waiting until *single-precision reduction* finishes. Problem (2) is solved because the computation between *single-precision multiplication* and *single-precision reduction* is the same. Problem (3) is solved because the right term of these equations never exceeds $2^{2w} - 1$ even if all single-precision variables in the right terms are $2^w - 1$, so that the lengths of c_1 and c_2 do not exceed w-bit.

Algorithm 4 shows an improved routine of Algorithm 3. Figure 2 shows the construction of the core loop in Algorithm 4.

Algorithm 4. *Proposed Montgomery multiplication algorithm.*
input: $A = (a_{g-1}, a_{g-2}, \ldots, a_0), B = (b_{g-1}, b_{g-2}, \ldots, b_0),$
$N' = (n'_{g-1}, n'_{g-2}, \ldots, n'_0), R = (2^w)^g.$
output: $Y = (y_g, y_{g-1}, \ldots, y_0) = ABR^{-1} \pmod{N}.$

401 $Y := 0$
402 *for* $j := 0$ *to* $g - 1$
403 $(c_1, tmp1) := y_0 + a_i \times b_j$
404 $m := tmp1 \times n'_0 \pmod{2^w}$
405 $(c_2, tmp1) := tmp1 + m \times n_0$
406 *for* $i := 1$ *to* $g - 1$
407 $(c_1, tmp1) := y_i + c_1 + a_i \times b_j$ *single-precision multiplication*
408 $(c_2, y_{i-1}) := tmp1 + c_2 + m \times n_i$ *single-precision reduction*

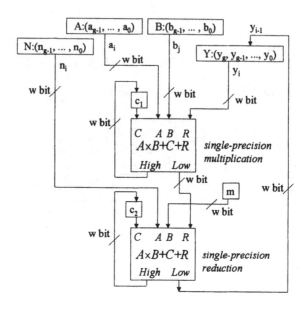

Fig. 2. Construction of the core loop in Algorithm 4.

409 next i
410 $(c_2, c_1) := c_1 + c_2 + y_g$
411 $y_{g-1} := c_1$
412 $y_g := c_2$
413 next j
414 if $Y \geq N$ then $Y := Y - N$
415 return Y

3 Fast elliptic doubling

We used a Weierstrass equation, $y^2 \equiv x^3 + ax + b \pmod{p}$ for the elliptic curve over prime fields where $4a^3 + 27b^2 \not\equiv 0 \pmod{p}$, and projective coordinate (X, Y, Z) which satisfies $(x, y) = (X/Z^2, Y/Z^3)$.

For exponentiation, such as m-ary [7] or window method [7], m elliptic doublings and 1 elliptic addition are processed alternatively. Remarking on this point, the m-*repeated* elliptic doublings method is proposed in [8] which is concerned with the computation on affine coordinates over binary fields. Compared to m times elliptic doublings, this method reduces the number of inverses by computing $2^m P$ for $P = (x, y)$ directly without computing intermediate points $2^i P (1 \leq i \leq m - 1)$.

We also remark this m-repeated elliptic doublings method, but take another approach to decrease the number of computation in terms of projective coordinates over prime fields. Our method is based on the m times elliptic doublings specified in the IEEE P1363 draft [17] and also reduces the number of additions and multiplications.

3.1 Reducing the number of multiplications

In this section, we describe our m-repeated elliptic doublings method which requires smaller multiplications than the m times elliptic doublings specified in the IEEE P1363 draft. In our method, the temporary value used in the t-th elliptic doubling is reused in the $(t+1)$-th elliptic doubling, and this eliminates 2 multiplications. Therefore, our method requires 10 multiplications in the first elliptic doubling, but requires only 8 multiplications from the second doubling to the m-th. Let $(X_m, Y_m, Z_m) = 2^m(X_0, Y_0, Z_0)$, Algorithm 5 shows m times elliptic doublings specified in the IEEE P1363 draft [17].

Algorithm 5. *m times elliptic doublings specified in the IEEE P1363 draft.*
input: Elliptic curve point (X_0, Y_0, Z_0), m and EC domain parameter a.
output: Elliptic curve point $(X_m, Y_m, Z_m) = 2^m(X_0, Y_0, Z_0)$.

501 for $i := 0$ to $m - 1$
502 $\quad W_i := aZ_i^4$
503 $\quad M_i := 3X_i^2 + aZ_i^4$
504 $\quad S_i := 4X_iY_i^2$
505 $\quad T_i := 8Y_i^4$
506 $\quad X_{i+1} := M_i^2 - 2S_i$
507 $\quad Y_{i+1} := M_i(S_i - X_{i+1}) - T_i$
508 $\quad Z_{i+1} := 2Y_iZ_i$
509 next i

If we consider $W_i = aZ_i^4$ and $Z_{i+1} = 2Y_iZ_i$ in line 502, 508, we notice that W_i can be computed from $W_i = 2T_{i-1}W_{i-1}$, which eliminates 2 multiplications. We show the improved routine of Algorithm 5 in Algorithm 6.

Algorithm 6. *Improved routine of Algorithm 5.*
input: Elliptic curve point (X_0, Y_0, Z_0), m and EC domain parameter a.
output: Elliptic curve point $(X_m, Y_m, Z_m) = 2^m(X_0, Y_0, Z_0)$.

601 $W_0 := aZ_0^4$
602 $M_0 := 3X_0^2 + W_0$
603 $S_0 := 4X_0Y_0^2$
604 $T_0 := 8Y_0^4$
605 $X_1 := M_0^2 - 2S_0$
606 $Y_1 := M_0(S_0 - X_1) - T_0$
607 $Z_1 := 2Y_0Z_0$
608 for $i := 1$ to $m - 1$
609 $\quad W_i := 2T_{i-1}W_{i-1}$
610 $\quad M_i := 3X_i^2 + W_i$
611 $\quad S_i := 4X_iY_i^2$

612 $T_i := 8Y_i^4$
613 $X_{i+1} := M_i^2 - 2S_i$
614 $Y_{i+1} := M_i(S_i - X_{i+1}) - T_i$
615 $Z_{i+1} := 2Y_i Z_i$
616 *next i*

3.2 Reducing the number of additions

Generally, an addition is regarded as much faster than a multiplication, and its running time is not considered. But on a DSP, multiplication can be computed efficiently with a fast hardware multiplier, and the running time of addition is not negligible. Table 1 shows a comparison of the running time of a modular multiplication and a modular addition based on our implementation on the DSP.

Table 1. Comparison of the running time of a modular multiplication and a modular addition @ 200 MHz.

	160-bit	192-bit	239-bit
Multiplication	1.36 μsec	1.76 μsec	2.68 μsec
Addition	0.250 μsec	0.254 μsec	0.291 μsec

In projective elliptic doubling, some computations such as modular multiplication by $2, 3, 4$, and 8 can be implemented by the combination of modular addition(s) and subtraction(s). Appending modular multiplication by $1/2$ to these computations, we define them "addition" in this paper. We estimate the computation amount of "addition" as follows:

- Modular addition and subtraction are "1 addition".
- Modular multiplication by 2 and 1/2 are "1 addition".
- Modular multiplication by 3 and 4 are "2 additions".
- Modular multiplication by 8 is "3 additions".

Now we consider reducing the number of additions in Algorithm 6 with this estimate. For example, computing $4Y^2$ as $(2Y)^2$ eliminates 1 addition compared with computing it as $4 \times (Y^2)$. Thus, additions in Algorithm 6 are reduced with $2Y$-based computation. With this technique, we can reduce the number of additions in Algorithm 6 by the following techniques:

(A) At the beginning, compute $Y_0' = 2Y_0$ as a base value, and compute $Y_i' (= 2Y_i)$ without computing Y_i for $i < m$.
(B) By reason of (A), compute $T_i = 16Y_i^4$ instead of $8Y_i^4$.
(C) Compute $S_i = 4X_i Y_i^2, Z_i = 2Z_{i-1}Y_{i-1}$ and $T_i = 16Y_i^4$ based on $Y_i' = 2Y_i$, viz. compute $S_i = X_i(Y_i')^2, Z_i = Z_{i-1}(Y_{i-1}')$ and $T = (Y_i')^4$ respectively.
(D) Finally, compute $Y_m = Y_m'/2$.

We show the improved routine of Algorithm 6 in Algorithm 7.

Algorithm 7. *Proposed m-repeated elliptic doublings routine.*
input: Elliptic curve point (X_0, Y_0, Z_0), m and EC domain parameter a.
output: Elliptic curve point $(X_m, Y_m, Z_m) = 2^m(X_0, Y_0, Z_0)$.

$701\ Y_0' := 2Y_0$

$702\ W_0 := aZ_0^4$

$703\ M_0 := 3X_0^2 + W_0$

$704\ S_0 := X_0(Y_0')^2$

$705\ T_0 := (Y_0')^4$

$706\ X_1 := M_0^2 - 2S_0$

$707\ Y_1' := 2M_0(S_0 - X_1) - T_0$

$708\ Z_1 := Y_0'Z_0$

$709\ for\ i := 1\ to\ m - 1$

$710\quad W_i := T_{i-1}W_{i-1}$

$711\quad M_i := 3X_i^2 + W_i$

$712\quad S_i := X_i(Y_i')^2$

$713\quad T_i := (Y_i')^4$

$714\quad X_{i+1} := M_i^2 - 2S_i$

$715\quad Y_{i+1}' := 2M_i(S_i - X_i) - T_i$

$716\quad Z_{i+1} := (Y_i')Z_i$

$717\ next\ i$

$718\ Y_m := Y_m'/2$

Table 2 shows the number of multiplications and additions required for the above algorithms. Our method eliminates $2m - 2$ multiplications and $5m - 2$ additions compared with the m times elliptic doublings specified in the IEEE P1363 draft.

Table 2. Number of multiplications and additions.

m-repeated elliptic doublings	Multiplication	Addition
Algorithm 5 (IEEE P1363 draft)	$10m$	$13m$
Algorithm 6	$8m + 2$	$14m - 1$
Algorithm 7 (Proposed)	$8m + 2$	$8m + 2$

4 Implementation

4.1 DSP and development tools

For the implementation, we used the DSP TMS320C6201 [18] (Texas Instruments). The DSP consists of eight parallel-operation functional units including two 16-bit multiplication units, and has a performance of 1600 MIPS at 200 MHz. The instruction processing system is of the VLIW/pipeline type and can execute conditional operations. And the maximum instruction code size is 64 Kbytes.

As the development tools, an assembler and C compiler are provided. We implemented arithmetic routines such as modular multiplication, addition, and subtraction in assembly language. Their performance greatly affects the total performance, because they are performed frequently. Other routines were written in C for easy implementation.

4.2 Implementation of RSA and DSA

We used the following methods:

- Modular multiplication with the Montgomery multiplication method [14] described in section 2.
- Modular exponentiation with m-ary method [7] for $m = 4$.

4.3 Implementation of ECC

We used following methods:

- Modular multiplication with the Montgomery multiplication method [14] described in section 2.
- Fast elliptic doubling with the method described in section 3, combined with the technique for increasing speed in case EC domain parameter $a = 0$.
- Elliptic addition based on IEEE P1363 draft [17].
- The base point exponentiation with fixed-base comb method [11], specified using two 5-bit precomputed tables.
- Random point exponentiation in combination with sliding-window exponentiation [11] with a 4-bit precomputed table and signed-binary [7] of the exponent.

4.4 Code size

We implemented RSA, DSA and ECDSA based on above method, and the total instruction code size was 41.1 Kbytes. Since TMS320C6201 allows a maximum instruction code size of 64 Kbytes, this implementation can deal with RSA, DSA and ECDSA without reloading.

4.5 Performance of RSA, DSA and ECC

Table 3 shows the performance of the RSA and DSA implementation. Table 4 shows the performance of the ECC implementation including the exponentiation on a random point. We measured the 100 times average clocks and figured the running time at 200 MHz.

In Table 3, we used $e = 2^{16} + 1$ for the RSA verification key, and Chinese remainder theorem for RSA signing.

In Table 4, the exponent of a random point has a same length as that of EC domain parameter p. The ECDSA scheme is based on the IEEE P1363 draft. Table 4 also shows the bit length of the order of the base point which affects the performance of ECDSA.

Table 3. Performance of RSA and DSA @ 200 MHz.

	RSA		DSA	
	1024bit	2048 bit	512 bit	1024 bit
Sign	11.7 msec	84.6 msec	2.62 msec	7.44 msec
Verify	1.2 msec	4.5 msec	4.82 msec	14.5 msec

Table 4. Performance of ECC @ 200 MHz.

	EC domain parameter p	160-bit	192-bit	239-bit
$a \neq 0$	Order of the base point	151-bit	192-bit	239-bit
	Exponentiation on a random point	3.09 msec	4.64 msec	8.47 msec
	ECDSA sign	1.13 msec	1.67 msec	2.85 msec
	ECDSA verify	3.97 msec	6.28 msec	11.2 msec
$a = 0$	Order of the base point	160-bit	185-bit	232-bit
	Exponentiation on a random point	2.88 msec	4.15 msec	7.60 msec
	ECDSA sign	1.09 msec	1.50 msec	2.66 msec
	ECDSA verify	3.78 msec	5.50 msec	9.78 msec

5 Conclusion

We proposed fast implementation methods of Montgomery multiplication and m-repeated elliptic doublings, which are efficient for any EC domain parameters and suitable for the server systems. Our methods are efficient not only for DSP, but also for any other processors.

Construction of our Montgomery multiplication method is suitable for the implementation on various pipeline processors. Furthermore, our method is also effective for the implementation on non-pipeline processors, because it computes all carries within a single-precision value.

Our m-repeated elliptic doublings method eliminates $2m - 2$ multiplications and $5m - 2$ additions compared with m times elliptic doublings specified in IEEE P1363 draft. This method is efficient on any processors. As the multiplication is faster in comparison with addition, our method is more effective.

We implemented RSA, DSA and ECC with our method on the latest DSP TMS320C6201(Texas Instruments). The performance is 11.7 msec for 1024-bit RSA signing, 14.5 msec for 1024-bit DSA verification and 3.97 msec for 160-bit ECDSA verification.

References

1. Paul Barrett, "Implementing the Rivest, Shamir, and Adleman Public-Key Encryption Algorithm on a Standard Digital Signal Processor", Advances in Cryptology-CRYTO'86(LNCS 263), pp.311-323, 1987.
2. E.F.Brickell, "A Survey of Hardware Implementations of RSA", Advances in Cryptology-CRYPTO'89(LNCS 435), pp.368-370, 1990.
3. D.Chudnovsky and G.Chudnovsky, "Sequences of numbers generated by addition in formal groups and new primality and factoring tests", Advances in Applied Mathematics, 7, pp.385-434, 1987.
4. Richard E.Crandall, "Method and apparatus for public key exchange in a cryptographic system", U.S. Patent, 5,159,632, 27 October 1992.
5. W.Diffie and M.Hellman, "New directions in cryptography", IEEE Transactions on Information Theory 22, pp.644-654, 1976.
6. S.R.Dusse and B.S.Kaliski Jr., "A Cryptographic Library for the Motorola DSP56000", Advances in Cryptology-Eurocrypt'90(LNCS 473), pp.230-244, 1991.

7. Daniel M.Gordon, "A Survey of Fast Exponentiation Methods", Journal of Algorithms 27, pp.129-146, 1998.
8. J.Guajardo and C.Paar, "Efficient Algorithms for Elliptic Curve Cryptosystems", Advances in Cryptology-CRYPTO'97(LNCS 1294), pp.342-356, 1997.
9. N.Koblitz, "Elliptic curve cryptosystems", Mathematics of Computation 48, pp.203-209, 1987.
10. Çetin Kaya Koç, Tolga Acar, B.S.Kaliski Jr., "Analyzing and Comparing Montgomery Multiplication Algorithms", IEEE Macro, Vol.16, No.3, pp.26-33, June 1996.
11. Alfred J.Menezes, Paul C.van Oorschot and Scott A.Vanstone, "HANDBOOK of APPLIED CRYPTOGRAPHY", CRC Press, 1997.
12. V.S.Miller, "Use of elliptic curves in cryptography", Advances in Cryptology-CRYPTO'85(LNCS 218), pp.417-426, 1986.
13. Atsuko Miyaji, "Method for Generating and Verifying Electronic Signatures and Privacy Communication Using Elliptic Curves", U.S. Patent, No.5,442,707, 15 August 1995.
14. P.L.Montgomery, "Modular Multiplication without Trial Division", Mathematics of Computation, Vol.44, No.170, pp.519-521, 1985.
15. R.L.Rivest, A.Shamir and L.Adleman, "A Method of obtaining digital signature and public key cryptosystems", Comm. of ACM, Vol.21, No.2, pp.120-126, Feb.1978.
16. FIPS 186, "Digital signature standard", Federal Information Processing Standards Publication 186, U.S.Department of Commerce/N.I.S.T., National Technical Information Service, Springfield, Virginia, 1994.
17. IEEE P1363/D9(Draft Version 9) Standard Specifications for Public Key, http://grouper.ieee.org/groups/1363/.
18. Texas Instruments, "Digital Signal Processing Solutions Products - TMS320C6x", http://www.ti.com/sc/docs/products/dsp/tms320c6201.html.

How to Implement Cost-Effective and Secure Public Key Cryptosystems

Pil Joong LEE[1], Eun Jeong LEE[2], and Yong Duk KIM[3]

[1] Dept. of Electronics and Electrical Eng., POSTECH, Pohang, Korea
[2] POSTECH Information Research Laboratories, Pohang, Korea
[3] Penta Security Systems Inc., 23-3 Yoido-dong, Yongdeungpo-ku, Seoul, Korea
pjl@postech.ac.kr, ejlee@postech.ac.kr, kyds@penta.co.kr

Abstract. The smart card has been suggested for personal security in public key cryptosystems. As the size of the keys in public key cryptosystems is increased, the design of crypto-controllers in smart cards becomes more complicated. This paper proposes a secure device in a terminal, "Secure Module", which can support precomputation technique for Schnorr-type cryptosystems such as Schnorr [7], DSA [1], KCDSA [5]. This gives a simple method to implement secure public key cryptosystems without technical efforts to redesign a cryptographic controller in $25mm^2$ smart card ICs.

Keywords : Secure module in terminal, smart cards, precomputation, digital signature/identification, public-key cryptosystem

1 Introduction

Public key cryptography proposed by Diffie and Hellman in 1976 allowed users to communicate securely without sharing secret information beforehand. These techniques can provide secure communication in today's open systems with privacy and message authentication. With public key techniques, a party has pairs of keys: a private key that is known only to the party and a public key available to all other users. A certificate for a public key is issued to a user so that any other party can verify the owner of the public key.

The user's personal security depends on the management of the private key. Not only must the private key be stored in a secure memory, but also the computation using the private key must be performed in a secure device. Smart cards, chip-embedded plastic cards with an MCU, make this possible with their tamper-resistant silicon chips. To store user-specific data securely against adversary, the EEPROM (Electrically Erasable Programmable Read Only Memory) in the chip is used.

The crypto-controller that uses the MCU and an additional arithmetic coprocessor for modular arithmetic provides the secure computation using the private key.

However, it is difficult to implement a secure and efficient public key cryptosystem due to the size of chip allowed in the smart card which is specified by

International Organization for Standardization (ISO) standard 7816. The parameters such as modulus and base integer in Schnorr-type protocol and the private key in RSA-type protocol are at least 128 bytes long, and may be more than 300 bytes, depending on the algorithm and security required [8]. Therefore, the smart card needs an additional arithmetic coprocessor optimized for modular exponentiation of long operands working on an 8-bit MCU [6]. Since the mechanical stability requirements of ISO standard 7816-1 limit the maximum chip size of the smart card ICs to 25 mm^2, the additional coprocessor for cryptographic algorithms must take up the space typically occupied by nonvolatile memory, specifically EEPROM.

Since the modular exponentiation of long operands is a much time-consuming job, Schnorr proposes to use precomputation for exponentiation using secret random numbers in idle time [7]. This idea can be applied to every schemes based on the discrete logarithm problem such as DSA [1], KCDSA [5], etc. With this precomputation, the signature generation can be computed by only one modular multiplication. However, since the ISO standard 7816 specifies the smart card is not supplied with electric power during its idle time, the precomputation technique cannot be applied to the smart card. In this paper, we propose a concept of "Secure Module (SM)" to design an efficient and secure public key cryptosystem without additional crypto-coprocessor. A normal smart card can be used as a SM when located in a terminal. With this SM and precomputation technique, we can efficiently generate Schnorr-type signatures.

Section 2 explains the environment of the proposed system and section 3 gives an example on how to implement the proposed system with Schnorr scheme. Then we consider the security of our proposed system in section 4.

2 The Environment of the Proposed System

A public key cryptosystem needs a secure place to store the user's private key. The nonvolatile memory, EEPROM, of the smart card is an appropriate space for the lengthy private key. To compute the exponentiation of a secret random number in signature/identification, we propose to outfit the terminal with a tamper-resistant device. We will call this device a Secure Module (SM).

The hardware specification is depicted in Fig. 1 and the requirements of the smart card (SC) and the terminal with secure module are as follows:

Smart Card (SC)
A normal IC card with 8-bit MCU, ROM, RAM, EEPROM, and COS (Card Operating System) according to ISO standard 7816 satisfies the hardware specification of the SC. The user data and cryptographic functions to be stored in EEPROM are as follows:

> **encryption algorithm,** $E( )$: symmetric key ciphering algorithm used in sending encrypted private key to SM.
> **random number generator,** $r_1( )$: algorithm to generate a secure random number used in mutual authentication with SM.

identifier of the SC, ID_{SC} : identifier of the smart card.
secret key, K_{SC} : symmetric key for $E(\ )$ used in mutual authentication. This is issued when the card is issued for a user and depends on the ID_{SC} and the master key K_M stored only in SM.
private key, X : user's private key for public-key cryptosystem.
certificate for public key, C : certificate for user's public key corresponding to the private key X.

The certificate C is optional, because, in some system, other users can obtain C from the trusted third party (TTP).

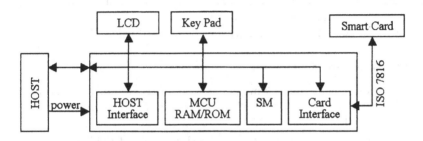

Fig. 1. The hardware specification of terminal with SM

Hardware Specification of Terminal

interface : LCD, keypad, host interface (RS232, modem, floppy, etc), and card interface (ISO 7816)
microprocessor
memory : RAM, ROM
secure module

Secure Module (SM)

A normal IC card with 8-bit MCU, ROM, RAM, EEPROM, and COS satisfies the hardware specification of SM. The user data and cryptographic functions to be stored in EEPROM are as follows:

encryption algorithm, $E(\ )$: symmetric key ciphering algorithm.
random number generator $r_2(\ )$: algorithm generating a secure random number used in mutual authentication with the SC and signature/identification schemes.
modular arithmetic functions : executable programs of modular multiplication, modular addition, and modular reduction.
precomputation functions : executable programs to precompute modular exponentiations for secret random numbers.
master key, K_M : symmetric key for $E(\ )$ used in mutual authentication. This is issued securely when the terminal is initialized.

3   Implementations of a Cryptographic Protocol

Every signature/identification scheme based on discrete log problem computes $W = g^k \pmod{p}$ for some random number $k \in \{1, 2, ..., q-1\}$, where p is a large prime and g is an element in $\{2, ..., p-1\}$ with order q, large prime divisor of $p-1$. Since k does not depend on the message to be signed or the identifier to be authenticated, these values can be precomputed in idle time, i.e. when the card is not in use. However, since the smart card is not supplied with power when the card is not in use, if there is the executable program of precomputation in smart card then we can not preprocess the exponentiation. Thus, it is the only time when the program of this algorithm is stored in the SM located in the terminal that we can preprocess the exponentiation.

An explanation for the Schnorr scheme is described below [7]:

Schnorr Signature Scheme for Message M:

(a) Choose a random number k in $\{1, ..., q-1\}$ and compute $W = g^k \pmod{p}$.
(b) Compute the first signature R, where $R = h(W\|M)$, where $h(\ )$ is a collision-resistant hash function.
(c) Compute the second signature S, where $S = k + X \cdot R \pmod{q}$.

Since (a) does not depend on the message, this step permits precomputation of the quantities needed to sign the next messages. We propose to precompute (k, W) values in the SM during idle time of the terminal. Since the SM is tamper-resistant, the random k's can be stored securely and the private key X saved in SC cannot be exposed. When the message is signed, with W precomputed, only one modular multiplication and one modular addition are needed. Since (c) uses the user's private key, the SM must bring the private key securely from the SC. A protocol of the Schnorr scheme using the proposed system is described below. This protocol is depicted in Fig. 2.

Protocol to Compute a Signature Using Schnorr Scheme:

Step 0. *Precompute values of (k, W) in SM.*
If there is any empty place in storage, pick a random number k in $\{1, ..., q-1\}$ and compute $W = g^k \bmod p$ and store the pair (k, W). Otherwise, wait for an interrupt signal that is sent by the terminal when a message arrives. If an interrupt signal arrives during the precomputation, SM quits the precomputation and dumps the intermediate result.

Step 1. *Establish mutual authentication and share a session key between the SC and the SM.*
The SC and the terminal must verify whether the other party is legitimate by carrying out mutual authentication algorithms. If the SC sends the ID_{SC} to the SM, then the SM generates the K_{SC} using ID_{SC} and K_M in a specified method. With this shared symmetric key K_{SC} and encryption algorithm $E(\ )$, the SC and the SM mutually authenticate and share a session key K_S using random numbers as in [2, 3].

Step 2. *Transfer the encrypted private key from the SC to the SM.*
    The SC encrypts the private key X with the session key K_S, denote $E_{K_S}(X)$, and sends the ciphertext to the SM. The SM decrypts the ciphertext $E_{K_S}(X)$ with K_S, and gets the private key X.

Step 3. *Compute signature values in SM.*
    The SM computes signature R and S using X, secret random number k, precomputed value W and message M as follows:

    Step 3-1. Retrieving a precomputed pair (k, W).
    Step 3-2. Compute $R = h(W \| M)$.
    Step 3-3. Compute $S = k + X \cdot R \pmod{q}$.
    Step 3-4. Output (R, S) and erase the used pair (k, W).

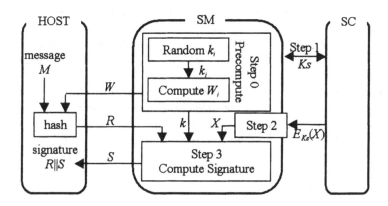

Fig. 2. The protocol to compute Schnorr-signature using SM

When the message is long, the hashing in Step 3-2 may be operated in the HOST like Fig. 2, because the channel between the terminal and the host has generally limited bandwidth.

We implemented this protocol with the digital signature algorithm, KCDSA [5] using the SM in a terminal with MCU 80C320 and the SC with Hitachi H8/3102 (8Kbyte EEPROM, 16Kbyte ROM, 512byte RAM). The cryptographic programs (modular operations, precomputation functions and random number generator) were coded using an 8051 assembler and their executable programs were stored in the EEPROM of the SM. We adapted algorithm [4] for exponent computation and this algorithm takes 30 seconds for one exponentiation, which is rather slow. However, thanks to the precomputation of this time-consuming job, the elapsed time to compute a digital signature is one second. Therefore, we could efficiently get a Schnorr-type digital signature of without using highly complex techniques for ICs.

4 Security Consideration

In our proposed system, each terminal has its own SM and all SM have the same master key K_M in common. The master key is issued when the terminal is initialized. Let us assume that an adversary knows the master key K_M. Since every one can get the identifier ID_{SC} of any smart card easily, the adversary can compute the session key K_{SC} shared between the smart card and the SM. In addition, if he succeeds in tapping or monitoring the transferred bit stream between the smart card and the terminal during mutual authentication and session key (K_S) establishment, he can get the session key K_S. Eventually he can get the user's private key X from the encrypted private key $E_{K_S}(X)$. Unfortunately, this reason will allown an adversary who knows the master key K_M to get the private key X stored in any smart card even though the private key is stored in the tamper resistant region of the smart card. Consequently, preventing the master key form being revealed is very important in our proposed system.

However, if there is no problem in initializing a terminal, i.e, the master key is stored in the tamper resistant region of the terminal without being revealed, and if computing of K_{SC} from the smart card identifier ID_{SC} and master key K_M is performed in tamper resistant region of SM, then we believe there will be no serious security hole in our system.

Even if an adversary knows the smart card identifier ID_{SC} but not the master key K_M, the attacks using differential analysis, timing attack or fault attack nearly do not work because the adversary knows only the values ID_{SC}, $E_{K_S}(X)$, $W = g^k \bmod p$, $R = h(W\|M)$ and $S = k + X \cdot R \bmod q$.

5 Conclusion

In this paper, we proposed to use a secure device in a terminal, "Secure Module", which can support precomputation technique for Schnorr-type cryptosystem. This gives an easy method to implement personal security without technical efforts to design cryptographic controller in $25mm^2$ smart card ICs. We could get KCDSA digital signature in one second using the proposed system.

References

1. NIST, Digital Signature standard, FIPS PUB 186,
   http://csrc.nist.gov/cryptval/dss.htm, 1994.
2. ISO/IEC IS 9798-2 Information technology - Security techniques - Entity Authentication, part 2: Mechanisms using a symmetric encipherment algorithm, 1995.
3. ISO/IEC IS 11770-2 Information technology - Security techniques - Key Management, part 2: Mechanisms using a symmetric techniques, 1996.
4. Chae Hoon Lim and Pil Joong Lee, "More flexible exponentiation with precomputation," *Advances in Cryptology-Crypto'94*, LNCS 839, Springer-Verlag, pp.95-107, 1994.

5. Chae Hoon Lim and Pil Joong Lee, "A study on the proposed Korean digital signature algorithm," *Advances in Cryptology-Asiacrypt'98*, LNCS 1514, pp.175-186, 1998.

6. David Naccache and David M'Raïhi, "Cryptographic Smart Cards," *IEEE Micro*, Vol. 16, No. 3, pp.14-24, 1996.

7. C. P. Schnorr, "Efficient signature generation by smart card," *J. of Cryptology*, vol. 4, pp.161-174, 1991.

8. A.M. Odlyzko, "The future of integer factorization," *Cryptobytes*, Vol. 1, No.2, pp. 5- 12, 1995.

Montgomery's Multiplication Technique: How to Make It Smaller and Faster

Colin D. Walter

Computation Department, UMIST
PO Box 88, Sackville Street, Manchester M60 1QD, UK
www.co.umist.ac.uk

Abstract. Montgomery's modular multiplication algorithm has enabled considerable progress to be made in the speeding up of RSA cryptosystems. Perhaps the systolic array implementation stands out most in the history of its success. This article gives a brief history of its implementation in hardware, taking a broad view of the many aspects which need to be considered in chip design. Among these are trade-offs between area and time, higher radix methods, communications both within the circuitry and with the rest of the world, and, as the technology shrinks, testing, fault tolerance, checker functions and error correction. We conclude that a linear, pipelined implementation of the algorithm may be part of best policy in thwarting differential power attacks against RSA.

Key Words: *Computer arithmetic, cryptography, RSA, Montgomery modular multiplication, higher radix methods, systolic arrays, testing, error correction, fault tolerance, checker function, differential power analysis, DPA.*

1 Introduction

An interesting fact is that the faster the hardware the more secure the RSA cryptosystem becomes. The effort of cracking the RSA code via factorization of the modulus M doubles for every 15 or so bits at key lengths of around 2^{10} bits [10]. However, adding the 15 bits only increases the work involved in decryption by $((1024+15)/1024)^2$ per multiplication and so by $((1024+15)/1024)^3$ per exponentiation, i.e. 5% extra! Thus speeding up the hardware by just 5% enables the cryptosystem to become about twice as strong without needing any other extra resources. Speed, therefore, seems to be everything. Indeed it is essential not just for cryptographic strength but also to enable real time decryption of the large quantities of data typically required in, for example, the use of compressed video.

On the other side of the Atlantic, the first electronic computer is generally recognised to be the Colossus, designed by Tommy Flowers, and built in 1943-4 at Bletchley Park, England. It was a dedicated processor to crack the Enigma code rather than a general purpose machine like the ENIAC, constructed slightly later by John Eckert and John Mauchly in Philadelphia. With the former view of history, cryptography has a fair claim to have started the (electronic) computer age. Breaking Enigma depended on a number of factors, particularly human weakness in adhering to strict protocols, but also on inherent implementation weaknesses.

Timing analysis and differential power analysis techniques [12] show that RSA cryptosystems mainly suffer not from lack of algorithmic strength but also from implementation weaknesses. No doubt governments worked on such techniques for many years before they appeared in the public domain and have developed sufficiently powerful techniques to crack any system through side channel leakage. Is it a coincidence that the US became less paranoid about the use of large keys when such techniques were first published? Now that there seems to be no significant gain to be made from further improvement of algorithms, the top priority must be to prevent such attacks by reducing or eliminating variations in timing, power and radiation from the hardware.

This survey provides a description of the main ideas in the hardware implementation of the RSA encryption process with an emphasis on the importance of Montgomery's modular multiplication algorithm [17]. It indicates the main publications where the significant contributions are to be found, but does not attempt to be exhaustive. The paper discusses the major problems associated with space- and time- efficient implementation and reviews their solution. Among the issues of concern are carry propagation, digit distribution, buffering, communication and use of available area. Finally, there are a few remarks on the reliability and cryptographic strength of such implementations.

2 Notation

An RSA cryptosystem [20] consists of a modulus M, usually of around 1024 bits, and two keys d and e with the property that $A^{de} \equiv A \bmod M$. Message blocks A satisfying $0 \leq A < M$ are encrypted to $C = A^e \bmod M$ and decrypted uniquely by $A = C^d \bmod M$ using the same algorithm for both processes. $M = PQ$ is a product of two large primes and e is often chosen small with few non-zero bits (e.g. a Fermat prime, such as 3 or 17) so that encryption is relatively fast. d is picked to satisfy $de \equiv 1 \bmod \phi(M)$ where ϕ is Euler's *totient* function, which counts the number of residue classes prime to its argument. Here $\phi(M) = (P-1)(Q-1)$ so that d usually has length comparable to M. The owner of the cryptosystem publishes M and e but keeps secret the factorization of M and the key d. Breaking the system is equivalent to discovering P and Q, which is computationally infeasible for the size of primes used.

The computation of $A^e \bmod M$ is characterised by two main processes: modular multiplication and exponentiation. Here we really only consider computing $(A \times B) \bmod M$. Exponentiation is covered in detail elsewhere, e.g. [9],[28]. To avoid potentially expensive full length comparisons with M, it is convenient to be able to work with numbers A and B which may be larger than the modulus. Assume numbers are represented with base (or radix) r which is a power of 2, say $r = 2^k$, and let n be the maximum number of digits needed for any number encountered. Here r is determined by the multiplier used in the implementation: an $r \times r$ multiplier will be used to multiply two digits together. The hardware also determines n if we are considering dedicated co-processors with a maximum register size of n digits. Generally, M must have fewer bits than the largest representable number; how much smaller will be determined by the algorithm used.

Except for the exponent e, each number X will have a representation of the form $X = \sum_{i=0}^{n-1} x_i r^i$. Here the ith digit x_i often satisfies $0 \le x_i < r$, yielding a *non-redundant* representation, i.e. one for which each representation is unique. The modulus M will always have such a representation. However, in order to limit the scope of interactions between neighbouring digits, a wider range of digits is very useful. Typically this is given by an extra (carry) bit so that digits lie in the range $0..2r-1$. For example, the output from a carry-save adder (with $r = 2$) provides two bits for each digit and so, in effect, is a *redundant* representation where digits lie in the range $0..3$ rather than the usual $0..1$. Here the k-bit architecture means that our adder will probably propagate carries up the length of a single digit, providing a "save" part in the range $0..r-1$ and a "carry" part representing a small multiple of r. A digit x split in this way is written $x = x_s + rx_c$. In fact, the addition cycle in our algorithms involves digit products, so that a double length result is obtained. Hence, with some notable exceptions, the carry component regularly consists of another digit and a further one or two more bits. In a calculation $X \leftarrow X+Y$, the digit slices can operate in parallel, with the jth digit slice computing

$$x_{j,s} + rx_{j,c} \leftarrow x_{j,s} + x_{j-1,c} + y_j$$

The extra range of x_j given through the carry component keeps x_j from having to generate a separate carry which would need propagating. Since only old values appear on the right, not new ones, carry propagation does not force sequential digit processing. The digit calculations can therefore be performed in parallel when there is sufficient redundancy.

3 Digit Multipliers and Their Complexity

Early modular multiplication designs treated radix 2, 4 or even 8 separately at a gate level. With rapidly advancing technology, these have had to be replaced by the generic radix r viewpoint which is now essential for a better understanding of the general principles as well as for a modular approach to design and for selecting from parametrised designs to make best use of available chip area. To-day's embedded cryptosystems are already using off-the-shelf 32-bit multipliers [21] where reduction of the critical path length by one gate makes virtually no difference to the speed – and would probably cost too much in terms of additional silicon area. These $r \times r$ multipliers form the core of an RSA processor, forming the digit-by-digit products.

In the absence of radical new algorithms we need to be aware of complexity results for multiplication but prepared to use pre-built state-of-the-art multipliers which contain years of experience in optimisation and which come with a guarantee of correctness. Practical planar designs are known for multipliers which are optimal with respect to some measure of time and area [3], [4], [15], [16], [19], [23]. A reasonable assumption which held until recently is that wires take area but do not contribute noticeably to time. Under such a model, $Area \times Time^2$ complexity for a k-bit multiplication is bounded below by k^2 [3] and this bound can be achieved for any time in the range $\log k$ to $\sqrt{k}$ [16]. Such designs tend to use the Discrete Fourier Transform and consequently involve large constants in their measures of time and area. There are more useful designs which are asymptotically poorer but perform better if k is not too large. The cross-over point is greater than the size of the digits here [15]. So classical multiplication methods are preferable. Indeed, for a chip area of around 10^7 transistors devoted entirely to RSA and containing hardware for a full length digit multiplication $a_i \times B$, k = 32 or 64 is the maximum practical since there must be space for registers and for other operations such as the modular reduction. In the latest technology wires have significant capacitance and resistance and there is a requirement from applications as diverse as cell phones and deep space exploration for low power circuitry. This requires a different model which is more sensitive to wire length and for which results are only just emerging [18].

Speed is most easily obtained by using at least n multipliers to perform a full length multiplication $a_i \times B$ (or equivalent) in one clock cycle. If we were not worried about modular reduction, the carry propagation problem could be taken care of by pipelining this row of multipliers (Fig. 1): $a_i b_j$ is then computed by the jth multiplier during the $i+j$th clock cycle, generating a carry which is fed into the next multiplier, which computes $a_i b_{j+1}$ in the next cycle.

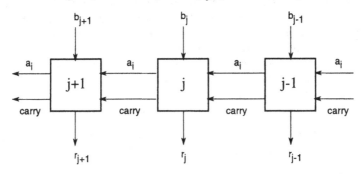

Figure 1. A Pipeline of Multipliers for $R \leftarrow a_i \times B$.

Is this set-up fast enough for real-time processing? A realistic measure of the speed required for real-time decryption is provided by an assumption that the internal bus speed is in the order of one k-bit digit per clock cycle. If the k-bit multiplier operates in one cycle with no internal pipelining then computing $A \times B$ takes n cycles using n multipliers in parallel in order to compute $a_i \times B$ in one cycle. The throughput is therefore one digit per cycle for a multiplication. Unfortunately, since RSA decryption requires $O(nk)$ multiplications, we may actually need a two dimensional array of multipliers rather than just a row of them to perform real-time decryption.

Of course, there is an immediate trade-off between time and area. Doubling the number of digit multipliers in an RSA co-processor allows the parallel processing of twice as many digits and so halves the time taken. This does not contradict the $Area \times Time^2$ measure being constant for non-pipelined multipliers, although it appears to require less area than expected for the speed-up achieved. Having two rows of digit multipliers with one row feeding into the other creates a pipeline (now with respect to digits of A) which doubles the throughput that the complexity rule expects. This indicates that choosing the largest r possible for the given silicon area may not be the best policy; a pipelined multiplier or several rows of smaller multipliers may yield better throughput for a given area.

Finally, despite a wish to use well-established multipliers, differential power analysis (DPA) attacks on cryptographic products [13] suggest that special purpose multipliers need to be designed for some RSA applications which contain the secret keys, such as smart cards. Briefly, switching a gate consumes more power than not doing so. Inputs for which Hamming weights are markedly more or less than average could therefore have a power consumption with measurable deviation from average and reveal useful information to an attacker. This is true of today's optimised multipliers.

4 Modular Reduction and the Classical Algorithm

The reduction of $A \times B$ to $(A \times B) \bmod M$ can be carried out in several ways [11]. Normally it is done through interleaving the addition of $a_i B$ with modular reductions instead of computing the complete product first. This makes some savings in hardware. In particular, it enables the partial product to be kept inside an n-digit register without overflow into a second such register. Each modular reduction involves choosing a suitable digit q and subtracting qM from the current result. The successive choices of digit q can be pieced together to form the integer quotient $Q = \lfloor (A \times B)/M \rfloor$ or a closely related quantity:

CLASSICAL MODULAR MULTIPLICATION ALGORITHM:
{ Pre-condition: $0 \le A < r^n$ }

```
R := 0 ;
For i := n-1 downto 0 do
Begin
    R   := r×R + aᵢ×B ;
    qᵢ  := R div M ;
    R   := R - qᵢ×M ;
End
```
{ Post-condition: $R = A \times B - Q \times M$
  and, consequently, $R \equiv (A \times B) \bmod M$ }

If we define $A_i = \sum_{j=i}^{n-1} a_j r^{j-i}$ so that $A_i = rA_{i+1}+a_i$ and use similar notation for Q_i then it is easy to prove by induction that at the end of the ith iteration $R = A_i \times B - Q_i \times M$. Hence the post-condition holds.

Brickell [5] observes that R div M need only be approximated using the top bits of R and M in order to keep R bounded above. In particular [24], suppose that $\underline{M}$ is the approximation to M given by setting all but the $k+3$ most significant bits of M to zero, and that $\underline{M}'$ is obtained from $\underline{M}$ by incrementing its $k+3$rd bit by 1. Then $\underline{M} \leq M < \underline{M}' \leq (1+2^{-2}r^{-1})\underline{M}$. Assume $\underline{R}$ is given similarly by setting the same less significant bits to zero and that the redundancy in the representation of R is small enough for $R < \underline{R}+M$ to hold. The approximation to q_i which is used is defined by the integer quotient $\underline{q_i} = \lfloor \underline{R}/\underline{M}' \rfloor$. Then

$$q_i = \lfloor \underline{R}/\underline{M}' \rfloor \leq \lfloor R/M \rfloor = q_i$$
$$\leq 1 + \underline{R}/M \leq 1 + (1+2^{-2}r^{-1})\underline{R}/\underline{M}' \leq 1 + (1+2^{-2}r^{-1})(\underline{q_i} + 1)$$

so that $q_i - \underline{q_i} \leq 1 + (1+2^{-2}r^{-1})^{-1} + q_i(1+4r)^{-1}$ from which, at the end of the loop, $R < (1+q_i-\underline{q_i})M < 2M + (1+4r)^{-1}q_iM$. Assume the (possibly redundant) digits a_i are bounded above by a. We will establish inductively that

$$R < 3M + a(1+3r)^{-1}B$$

at each end of every iteration of the loop. Using the bound on the initial value of R in the loop yields $q_iM \leq rR+aB < 3rM + a(1+r(1+3r)^{-1})B$ so that, from the above, the value for R at the end of the loop satisfies $R < 2M + (1+4r)^{-1}(3rM + a(1+r(1+3r)^{-1})B) < 3M + a(1+3r)^{-1}B$, as desired. Naturally, more bits for a better approximation to q_i will yield a lower bound on R, whilst fewer will yield a worse bound or even divergence.

The output from such a multiplication can be fed back in as an argument to a further such multiplication without any further adjustment by a multiple of M providing the bound on R does not grow too much. Assuming, reasonably, that redundancy in A is bounded by $a \leq 2r$ we obtain $R < 3M + \frac{2}{3}B$ from which i) if $B \geq 9M$ then B is an upper bound for successive modular multiplications and indeed the bound decreases towards a limit of $9M$, and ii) if $9M$ is an upper bound for the input it is also an upper bound for the output. Hence only a small number of extra subtractions at the end of the exponentiation yields M as a final upper bound.

There are no communication or timing problems when there is just one multiplier. So, for the rest of this article, we will assume that the hardware for the algorithm consists of an array of cells, each one for computing a digit of R, and so each containing two multipliers. The main difficulty is that q_i needs to be computed before any progress can be made on the partial product R. Scaling M to make its leading two digits 10_r makes the computation easier [25],[27], as does shifting B downwards to remove the dependency on B. However, $\underline{q_i}$ still has to be broadcast simultaneously to every digit position (Fig. 2) and redundancy has to be employed in R so that digit operations can be performed in parallel [5], [27]. These severe drawbacks make Montgomery's algorithm for modular multiplication appear more attractive.

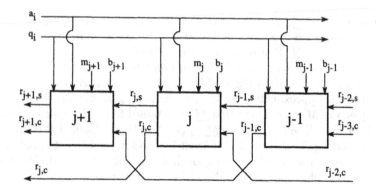

Figure 2. Classical Algorithm with Shift Up, Carries and Digit Broadcasting

5 Montgomery's Algorithm

Peter Montgomery [17] has shown how to reverse the above algorithm so that carry propagation is away from the crucial bits, redundancy is avoided and simultaneous broadcasting of digits is no longer required. This noticeably reduces silicon area and shortens the critical path. Although the complexities of the two algorithms seem to be identical in time and space, the constants for Montgomery's version are better in practice. Montgomery uses the *least* significant digit of an accumulating product R to determine a multiple of M to *add* rather than subtract. He chooses multiplier digits in the opposite order, from *least* to *most* significant and shifts *down* instead of up on each iteration:

MONTGOMERY'S MODULAR MULTIPLICATION ALGORITHM:
{ Pre-condition: $0 \leq A < r^n$ }

```
R := 0 ;
For i := 0 to n-1 do
Begin
      R  := R + a_i×B ;
      q_i := (-r_0 m_0^{-1}) mod r ;
      R  := (R + q_i×M) div r ;
      { Invariant: 0 ≤ R < M+B }
End
{ Post-condition: Rr^n = A×B + Q×M
    and, consequently, R ≡ (A×B×r^{-n}) mod M }
```

Here m_0^{-1} is a residue mod r satisfying $m_0 \times m_0^{-1} \equiv 1 \bmod r$. Since r is a power of 2 and M is odd (because it is a product of two large primes) r and M are coprime, which is enough to guarantee the existence of m_0^{-1}. The digit q_i is chosen so that the expression $R + q_i \times M$ is exactly divisible by r. Its lowest digit is clearly 0. If we define $A_i = \sum_{j=0}^{i} a_j r^j$ and Q_i analogously then $A_i = A_{i-1} + a_i r^i$ and

$A_n = A$. The value of R at the end of the iteration whose control variable has value i is easily shown by induction to satisfy $Rr^{i+1} = A_i \times B + Q_i \times M$ because the division is exact. Hence the post-condition holds. The digits of A are required in ascending order. Thus, they can be converted on-line into a non-redundant form and so we may assume $a_i \leq r-1$. This enables the loop invariant bounds to be established by induction.

The extra power of r factor in the output R is easily cleared up by minor pre- and post-processing [6]. The easiest way to explain this is to associate with every number its *Montgomery class* mod M, namely

$$\overline{A} \equiv Ar^n \bmod M$$

and to use $\overline{\times}$ to denote the Montgomery modular multiplication. The *Montgomery product* of $\overline{A}$ and $\overline{B}$ is $\overline{A} \times \overline{B} \equiv \overline{A}\,\overline{B}\,r^{-n} \equiv ABr^n \equiv \overline{AB} \bmod M$. So applying Montgomery multiplication to $\overline{A}$ in an exponentiation algorithm is going to produce $\overline{A^e}$ rather than $(\overline{A})^e$. Introduction of the initial power of r to obtain $\overline{A}$ is performed using the precomputed value

$$R_2 = \overline{r^n} \equiv r^{2n} \bmod M$$

and a Montgomery multiplication thus [7]: $A \overline{\times} R_2 \equiv Ar^n \equiv \overline{A} \bmod M$. Removal of the final extra power of r is also performed by a Montgomery multiplication: $\overline{A^e} \overline{\times} 1 \equiv A^e \bmod M$.

Throughout the exponentiation, an output from one multiplication is used as an input to a subsequent multiplication. Without care the outputs will slowly increase in size. However, suppose $a_{n-1} = 0$. Then the bound $R < M+B$ at the end of the second last loop iteration is reduced to $M+r^{-1}B$ on the final round, which prevents unbounded growth when outputs are used as inputs. In particular, if the second argument satisfies $B < 2M$ then the output also satisfies $R < 2M$. Thus, suppose $2rM < r^n$. It is reasonable to assume that A and R_2 are less than $2M$, even less than M, so that their topmost digits are both zero. Then the scaling of A to $\overline{A}$ by Montgomery multiplication yields $\overline{A} < 2M$ and this bound is maintained as far as the final output A^e. So only a single extra subtraction of M may be necessary at the very end to obtain a least non-negative residues.

However, when all the I/O is bounded by $2M$, an interesting and useful observation about the output R of the final multiplication $\overline{A^e} \overline{\times} 1 \equiv A^e \bmod M$ can be derived from the post-condition of the modular multiplication, namely $Rr^n = \overline{A^e} + QM$. Q has a maximum value of r^n-1. Hence, $\overline{A^e} < 2M$ would lead to the output satisfying $Rr^n < (r^n+1)M$ and so to $R \leq M$, whilst a sub-maximal value for Q immediately yields $R < M$ in the same way. Hence a final subtraction is *only* necessary when $R = M$, i.e. when $\overline{A^e} \equiv 0 \bmod M$, that is, for $A \equiv 0 \bmod M$. It is entirely reasonable to assume that this never occurs in the use of the RSA cryptosystem as it would require starting with $A = M$, whereas invariably the initial value should be *less* than M. Moreover, each modular multiplication in the exponentiation would also have to return M rather than 0 to prevent all subsequent operands becoming 0. Hence the final subtraction need never occur after the final re-scaling of an exponentiation.

Computation of q_i is a potential bottleneck here. Simplification may be achieved in the same two ways that applied to the classical algorithm above: this time scale M to make $m_0 = -1$ and shift B up so that $b_0 = 0$. These would make $q_i = r_0$, avoiding any computation at all. We consider the shift first, supposing initially that $B < 2M$. If the shift $B \leftarrow rB$ is added to the start of the multiplication code, then the loop invariant becomes $R < M+rB$. Hence we require the top *two* digits of A to be zero in order for the bound on R to be reduced first to $M+B$ and then to $M+r^{-1}B$, so that $R < 2M$ is output. If we always have $A < 2M$ then this is achieved if $2r^2M$ fits into the hardware registers of n digits. The cost of this shift is hidden by the definition of n. The shift can be hardwired and counted as free, as can a balancing adjustment of the output through a higher power of r in R_2. However, there is an extra iteration of the loop: before we had one more iteration than digits in M, now we have two more. Fortunately, the cost of one more iteration per multiplication is low compared with the delay on every iteration which computing q_i may cause. Apart from the extra storage cost of the scaled number, scaling M has a similar cost, namely one more iteration per multiplication: M is replaced by $(r-m_0^{-1})M$, which increases its number of digits by 1 [26]. However, at the end of the exponentiation, the original M also needs to be loaded into the hardware and some extra subtractions of M may be necessary to reduce the output from a bound of $2rM$.

6 Digit-Parallel and Digit-Serial Implementations

To overcome the problems of carry propagation in the classical algorithm, redundancy and extra hardware for digit broadcasting were required. Here, too, the same methods enable parallel digit processing. Indeed, if the shift direction were reversed, the diagram of Fig. 2 would cover Montgomery's algorithm also.

For both algorithms, define the ith value of R, written $R^{(i)} = \sum_{j=0}^{n-1} r_j^{(i)} r^j$, to be the value immediately before the shift is performed. This is calculated at time $2i$ in the parallel digit implementations, with q_i computed and broadcast at time $2i+1$, say. The jth cell operates on the jth digits, transforming the $i-1$st value of R into the ith value. A common view of this process is:

$$r_j^{(i)} \leftarrow r_{j\pm1,s}^{(i-1)} + r_{j\pm1-1,c}^{(i-1)} + a_i b_j \pm q_i m_j$$

where the choice of signs is $-$ for the classical algorithm and $+$ for Montgomery's. (The input values of R on the right are partitioned into save and carry/borrow parts.)

A restriction to only nearest neighbour communication is desirable because of the delays and wiring associated with global movement of data. For Montgomery's algorithm, a systolic array makes this possible [26]. In this, the cells are transformed into a pipeline in which the jth cell computes $r_j^{(i)}$ at time $2i+j$ (Fig. 3). The input $r_{j+1}^{(i-1)}$ is calculated on the preceding cycle by cell $j+1$ and a carry $c_{j-1}^{(i)}$ from $r_{j-1}^{(i)}$ is computed in cell $j-1$, also in that cycle. This means that carries can be propagated and so the cell function can become:

$$r_j^{(i)} + rc_j^{(i)} \leftarrow r_{j+1}^{(i-1)} + c_{j-1}^{(i)} + a_i b_j + q_i m_j$$

where the digits $r_j^{(i)}$ are now in the standard, non-redundant range $0\mathinner{..}r-1$. (The different notation for the carries recognises that they do not form part of the value of $R^{(i)}$, unlike in the carry-save view.) If the digit q_i is produced at time $2i$ in cell 0, it can be pipelined and received from cell $j-1$ at time $2i+j-1$ ready for the calculation. This pipeline can be extended to part or all of a 2-dimensional array with n rows which computes iterations of the loop in successive rows.

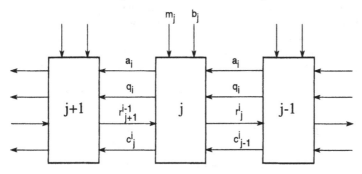

Figure 3. Pipelined Montgomery Multiplication: I/O for cell j at time $2i+j$.

The lowest cell, cell 0, computes only the quotient digit q_i. Digits of index 0 are always discarded by the shift down and so do not need computing; the lowest digit of the final output is shifted down from index 1. We have $q_i = -(r_1^{(i-1)} + a_i \times b_0){m_0}^{-1} \bmod r$. This can indeed be calculated at time $2i$ because its only timed input $r_1^{(i-1)}$ is computed at time $2i-1$. Observe that pre-computation of $b_0 \times {m_0}^{-1}$ reduces the computation of q_i to a single digit multiplication and an addition, giving lower complexity for cell 0 than for the other cells. Hence, computing q_i no longer holds up the multiplication. Instead, the critical path lies in the repeated, standard cell.

The communication infrastructure is less here than for the parallel digit operations illustrated in Fig. 2. Although the number of bits transmitted is almost the same in both cases and is independent of n, the parallel digit set-up requires an additional $O(\log n)$ depth network of multiplexers to distribute the digit q_i. Here the inputs and output are consumed, resp. generated, at a rate of one digit every other cycle for A and R, and one digit every cycle in the case of B. Unlike the parallel digit model, this is very convenient for external communication over the bus, reducing the need for buffering or increased bandwidth.

When one multiplication has completed, another one can start without any pause. However, the opposing directions of carry propagation and shift mean that each cell is idle on alternate cycles. Thus, full use of the hardware requires two modular multiplications to be processed in parallel. The normal square and multiply algorithm for exponentiation can be programmed to compute squares nose-to-tail starting loop iterations on the even cycles and interleave any necessary multiplications to start on the odd cycles. This enables an average 75% take-up of the processing power, but has some overhead in storage and switching between the two concurrent multiplications. Overall, with this added complexity, the classical, parallel digit, linear array might be faster for small n, but for larger n and/or smaller r the broadcasting problem for q_i means that a pipelined implementation of Montgomery's algorithm should be faster.

In [14] Peter Kornerup modified this arrangement, pairing or grouping digits in order to reduce the waiting gap. In effect he alters the angle of the timing front in the data dependency graph and, in the case he illustrates, he uses half the number of cells with twice the computing power. This can be advantageous in some circumstances.

An idea of the current speed of such array implementations is given in Blum and Paar [1] and amongst those actually constructed is one by Vuillemin *et al.* [22]

7 Data Integrity

Correct functioning is important not only for critical applications but as a protection against, for example, attacks on RSA signature schemes through single fault analysis [2]. Moreover, it is difficult and expensive to check all gate combinations for faults at fabrication time because of the time need to load sufficiently many different moduli [29] and, when smart cards are involved, a low unit price may only be possible by using tests which occasionally allow sub-standard products through to the market.

However, run-time checker functions are possible. These can operate in a similar way to those for multiplication in current chips. For example, results there are checked $\bmod 3$ and $\bmod 5$ in one case [8]. Here the cost of a similar check is minimal compared to that of the total hardware. The key observation is that the output from the modular multiplication algorithm satisfies an arithmetic equation:

$$R = A \times B - Q \times M \quad \text{or} \quad Rr^n = A \times B + Q \times M$$

These are easily checked $\bmod m$ for some suitable m by accumulating partial results for both sides on a digit-by-digit basis as the digits become available. A particularly good choice for m is a prime just above the maximum cell output value, but smaller m prime to r are also reasonable. The hardware complexity for this is then equivalent to about one cell in the linear array and so the cost is close to that of increasing n by 1.

If a discrepancy is found by the checker function, the computation can be aborted or re-computed by a different route. For example, to avoid the problem, M might be replaced by dM for a digit d prime to r and combined as necessary with some extra subtractions of the original M at the end.

8   Timing and Power Attacks

The literature contains descriptions of a number of attacks on the RSA cryptosystem which use timing or power information from hardware implementations which contain secret keys [12], [13]. Experience of implementing both the classical and Montgomery algorithms for modular multiplication suggests that most optimisation techniques which work with one also apply to the other. This suggests that attacks which succeed on well-designed implementations of the classical algorithm will have equivalents which apply to implementations of Montgomery's algorithm.

However, an important difference arises when the pipelined linear array is used since, judging from the data dependency graph, there seems to be no equivalent for the classical algorithm. With parallel digit processing of the multiplication $A \times B \bmod M$, the same digits of A and Q are used in every digit slice, opening up the possibility of extracting information about both by averaging the power consumption over all cells. However, during the related Montgomery multiplication in a pipelined array, many digits of A and Q are being used simultaneously for forming digit products. This should make identification of the individual digits much more difficult, and certainly increases the difficulty of any analysis.

9   Conclusion

We have reviewed and compared the main bottlenecks which may arise in hardware for implementing the RSA cryptosystem using both the classical algorithm for modular multiplication and Montgomery's version, and shown how these are solved. The hardware still suffers from broadcasting problems with the classical algorithm and scheduling complications with Montgomery's. However, as far as implementation attacks using power analysis are concerned, the pipelined array for the latter seems to offer considerable advantages over any other implementations.

References

1. T. Blum & C. Paar, "Montgomery Modular Exponentiation on Reconfigurable Hardware", *Proc. 14th IEEE Symp. on Computer Arithmetic*, Adelaide, 14-16 April 1999, IEEE Press (1999) 70-77
2. D. Boneh, R. DeMillo & R. Lipton, "On the Importance of Checking Cryptographic Protocols for Faults", *Eurocrypt '97*, Lecture Notes in Computer Science, vol. 1233, Springer-Verlag (1997) 37-51
3. R. P. Brent & H. T. Kung, "The Area-Time Complexity of Binary Multiplication", *J. ACM* **28** (1981) 521-534
4. R. P. Brent & H. T. Kung, "A Regular Layout for Parallel Adders", *IEEE Trans. Comp.* **C-31** no. 3 (March 1982) 260-264

5. E. F. Brickell, "A Fast Modular Multiplication Algorithm with Application to Two Key Cryptography", *Advances in Cryptology - CRYPTO '82*, Chaum et al. (eds.), New York, Plenum (1983) 51-60

6. S. E. Eldridge, "A Faster Modular Multiplication Algorithm", *Intern. J. Computer Math.* **40** (1991) 63-68

7. S. E. Eldridge & C. D. Walter, "Hardware Implementation of Montgomery's Modular Multiplication Algorithm", *IEEE Trans. Comp.* **42** (1993) 693-699

8. G. Gerwig & M. Kroener, "Floating Point Unit in Standard Cell Design with 116 bit Wide Dataflow", *Proc. 14th IEEE Symp. on Computer Arithmetic*, Adelaide, 14-16 April 1999, IEEE Press (1999) 266-273

9. D. E. Knuth, *The Art of Computer Programming*, vol. 2, *Seminumerical Algorithms*, 2nd Edition, Addison-Wesley (1981) 441-466

10. N. Koblitz, *A Course in Number Theory and Cryptography*, Graduate Texts in Mathematics **114**, Springer-Verlag (1987)

11. Ç. K. Koç, T. Acar & B. S. Kaliski, "Analyzing and Comparing Montgomery Multiplication Algorithms", *IEEE Micro* **16** no. 3 (June 1996) 26-33

12. P. Kocher, "Timing Attacks on Implementations of Diffie-Hellman, RSA, DSS, and Other Systems", *Advances in Cryptology, Proc Crypto 96*, Lecture Notes in Computer Science **1109**, N. Koblitz editor, Springer-Verlag (1996) 104-113

13. P. Kocher, J. Jaffe & B. Jun, *Introduction to Differential Power Analysis and Related Attacks* at www.cryptography.com/dpa

14. P. Kornerup, "A Systolic, Linear-Array Multiplier for a Class of Right-Shift Algorithms", *IEEE Trans. Comp.* **43** no. 8 (1994) 892-898

15. W. K. Luk & J. E. Vuillemin, "Recursive Implementation of Optimal Time VLSI Integer Multipliers", *VLSI '83*, F. Anceau & E.J. Aas (eds.), Elsevier Science (1983) 155-168

16. K. Mehlhorn & F. P. Preparata, "Area-Time Optimal VLSI Integer Multiplier with Minimum Computation Time", *Information & Control* **58** (1983) 137-156

17. P. L. Montgomery, "Modular Multiplication without Trial Division", *Math. Computation* **44** (1985) 519-521

18. S. F. Obermann, H. Al-Twaijry & M. J. Flynn, "The SNAP Project: Design of Floating Point Arithmetic Units", *Proc. 13th IEEE Symp. on Computer Arith.*, Asilomar, CA, USA, 6-9 July 1997, IEEE Press (1997) 156-165

19. F. P. Preparata & J. Vuillemin, "Area-Time Optimal VLSI Networks for computing Integer Multiplication and Discrete Fourier Transform", *Proc. ICALP*, Haifa, Israel, 1981, 29-40

20. R. L. Rivest, A. Shamir & L. Adleman, "A Method for obtaining Digital Signatures and Public-Key Cryptosystems", *Comm. ACM* **21** (1978) 120-126

21. A. van Someren & C. Attack, *The ARM RISC Chip: a programmer's guide*, Addison-Wesley (1993)

22. J. Vuillemin, P. Bertin, D. Roncin, M. Shand, H. Touati & P. Boucard, "Programmable active memories: Reconfigurable systems come of age", *IEEE Trans. on VLSI Systems* **5** no. 2 (June 1997) 211-217

23. C. S. Wallace, "A Suggestion for a Fast Multiplier", *IEEE Trans. Electronic Computers* **EC-13** no. 2 (Feb. 1964) 14-17

24. C. D. Walter, "Fast Modular Multiplication using 2-Power Radix", *Intern. J. Computer Maths.* **39** (1991) 21-28

25. C. D. Walter, "Faster Modular Multiplication by Operand Scaling", *Advances in Cryptology - CRYPTO '91*, J. Feigenbaum (ed.), Lecture Notes in Computer Science **576**, Springer-Verlag (1992) 313-323

26. C. D. Walter, "Systolic Modular Multiplication", *IEEE Trans. Comp.* **42** (1993) 376-378
27. C. D. Walter, "Space/Time Trade-offs for Higher Radix Modular Multiplication using Repeated Addition", *IEEE Trans. Comp.* **46** (1997) 139-141
28. C. D. Walter, "Exponentiation using Division Chains", *IEEE Trans. Comp.* **47** no. 7 (July 1998) 757-765
29. C. D. Walter, "Moduli for Testing Implementations of the RSA Cryptosystem", *Proc. 14th IEEE Symp. on Computer Arithmetic*, Adelaide, 14-16 April 1999, IEEE Press (1999) 78-85

A Scalable Architecture for Montgomery Multiplication *

Alexandre F. Tenca and Çetin K. Koç

Electrical & Computer Engineering
Oregon State University, Corvallis, Oregon 97331
{tenca,koc}@ece.orst.edu

Abstract. This paper describes the methodology and design of a scalable Montgomery multiplication module. There is no limitation on the maximum number of bits manipulated by the multiplier, and the selection of the word-size is made according to the available area and/or desired performance. We describe the general view of the new architecture, analyze hardware organization for its parallel computation, and discuss design tradeoffs which are useful to identify the best hardware configuration.

1 Introduction

The Montgomery multiplication algorithm [10] is an efficient method for modular multiplication with an arbitrary modulus, particularly suitable for implementation on general-purpose computers (signal processors or microprocessors). The method is based on an ingenious representation of the residue class modulo M, and replaces division by M operation with division by a power of 2. This operation is easily accomplished on a computer since the numbers are represented in binary form. Various algorithms [11, 7, 1] attempt to modify the original method in order to obtain more efficient software implementations on specific processors or arithmetic coprocessors, or direct hardware implementations. In this paper we are interested in hardware implementations of the Montgomery multiplication operation.

Several algorithms and hardware implementations of the Montgomery multiplication for a limited precision of the operands were proposed [1, 11, 3]. In order to get improved performance, high-radix algorithms have also been proposed [11, 8]. However, these high-radix algorithms usually are more complex and consume significant amounts of chip area, and it is not so evident whether the complex circuits derived from them provide the desired speed increase. A theoretical investigation of the design tradeoffs for high-radix modular multipliers is given in [15]. An example of a design in radix-4 is shown in [13]. The increase in the radix forces the use of digit multipliers, and therefore more complex designs and longer

---

* Readers should note that Oregon State University has filed or will file a patent application containing this work to the US Patent and Trademark Office.

clock cycle times. For this reason, low-radix designs are usually more attractive for hardware implementation.

The Montgomery multiplication is the basic building block for the modular exponentiation operation [4, 5] which is required in the Diffie-Hellman and RSA public-key cryptosystems [2, 12]. Currently, most modular exponentiation chips perform the exponentiation as a series of modular multiplications, which is the most compelling reason for the research of fast and inexpensive modular multipliers for long integers. Recent implementations of the Montgomery multiplications are focused on elliptic key cryptography [9] over the finite field $GF(p)$. The introduction [6] of the Montgomery multiplication in $GF(2^k)$ opened up new possibilities, most notably in elliptic key cryptography over the finite field $GF(2^k)$ and discrete exponentiation over $GF(2^k)$ [5].

In this paper, we propose a *scalable* Montgomery multiplication architecture, which allows us to investigate different areas of the design space, and thus, analyze the design tradeoffs for the best performance in a limited chip area. We start with a short discussion of the scalability requirement which we impose in our design, and then give a presentation of the general theoretical issues related to the Montgomery multiplication. We then propose a word-based algorithm, and show the parallel evaluation of its steps in detail. Using this analysis, we derive an architecture for the modular multiplier and present the design of the module. We also perform simulations in order to provide area/time tradeoffs and give a first order evaluation of the multiplier performance for various operand precision.

2 Scalability

We consider an arithmetic unit as scalable if

> the unit can be reused or replicated in order to generate long-precision results independently of the data path precision for which the unit was originally designed.

For example, a multiplier designed for 768 bits [13] cannot be immediately used in a system which needs 1,024 bits. The functions performed by such designs are not consistent with the ones required in the larger precision system, and the multiplier must be redesigned. In order to make the hardware scalable, the usual solution is to use software and standard digit multipliers. The algorithms for software computation of Montgomery's multiplication are presented in [6, 7]. The complexity of software-oriented algorithms is much higher than the complexity of the radix-2 hardware implementation [1], making a direct hardware implementation not attractive. In the following, we propose a hardware algorithm and design approach for the Montgomery multiplication that are attractive in terms of performance and scalability.

3   Montgomery Multiplication

Given two integers X and Y, the application of the radix-2 Montgomery multiplication (MM) algorithm with required parameters for n bits of precision will result in the number

$$Z = \text{MM}(X, Y) = XYr^{-1} \bmod M , \tag{1}$$

where $r = 2^n$ and M is an integer in the range $2^{n-1} < M < 2^n$ such that $\gcd(r, M) = 1$. Since $r = 2^n$, it is sufficient that the modulus M is an odd integer. For cryptographic applications, M is usually a prime number or the product of two primes, and thus, the relative primality condition is always satisfied. The Montgomery algorithm transforms an integer in the range $[0, M-1]$ to another integer in the same range, which is called the image or the M-residue of the integer. The image or the M-residue of a is defined as $\bar{a} = ar \bmod M$. It is easy to show that the Montgomery multiplication over the images $\bar{a}$ and $\bar{b}$ computes the image $\bar{c} = \text{MM}(a, b)$ which corresponds to the integer $c = ab \bmod M$ [7]. The transformation between the image and the integer set is accomplished using the MM as follows.

- From the integer value to the M-residue: $\bar{a} = \text{MM}(a, r^2) = ar^2r^{-1} \bmod M = ar \bmod M$.
- From the M-residue to the integer value: $a = \text{MM}(\bar{a}, 1) = arr^{-1} \bmod M = a \bmod M$.

Provided that r (mod M) and r^2 (mod M) are precomputed and saved, we need only a single MM to perform either of these transformations. The tradeoff is the lower complexity of the MM algorithm when compared to the conventional modular multiplication which requires a division operation. Another important aspect of the advantage of the MM over the conventional multiplication is exposed in modular exponentiation, when multiple MMs are computed over the M-residues before the result is translated back to the original integer set.

The radix-2 Montgomery multiplication algorithm for m-bit operands $X = (x_{m-1}, ..., x_1, x_0)$, Y, and M is given as:

> The Radix-2 Algorithm
> _____
> $S_0 = 0$
> for $i = 0$ to $m - 1$
>   if $(S_i + x_iY)$ is even
>   then $S_{i+1} := (S_i + x_iY)/2$
>   else $S_{i+1} := (S_i + x_iY + M)/2$
> if $S_m \geq M$ then $S_m := S_m - M$ — the final correction step

This algorithm is adequate for hardware implementation because it is composed of simple operations: word-by-bit multiplication, bit-shift (division by 2), and addition. The test of the *even* condition is also very simple to implement, consisting on checking the least significant bit of the partial sum S_i to decide if the addition of M is required. However, the operations are performed on full precision of the operands, and in this sense, they will have an intrinsic limitation on the operands' precision. Once a hardware is defined for m bits, it cannot work with more bits.

4 A Multiple-Word Radix-2 Montgomery Multiplication Algorithm

The use of short precision words reduces the broadcast problem in the circuit implementation. The broadcast problem corresponds to the increase in the propagation delay of high-fanout signals. Also, a word-oriented algorithm provides the support we need to develop scalable hardware units for the MM. Therefore, an algorithm which performs bit-level computations and produces word-level outputs would be the best choice. Let us consider w-bit words. For operands with m bits of precision, $e = \lceil (m+1)/w \rceil$ words are required. The extra bit used in the calculation of e is required since it is known that S (internal varible of the radix 2 algorithm) is in the range $[0, 2M - 1]$, where M is the modulus. Thus the computations must be done with an extra bit of precision. The input operands will need an extra 0 bit value at the leftmost bit position in order to have the precision extended to the correct value.

We propose an algorithm in which the operand Y (multiplicand) is scanned word-by-word, and the operand X (multiplier) is scanned bit-by-bit. This decision enables us to obtain an efficient hardware implementation. We call it *Multiple Word Radix-2 Montgomery Multiplication algorithm (MWR2MM)*. We make use of the following vectors:

$$M = (M^{(e-1)}, ..., M^{(1)}, M^{(0)}) ,$$
$$Y = (Y^{(e-1)}, ..., Y^{(1)}, Y^{(0)}) ,$$
$$X = (x_{m-1}, ..., x_1, x_0) ,$$

where the words are marked with superscripts and the bits are marked with subscripts. The concatenation of vectors a and b is represented as (a, b). A particular range of bits in a vector a from position i to position j, $j > i$ is represented as $a_{j..i}$. The bit position i of the kth word of a is represented as $a_i^{(k)}$. The details of the MWR2MM algorithm are given below.

The MWR2MM Algorithm
$S = 0$ — initialize all words of S
for $i = 0$ to $m - 1$
     $(C, S^{(0)}) := x_i Y^{(0)} + S^{(0)}$
     if $S_0^{(0)} = 1$ then
         $(C, S^{(0)}) := (C, S^{(0)}) + M^{(0)}$
         for $j = 1$ to $e - 1$
             $(C, S^{(j)}) := C + x_i Y^{(j)} + M^{(j)} + S^{(j)}$
             $S^{(j-1)} := (S_0^{(j)}, S_{w-1..1}^{(j-1)})$

$$S^{(e-1)} := (C, S^{(e-1)}_{w-1..1})$$
else
$$\text{for } j = 1 \text{ to } e - 1$$
$$(C, S^{(j)}) := C + x_i Y^{(j)} + S^{(j)}$$
$$S^{(j-1)} := (S^{(j)}_0, S^{(j-1)}_{w-1..1})$$
$$S^{(e-1)} := (C, S^{(e-1)}_{w-1..1})$$

The MWR2MM algorithm computes a partial sum S for each bit of X, scanning the words of Y and M. Once the precision is exhausted, another bit of X is taken, and the scan is repeated. Thus, the algorithm imposes no constraints to the precision of operands. The arithmetic operations are performed in precision w bits, and they are independent of the precision of operands. What varies is the number of loop iterations required to accomplish the modular multiplication. The carry variable C must be in the set $\{0, 1, 2\}$. This condition is imposed by the addition of the three vectors S, M, and $x_i Y$. To have containment in the addition of 3 w-bit words and a maximum carry value C_{max} (generated by previous word addition), the following equation must hold:

$$3(2^w - 1) + C_{max} \le C_{max} 2^w + 2^w - 1$$

which results in $C_{max} \ge 2$. Thus, choosing $C_{max} = 2$ is enough to satisfy the containment condition. The carry variable C is represented by two bits.

5 Parallel Computation of the MWR2MM

In this section we analyze the data dependencies on the proposed algorithm (MWR2MM) giving more information on the its potential parallelism and investigating parallel organizations suitable for its implementation.

The dependency between operations within the loop for j restricts their parallel execution due to dependency on the carry – C. However, parallelism is possible among instructions in different i loops. The dependency graph for the MWR2MM algorithm is shown in Figure 1. Each circle in the graph represents an atomic computation and is labeled according to the type of action performed. Task A corresponds to three steps: (1) test the least significant bit of S to determine if M should be added to S during this and next steps, (2) addition of words from S, $x_i Y$, and M (depending on the test performed), and (3) one-bit right shift of a S word. Task B corresponds to steps (2) and (3). We observe from this graph that the degree of parallelism and pipelining can be very high.

Each column in the graph may be computed by a separate processing element (PE), and the data generated from one PE may be passed to another PE in a pipelined fashion. An example of the computation executed for 5-bit operands is shown in Figure 2 for the word size of $w = 1$ bit. Since the jth word of each input operand is used to compute word $j - 1$ of S, the last B task in each column must receive $M^{(e)} = Y^{(e)} = 0$ as inputs. This condition is enough to guarantee that $S^{(e-1)}$ will be generated based only on the internal PE information. Note also that there is a delay of 2 clock cycles between processing a column for x_i

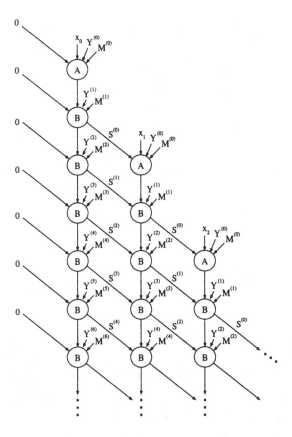

Fig. 1. The dependency graph for the MWR2MM Algorithm.

and a column for x_{i+1}. The total execution time for the computation shown in Figure 2 is 14 clock cycles.

Tasks A and B are performed on the same hardware module. The local control circuit of the module must be able to read the least significant bit of $S^{(0)}$ at the beginning of the operation, and keep this value for the entire operand scanning. Recall that the *even* condition of $S_0^{(0)}$ determines if the processing unit should add M to the partial sum during the pipeline cycle. The *pipeline cycle* is the sequence of steps that a PE needs to execute to process all words of the input operands.

The maximum degree of parallelism that can be attained with this organization is found as

$$p_{max} = \left\lceil \frac{e+1}{2} \right\rceil . \tag{2}$$

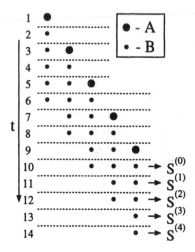

Fig. 2. An example of computation for 5-bit operands, where $w = 1$ bit.

It is easy to see from Figure 2 that $p_{max} = 3$. When less than p_{max} units are available, the total execution time will increase, but it is still possible to perform the full precision computation with the smaller circuit. Figure 3 shows what happens when only 2 processing modules are used for the same computation shown in Figure 2. In this case, the computation during the last pipeline cycle wastes one of the stages, because m is not a multiple of 2.

The total computation time T (in clock cycles) when $n \leq p_{max}$ modules (stages) are used in the pipeline is

$$T = \left\lceil \frac{m+1}{n} \right\rceil (e + 1) - 1 + 2(n - 1) \tag{3}$$

where the first term corresponds to the number of pipeline cycles ($\lceil (m + 1)/n \rceil$) times the number of clock cycles required by a pipeline stage to compute one full-precision operand, and the last term corresponds to the latency of the pipeline architecture. With n units, the average utilization of each unit is found as

$$U = \frac{\text{Total number of time slots per bit of } X \times m}{\text{Total number of time slots} \times n} = \frac{m(e+1)}{Tn} . \tag{4}$$

Figure 4 shows the hardware utilization U, total computation time T, and speedup of 2 or 3 units over one unit, for a small range of the precision, and word size $w = 8$ bits. We can see that the overhead of the pipelined organization becomes insignificant for precision $m \geq 3w$. We can attain a speedup very close to the optimum for even small number of operand bits.

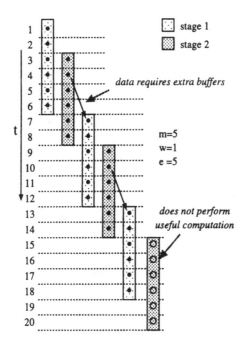

Fig. 3. An example of computation for 5-bit operands with two pipeline stages.

6 Design of the Scalable Architecture

A pipeline with 2 computational units is shown in Figure 5. One aspect of this organization is the register file design. Since the data is received word-serially by the kernel, the registers must work as rotators (circular shift registers) in some cases and shifters in other cases. The registers which store Y and M work as rotators. The processing elements itself must relay the received digits to the next unit in the pipeline. All paths are w bits wide, except for the x_i inputs (only 1 bit). The values of x_i come from a p-shift register, where p equals to the number of processing elements in the pipeline. The register for S must be a shift register, since its contents is not reused. The length (L) of the shift register for S values depends on the number of words (e) and the number of stages (n) in the pipeline. This length is determined as:

$$L = \begin{cases} e + 2 - 2n & \text{if } (e+2) > 2n \\ 0 & \text{otherwise} \end{cases} \tag{5}$$

Observe that these registers will not consume more than what is normally used in a conventional radix-2 design of the MM. These registers can be easily implemented by connecting one memory element to another in a chain (or loop), which will not impact the clock cycle of the whole system. Since we also need

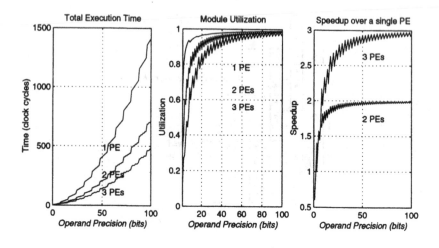

Fig. 4. The performance figures for multiple units with $w = 8$ bits.

loading capability for the rotator, multiplexers (MUXes) should be used between certain memory elements in the chain. The delay imposed by these MUXes will not create a critical path in the final circuit. To avoid having too many MUXes, we may load M and Y serially, during the last pipeline cycle of the algorithm. In this case, MUXes are required between two memory elements of the rotator only (not between all of the memory elements).

The global control block was not included in this figure for simplicity. Its function can be inferred from the dependency graph and the algorithm already presented. The shaded box represent flip flops.

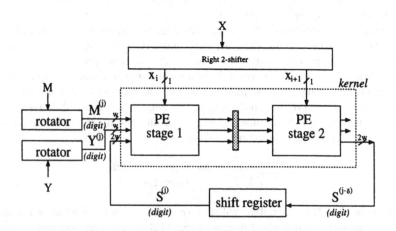

Fig. 5. Pipelined organization with 2 units.

6.1 Processing Element

The block diagram of the processing element is shown in Figure 6. The data path receives the inputs from the previous stage in the pipeline, and computes the next $S^{(j)}$ digits serially. The inputs are delayed one extra clock cycle before they are sent to the next stage.

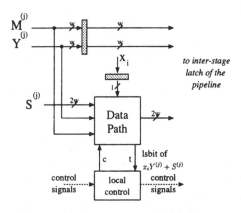

Fig. 6. The block diagram of the processing unit.

To reduce storage and arithmetic hardware complexity, we consider that M, X, and Y are available in non-redundant form. The internal sum S is received and generated in the redundant Carry-Save form. In this case, $2w$ bits per word are transferred between units in each clock cycle. The data path also makes available the information on the least significant bit (t) of the computation $S^{(j)} + x_i Y^{(j)}$ which is the first computation step performed by the data path in each pipeline cycle. Only the value t obtained when the least significant digits of Y and S come into the unit should be used to control the addition of M (control signal c). The local control is responsible for storing the t value during the pipeline cycle, and also relay some control signals to the downstrem modules.

The design of the data path follows the idea presented in [14] modified for least-significant-digit-first type of computation. The basic organization of the data path consists of two layers of carry-save adders (CSA). Assuming a full-precision structure as in Figure 7(a), we propose the retiming process shown for the case $w = 1$ to generate the serial circuit design presented in Figure 7(b). For $w > 1$, larger groups of adders are considered, based on the same approach. Notice that the cycle time may increase for larger w as a result of the broadcast problem only; it will not depend on the arithmetic operation itself. The high-fanout signals in the design are x_i and c, and both change value only once for each pipeline cycle. Observe that the bit-right-shift that must be performed by the data path is already included in the CSA structure shown in the Figure.

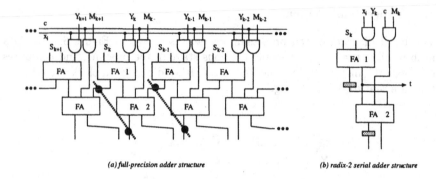

(a) full-precision adder structure

(b) radix-2 serial adder structure

Fig. 7. The serial computation of the MM operations.

The data path design for the case $w = 3$ is shown in Figure 8. It has a more complicated shift and alignment section to generate the next S word. When computing the bits of word j (step j), the circuit generates $w - 1$ bits of $S^{(j)}$, and the most significant bit of $S^{(j-1)}$. The bits of $S^{(j-1)}$ computed at step $j - 1$ must be delayed and concatenated with the most significant bit generated at step j (alignment).

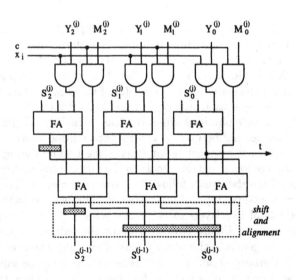

Fig. 8. PE's data path for $w = 3$ bits.

7 Area/Time Tradeoffs

After describing the general building block for the implementation of our scalable MM architecture, we discuss the area/time tradeoffs that arise for different values of operand precision m, word size w, and the pipeline organization. The area A is given as a design constraint. In this analysis, we do not consider the wiring area. For a first order approximation we consider that the propagation delay of the processing element is independent of w (this hypothesis is reasonable when w is small). This assumption implies that the clock cycle is the same for all cases and the comparison of speed among different designs can be made based on clock cycles. The area used by registers for the intermediate sum, operands and modulus is the same for all designs.

It is clear that the proposed scheme has the worst execution time for the case $w = m$, since some extra cycles were introduced by the computational unit in order to allow word-serial computation, when compared to other full-precision designs. Thus, we will consider the case when the available chip area is not sufficient to implement a full-precision conventional design. The performance evaluation resumes to the question:

What is the best organization for the scalable architecture for a given area?

We used VHDL on the Mentor graphics tools to synthesize the circuit with the $1.2\mu m$ CMOS technology. The cell area for a given word size w is obtained as

$$A_{cell}(w) = 47.2w ,$$

where the value 47.2 is the area cost provided by the tool (a 2-input NAND gate corresponds to 0.94). When using the pipelined organization, the area of each inter-stage latch is important, and was measured as $A_{latch}(w) = 8.32w$. The area of a pipeline with n units is given as

$$A_{pipe}(n, w) = (n - 1)A_{latch}(w) + nA_{cell}(w) = 55.52nw - 8.32w . \quad (6)$$

The maximum word size that can be used in the particular design (w_{max}) is a function of the available area A and the number of pipeline stages n. It is found as

$$A_{pipe}(n, w) \leq A$$
$$55.52nw - 8.32w \leq A$$
$$w \leq \frac{A}{55.52n - 8.32}$$
$$w_{max}(A, n) = \left\lfloor \frac{A}{55.52n - 8.32} \right\rfloor . \quad (7)$$

Based on w_{max}, we obtain the total execution time (in clock cycles) for operands with precision m from Equation 3, as follows:

$$T(m, A, n) = \left\lceil \frac{m + 1}{n} \right\rceil \left(\left\lceil \frac{m + 1}{w_{max}(A, n)} \right\rceil + 1 \right) - 1 + 2(n - 1) . \quad (8)$$

For a given area A, we are able to try different organizations and select the faster one. The graph given in Figure 9 shows the computation time for various pipeline configurations for $A = 20,000$. The number of stages that provides the best performance varies with the precision required in the computation. For the cases shown, 5 stages would provide good performance. We don't want to have too many stages for two reasons: (1) high utilization of the processing elements will be possible only for very high precision and (2) the execution time may have undesirable oscillations (as shown in the rightmost part of the curve for $m = 1024$). The behavior mentioned in (2) is the result of (i) word size w is not a good divisor for m, producing one word (most significant) with few significant bits, and (ii) there is not a good match between the number of words e and n, causing a sub-utilization of stages in the pipeline.

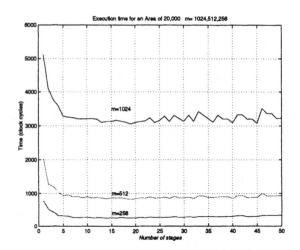

Fig. 9. The execution time of the MM hardware for various precision and configurations.

For a fixed area, the word size becomes a function of the number of stages only. The word size decreases as the number of stages in the pipeline increases. The word size for some values of n is given on Table 1.

n (stages)	1	2	3	4	5	6	7	8	9	10
w (bits)	423	194	126	93	74	61	52	45	40	36

Table 1. The number of pipeline stages versus the word size, for a fixed chip area.

From the synthesis tools we also obtained a minimum clock cycle time of 11 ns (clock frequency of 90MHz). For the case $m = 1024$ bits, $n = 10$ stages, and $w = 36$ bits, the total execution time is $3107 * 11 = 34,177$ nanoseconds. The correction step was not included in these estimates, but it would require another pipeline cycle to be performed.

8 Conclusions

We presented a new architecture for implementing the Montgomery multiplication. The fundamental difference of our design from other designs described in the literature is that it is scalable to any operand size, and it can be adjusted to any available chip area. The proposed architecture is highly flexible, and provides the investigation of several design tradeoffs involved in the computation of the Montgomery multiplication. Our analysis shows that a pipeline of several units is more adequate than a single unit working with a large word length. This is an interesting result since using more units we can reduce the word size and consequently the data paths in the final circuit, reducing the required bandwidth. The proposed data path for the multiplier was synthesized to a circuit that is able to work with clock frequencies up to 90MHz (for the CMOS technology considered in this work). The total time to compute the Montgomery multiplication for a given precision of the operands will depend on the available area and the chosen pipeline configuration. The upper limit on the precision of the operands is dictated by the memory available to store the operands and internal results.

Acknowledgements

This research is supported in part by Secured Information Technology, Inc. The authors would like to thank Erkay Savaş (Oregon State University) for his comments on the algorithm definition.

References

1. A. Bernal and A. Guyot. Design of a modular multiplier based on Montgomery's algorithm. In *13th Conference on Design of Circuits and Integrated Systems*, pages 680–685, Madrid, Spain, November 17–20 1998.
2. W. Diffie and M. E. Hellman. New directions in cryptography. *IEEE Transactions on Information Theory*, 22:644–654, November 1976.
3. S. E. Eldridge and C. D. Walter. Hardware implementation of Montgomery's modular multiplication algorithm. *IEEE Transactions on Computers*, 42(6):693–699, June 1993.
4. T. Hamano, N. Takagi, S. Yajima, and F. P Preparata. $O(n)$-Depth circuit algorithm for modular exponentiation. In S. Knowles and W. H. McAllister, editors, *Proceedings, 12th Symposium on Computer Arithmetic*, pages 188–192, Bath, England, July 19–21 1995. Los Alamitos, CA: IEEE Computer Society Press.

5. Ç. K. Koç and T. Acar. Fast software exponentiation in GF(2^k). In T. Lang, J.-M. Muller, and N. Takagi, editors, *Proceedings, 13th Symposium on Computer Arithmetic*, pages 225–231, Asilomar, California, July 6–9, 1997. Los Alamitos, CA: IEEE Computer Society Press.

6. Ç. K. Koç and T. Acar. Montgomery multiplication in GF(2^k). *Designs, Codes and Cryptography*, 14(1):57–69, April 1998.

7. Ç. K. Koç, T. Acar, and B. S. Kaliski Jr. Analyzing and comparing Montgomery multiplication algorithms. *IEEE Micro*, 16(3):26–33, June 1996.

8. P. Kornerup. High-radix modular multiplication for cryptosystems. In E. Swartzlander, Jr., M. J. Irwin, and G. Jullien, editors, *Proceedings, 11th Symposium on Computer Arithmetic*, pages 277–283, Windsor, Ontario, June 29 – July 2 1993. Los Alamitos, CA: IEEE Computer Society Press.

9. A. J. Menezes. *Elliptic Curve Public Key Cryptosystems*. Boston, MA: Kluwer Academic Publishers, 1993.

10. P. L. Montgomery. Modular multiplication without trial division. *Mathematics of Computation*, 44(170):519–521, April 1985.

11. H. Orup. Simplifying quotient determination in high-radix modular multiplication. In S. Knowles and W. H. McAllister, editors, *Proceedings, 12th Symposium on Computer Arithmetic*, pages 193–199, Bath, England, July 19–21 1995. Los Alamitos, CA: IEEE Computer Society Press.

12. R. L. Rivest, A. Shamir, and L. Adleman. A method for obtaining digital signatures and public-key cryptosystems. *Communications of the ACM*, 21(2):120–126, February 1978.

13. A. Royo, J. Moran, and J. C. Lopez. Design and implementation of a coprocessor for cryptography applications. In *European Design and Test Conference*, pages 213–217, Paris, France, March 17-20 1997.

14. A. F. Tenca. *Variable Long-Precision Arithmetic (VLPA) for Reconfigurable Coprocessor Architectures*. PhD thesis, Department of Computer Science, University of California at Los Angeles, March 1998.

15. C. D. Walter. Space/Time trade-offs for higher radix modular multiplication using repeated addition. *IEEE Transactions on Computers*, 46(2):139–141, February 1997.

Arithmetic Design for Permutation Groups

Tamás Horváth

Secunet AG, Im Teelbruch 116, 45219 Essen, Germany, email: horvath@secunet.de

Abstract. This paper investigates the hardware implementation of arithmetical operations (multiplication and inversion) in *symmetric* and *alternating groups*, as well as in *binary permutation groups* (permutation groups of order 2^r). Various fast and space-efficient hardware architectures will be presented. High speed is achieved by employing switching networks, which effect multiplication in one clock cycle (full parallelism). Space-efficiency is achieved by choosing, on one hand, proper network architectures and, on the other hand, the proper representation of the group elements. We introduce a non-redundant representation of the elements of binary groups, the so-called *compact representation*, which allows low-cost realization of arithmetic for binary groups of large degrees such as 128 or even 256. We present highly optimized multiplier architectures operating directly on the compact form of permutations. Finally, we give complexity and performance estimations for the presented architectures.

Keywords: permutation multiplier, switching network, destination-tag routing, sorting network, separation network, binary group, compact representation, secret-key cryptosystem, PGM.

1 Introduction

Several cryptosystems, such as RSA, elliptic curve systems, IDEA or SAFER, utilize operations in algebraic domains like polynomial rings or Galois-fields. Efficient implementations of the basic arithmetical operations in those domains have been extensively studied but not much attention has been spent to simpler constructs like permutation groups. Our research on permutation group arithmetic has been motivated by the implementation of a secret-key cryptosystem called PGM (Permutation Group Mapping) [11, 12], which utilizes some generator sets, called *group bases*, for encryption.

Briefly, a *basis* for a permutation group G is an ordered collection $\beta = (B_0, B_1, \ldots, B_{w-1})$ of ordered subsets (so-called *blocks*) $B_i = (b_{i,0}, \ldots, b_{i,r_i-1})$ of G, such that each element $g \in G$ has a unique representation in form of a product $g = b_{0,x_0} \cdot b_{1,x_1} \cdots b_{w-1,x_{w-1}}$, where $b_{i,x_i} \in B_i$ and $0 \leq x_i \leq r_i-1$. Thus, β defines a bijective mapping $\hat{\beta} : G \to X$ which assigns to each element $g \in G$ a unique vector $x = (x_0, \ldots, x_{w-1}) \in X$, where $X = \mathbb{Z}_{r_0} \times \mathbb{Z}_{r_1} \times \cdots \times \mathbb{Z}_{r_{w-1}}$. Clearly, $|G| = |X| = r_0 r_1 \cdots r_{w-1}$. $\hat{\beta}^{-1}$ can be effected by means of permutation multiplications, whereas $\hat{\beta}$ involves finding the proper factors in β, and is hence called *factorization*. There is a huge amount of so-called *transversal* bases, for which factorization can be effected very efficiently by means of permutation multiplications and inversions. A pair of such, randomly chosen bases β_1 and β_2 for some group G (called the *carrier group*) form the key for PGM. The encryption of a cleartext message m is $c = \hat{\beta}_2(\hat{\beta}_1^{-1}(m))$. Decryption is performed in the same manner by exchanging the roles of β_1 and β_2, i.e. $m = \hat{\beta}_1(\hat{\beta}_2^{-1}(c))$. To accomodate the cryptosystem to some binary cleartext and ciphertext space $\mathcal{M} = \mathcal{C} = \mathbb{Z}_{2^k}$, an additional, fixed mapping $\lambda : \mathcal{M} \rightarrow X$ has to be effected prior to and λ^{-1} after the actual encryption.

It is very natural to represent permutations in a computer in the so-called *Cartesian* form. Section 2 introduces the basic principle of multiplying two Cartesian permutations in a switching network. Section 3 presents different, mostly novel multiplier architectures operating in the symmetric group. Unfortunately, a symmetric carrier group S_n has a serious drawback. It is namely that any basis for S_n ($n > 2$) has several blocks with length $r_i \neq 2^{k_i}$. It follows that λ, which can be seen as a conversion from a binary to a *mixed* radix $r = (r_0, r_1, \ldots, r_{w-1})$, is computationally rather intensive [14].

As oppesed to a symmetric group, any basis for a *binary group* (a permutation group of order 2^r) has block lengths $r_i = 2^{k_i}$, and thus the mapping λ for such a carrier group is trivial. Since a binary group of degree n is only a small subgroup of the symmetric group S_n, the use of some large degree ($n = 128 \ldots 256$) is indicated, which makes the use of use of the multipliers, proper proper for symmetric groups, infeasible. The problem lies in the fact that the Cartesian representation of binary group elements contains a large amount of redundancy. In Sec. 4 we introduce a novel, non-redundant representation, the so-called *compact representation*, and present various multiplier architectures operating directly on the compact form of permutations. Finally, Sec. 5 gives complexity and performance estimations for the presented architectures. It turns out that a PGM system with a binary carrier group is indeed much more efficient than one based on a symmetric group of similar order.

2 Multiplication in Permutation Networks

To briefly recall, a permutation p of degree n is a bijection $p : L \rightarrow L$, where L is a set of n arbitrary symbols or *points*. In the arithmetic we propose, elements of the symmetric group S_n are represented in the so-called *Cartesian* form. In this representation, $L = \{0, 1, \ldots, n-1\}$ and a permutation $p \in S_n$ is a vector $p = (p(0), p(1), \ldots, p(n-1))$ of the n function values. Suppose now, elements of vector p are physically stored in their natural order in a block P of registers, i.e. $P[i] = p(i)$ for $0 \leq i \leq n-1$, where $P[i]$ denotes the content of the ith register.

By definition, the product of two permutations a and b is permutation $q = a \cdot b : q(i) = b(a(i))$, for $0 \leq i \leq n-1$. Using the Cartesian representation, the product q can be computed by means of n memory transfer operations: $Q[i] := B[A[i]]$ for $0 \leq i \leq n-1$, where A, B and Q are the memory blocks storing a, b and q, respectively. For simplicity of the notation, we are not going to distinguish register blocks from their content, but simply write $Q = A \cdot B$ to denote the product.

By definition, the inverse of a permutation a is permutation $q = a^{-1}$, for which $a \cdot a^{-1} = a^{-1} \cdot a = \iota$, where ι denotes the *identity* permutation (i.e. $\iota(i) = i$). The inverse a^{-1} can be obtained in block Q by applying n memory transfers $Q[A[i]] := i$. We denote the inverse simply as $Q = A^{-1}$.

The memory transfers can be carried out either sequentially or in parallel. The former is the typical software implementation. The parallel implementation exploits the fact that the n memory transfers are completely independent and can thus be carried out simultaneously in *switching networks*, as follows. For multiplication, the $A[i]^{\text{th}}$ register of source block B is connected to the ith register of the destination block Q, i.e. A is interpreted during *routing* as a vector of source addresses. After setting up the network, the content of B is copied to Q via the established connections, forming the product $A \cdot B$ in Q. Fig. 1a illustrates this principle on a small example.

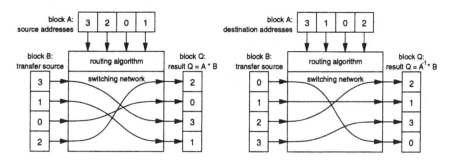

Fig. 1. a, Multiplication and   **b,** inversion in a switching network

When interpreting A as a vector of destination addresses, the reverse connections are established. By copying then the content of B to Q, the product $A^{-1} \cdot B$ is obtained in Q. By substituting $B = \iota$, the network delivers the inverse of A, as shown in Fig. 1b.

3 Arithmetic in Symmetric Groups

According to the computing principles introduced above, a multiplier network for S_n should be an n-input, n-output network (briefly (n, n) network). For the sake of full parallelism, the network must be able to connect the n inputs to the n outputs simultaneously, that is without collision (or *blocking*) at any of the links. Since the network has to be completely re-routed after each multiplication, *rearrangeable networks* are favourable compared to the more complex *non-blocking* networks [3]. Moreover, since the routing operand may come from the entire symmetric group S_n, the network must be able to realize all possible n-to-n connections. Exactly these characteristics describe a specific class of switching networks, called *permutation networks*.

3.1 Crossbar Networks

The *crossbar network* is the most fundamental *single stage* permutation network [3]. As depicted in Fig. 2, it consists of $n \times n$ switches in a matrix form, which pass the signals from input port C to output port Q. In the following, we propose three different schemes for fast routing of the network.

In the first routing scheme the pure crossbar network is equipped with a routing port A and with corresponding horizontal lines, each controlling an n-to-1 multiplexer (MUX), as shown in Fig. 2a. According to the control signal, each MUX selects one of the n signals of port C, and forwards it to port Q. Note in this mechanism that the data items entered at port A are used as source addresses. Hence, when entering Cartesian permutations at A and C, the network computes the product $Q = A \cdot C$.

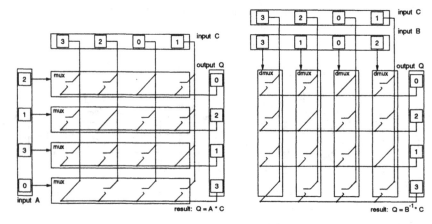

Fig. 2. a, MUX-type multiplier   **b,** DMUX-type multiplier

The second routing scheme in Fig. 2b adds routing port B and corresponding vertical lines to the pure crossbar network. Each of these lines controls a 1-to-n demultiplexer (DMUX) which transmits the input signal through one of the n output lines towards Q, while disconnecting from all other output lines. Note that data items of A are interpreted in this mechanism as destination addresses. Accordingly, when entering Cartesian permutations at B and C, the network computes the product $Q = B^{-1} \cdot C$.

A combination of the above two routing schemes yields the third one of Fig. 3. Both routing ports A and B are included here, and are connected to horizontal and, respectively, vertical addressing lines. In addition, each switching cell is equipped with an **equivalence** comparator, which compares the addresses received from the neighboring addressing lines. If the addresses are equal, the comparator closes the switch, otherwise opens it. By entering Cartesian permutations at ports A, B and C respectively, the result obtained at port Q is the product $Q = A \cdot B^{-1} \cdot C$. A more detailed description of this architecture and of a bit-parallel realization has been published in [13].

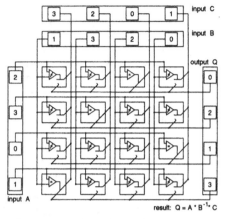

Fig. 3. A 3-operand crossbar multiplier

3.2 Sorting Networks

The *Beneš-network* [3, 5] is known to be the most efficient rearrangeable multi-stage network topology, based on elementary (2,2) switching cells. However, its routing algorithm, the so-called *looping algorithm* [5], is intrinsically sequential and can only be effected in a *centralized control* unit. Accordingly, the mechanism is rather slow and thus not suitable for a multiplier network. On the other hand, there exists a large class of multistage networks, the so-called *digit-controlled* (or delta) *networks*, which possess a very convenient routing algorithm, the so-called *destination-tag routing* (or *self-routing*) [5, 7, 8, 10]. This *distributed control* mechanism is very fast, since the individual cells decide independently and simultaneously. Unfortunately, delta networks are blocking ones.

A *sorting network* is an (n, n) multistage network effecting some deterministic sorting algorithm [2, 4, 6, 9]. The network is built from elementary (2,2) *compare-exchange modules*. Each module compares the two incoming numbers and routes them according to their magnitudes. No matter in which order input numbers are entered at the input, the network applies the proper permutation to them and delivers the sorted sequence. Hence, any sorting network can be regarded as a rearrangeable permutation network.

Though all known sorting networks are more complex than the Beneš network, they offer a way for destination-tag routing, as follows. If entering elements $a(i)$ of a Cartesian permutation a at input $A[i]$ $(0 \le i \le n - 1)$, the network forwards each $a(i)$ to output $Q[a(i)]$. Put another way, vector a carries routing information and designates n parallel paths from $A[i]$ to $Q[a(i)]$. When now attaching the elements of a permutation b to a, i.e. entering packets of the form $(a(i), b(i))$ at input line i, destination tag $a(i)$ will route $b(i)$ through the network towards $Q[a(i)]$, i.e. eventually $Q[a(i)] = b[i]$ is obtained. It is seen that the Cartesian permutation q obtained at Q is $q(a(i)) = b(i)$, or equivalently, $a \cdot q = b$, and thus $q = a^{-1} \cdot b$, which fact amounts to the general multiplication principle of Sec. 2. Figure 4 illustrates the method on a small example.

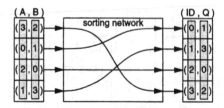

Fig. 4. Multiplication in a sorting network

In Fig. 5 we introduce two classical sorting networks. The *odd-even transposition sorter* is the parallel implementation of the insertion sort and, at the same time, of the selection sort algorithms. The n input numbers are sorted in n stages, comprising $n(n-1)/2$ modules. Accordingly we say that the network has depth n and complexity $n(n-1)/2$. The arrows in the symbols of the compare-exchange modules indicate the direction which the larger numbers are forwarded to. Note that this network has a completely "straight" wiring topology, which is advantageous in view of wiring area. The *bitonic sorter* as well as *Batcher's odd-even sorter* [9] are known to be the most efficient regular topologies, having $O(\log^2 n)$ stages in a recursive structure. Note that many lines cross between certain stages, which is in direct correspondence with the wiring area.

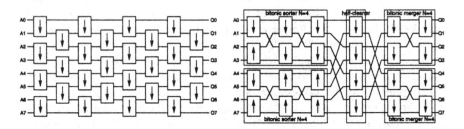

Fig. 5. a, The odd-even transposition and **b,** the bitonic sorters

In a straightforward realization of the compare-exchange modules, comparison is carried out first, and the result is then used to set the switches. In this method, no data can be transferred until the comparison is completed. Considerable acceleration can be achieved by recognizing that comparison can be performed sequentially, scanning from the MSBs towards the LSBs of the input numbers X and Y, according to the following algorithm:

1. As long as $X_i = Y_i$ while scanning bits in decreasing order of i, it does not matter how the switch is set, and thus X_i and Y_i can be passed to the next stage;
2. as soon as difference is noticed at bit j, i.e. $X_j \neq Y_j$, all switches for bits $i \leq j$ can be set to the same state, which is determined by the relation of X_j and Y_j.

In the improved scheme, corresponding bits X_i and Y_i are transmitted to the next stage immediately after their comparison, that is before the comparison of lower bits is completed. The higher order bits reach the next stage therefore earlier than the lower order ones, where their comparison starts immediately. In this way, each comparator stage delays the destination tag effectively only by the time of a single bit-comparison. For implementation details we refer to [14].

3.3 Separation Networks

Sorting networks are able to sort arbitrary number sequences. Note however that Cartesian permutations are special sequences, such that each number of the range $0 \ldots n-1$ occurs exactly once. This kind of sequence we call a *permutation sequence*. In the following, we introduce a class of novel permutation network architectures, which exploit this property to reduce hardware complexity. The new networks employ the *radix sorting* algorithm [1] for routing: destination addresses are represented as binary strings, starting with the MSB as first letter. Sorting proceeds as follows: first the strings starting with a '1' as first letter are separated from those starting with a '0'. As second step, both of the resulting subsequences are further be split up so that strings having '1' as second letter get separated from those having '0' at the same position. The 'divide-and-conquer' principle is followed in this way till the last step, where strings with trailing '1' are separated from those with trailing '0'.

Since a permutation sequence contains a predetermined set of strings, the number of strings with '1' and respectively '0' at any particular position is constant, irrespective of the actual sequence. Due to this fact, the length of the separated subsequences is known and constant for all separation steps. In the specific case of $n = 2^m$, all separated subsequences are *balanced*, i.e. contain exactly as many 1's as 0's at any particular position. This property is the basis for the design of *separator networks*. Each separation step is effected in a dedicated *separator stage*. The first separator stage splits the input sequence in two halves of length $n/2$ (without actually achieving perfect ordering), the next stage produces subsequences of length $n/4$, and so on. Networks of degree $n \neq 2^m$ can be constructed by omitting parts of a network of degree 2^m, where $n < 2^m$.

The strength of the technique lies in the fact that any particular stage can achieve the separation by looking at corresponding single bits of the destination tags. The method can thus be considered as the generalization of the bit-controlled self-routing algorithm for permutation networks. Interestingly, comparing corresponding bits X and Y of two destination tags and routing them towards the proper output H ("higher" value) and respectively L ("lower" value) requires no logic at all. To see this, consider the truth-table of the "binary" compare-exchange module:

X	Y	switch state required	switch state chosen	H	L
0	0	don't care	across	0	0
0	1	across	across	1	0
1	0	straight	straight	1	0
1	1	don't care	straight	1	1

By choosing the switch state for *don't care*'s as shown above, the switches can be controlled directly by input bit X, whereas H and L can be formed by a single OR- and respectively AND-gate. See [14] for implementation details.

In the following, we present a couple of novel separator network architectures. The first scheme is related to the bitonic sorter (Fig. 5b). The two bitonic sorters of length $n/2$ and a *half-cleaner stage* of this network form a so-called *selection network* [9], which separates the $n/2$ largest from the $n/2$ smallest elements. If entering a sequence of $n/2$ 1's and $n/2$ 0's, the selection network separates the 1's from the 0's. Clearly, this is also achieved when the magnitude comparator modules are replaced with "binary" comparator modules. Such *bitonic separator* stages can be used to build a *bitonic separator network*, as illustrated in Fig. 6 for $n = 8$. The network has depth of order $O(\log^3 n)$.

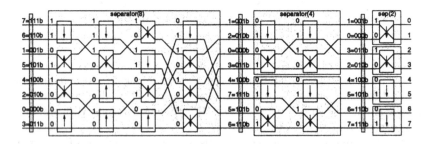

Fig. 6. A separator network based on bitonic separators

Note that the $(n/2,n/2)$-sorters in front of the half-cleaner can actually be replaced by any kind of sorting network, for instance, by odd-even transposition sorters. We call the network obtained in this way the *linear odd-even separator network*. As the name suggests, it has depth of order $O(n)$.

Separator stages can rely on other principles, too. The sorter depicted in Fig. 7 employs a novel separator type of depth $O(n)$, which we call a *diamond separator*. The underlying sorting principle is similar to that of the odd-even transposition sorter. The advantage of this architecture is the completely straight wiring pattern. Its drawbacks are that the network is rather deep and hard to lay-out in a rectangular form.

The *rotation separator* offers lower depth, rectangular layout and still a "nearly" straight wiring topology. The separation principle can be followed in Fig. 8. Links running across the network are considered to be of two types: *0-lines*, which are expected to deliver 0's at the output, and *1-lines*, that should deliver 1's. A '1' on a 0-line (and a '0' on a 1-line, respectively) is considered as a *1-error* (a *0-error*, respectively). Due to the balance in a permutation sequence of length $n = 2^m$, 1-errors are present in the same number as 0-errors at any particular stage. Each compare-exchange module receives input from a 0-line and a 1-line, and outputs to a 0-line and a 1-line. When a 1-error and a 0-error are received, they "neutralize" each other, i.e. both errors disappear.

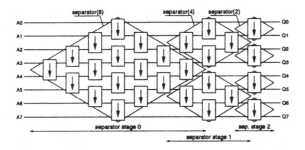

Fig. 7. The "diamond" separator network with 8 inputs

The topology of the network implements the following strategy for eliminating all errors in the input sequence: 0-lines (carrying potentially 1-errors) are iteratively "rotated around" and combined pairwise with 1-lines (carrying potentially 0-errors). In order for all 0-lines to be combined with all 1-lines, $n/2$ rotation steps are needed, and hence the separator stage has depth $n/2$. The total depth of the entire network is $n-1$.

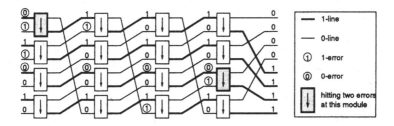

Fig. 8. A "rotation" separator stage with 8 inputs

4 Arithmetic in Binary Groups

As mentioned, any binary group of degree n is a subgroup of S_n. Unfortunately, even a so-called Sylow-2 subgroup $\mathcal{H}_s$ of S_n, which is of maximal order, is rather small; it has order $|\mathcal{H}_s| = 2^{n-1}$ if $n = 2^s$. Hence, if a certain group size is required, the usage of a binary group of some large degree is indicated. For instance, if a group order of at least 2^{127} is required, not unusual in cryptographic applications, either a symmetric group of degree $n = 34$ or a Sylow-2 subgroup of degree $n = 128$ may be chosen. Unfortunately, the storage of Cartesian permutations (7*127=896 bits in the above example) as well as the multipliers based on permutation networks are very extensive for binary groups of such large degrees. Note however that the Cartesian form is very redundant for representing binary group elements, and that the multiplier networks would be used rather inefficiently too, because most of the possible permutation patterns, namely thus in S_n but not in $\mathcal{H}_s$, would never be configured.

A study of the *indirect binary cube* (IBC) network for $n = 2^s$ has shown that though the set of permutations realized by the network is not a group, it embeds a Sylow-2 subgroup $\mathcal{H}_s$ of $\mathcal{S}_n$. Similar results can be obtained for the "inverse" of the IBC network, the so-called *generalized cube* (also called *butterfly* or *SW-Banyan*) network, as well as for other (n, n) delta networks, such as the *omega*, the *baseline*, the *modified data manipulator* MDM and their respective "inverses", the *reverse omega* (also called *flip*), the *reverse baseline* and the *inverse* MDM networks [5, 10]. The different delta networks realize various instances of $\mathcal{H}_s$, while it is known that all Sylow-2 subgroups $\mathcal{H}_s$ of $\mathcal{S}_n$ are isomorphic.

A delta network of degree $n = 2^s$ comprises $s * n/2$ switches in s stages, and is thus considerably more efficient for $\mathcal{H}_s$ then any of the permutation networks. The construction of a multiplier we illustrate on the IBC network of degree $n = 8$, depicted on the left of Fig. 9. The network contains 12 binary switches and since it is a banyan network (i.e. there is one unique path from each input to each output), all of the 2^{12} different configurations realize different permutations. This permutation set of size 2^{12} is not a group, but it contains $\mathcal{H}_3$, which is of order 2^7. Clearly, some configurations will never be used if working in $\mathcal{H}_3$.

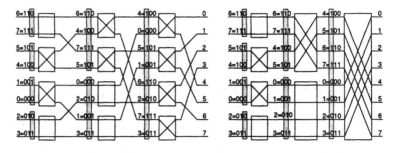

Fig. 9. The indirect binary cube network for $n = 8$ in two presentations

As illustrated in the figure, the IBC network can be configured by the bit-controlled self-routing algorithm, where switches in the first stage are controlled by the LSBs, while subsequent stages by succeeding bits of the destination tags. If first routing with a Cartesian permutation $a \in \mathcal{H}_3$ and then transferring another permutation $b \in \mathcal{H}_3$, the network delivers, according to the general multiplication principle of Sec. 2, the product $q = a^{-1} \cdot b$.

It turns out that if working in $\mathcal{H}_3$, switches of certain switch groups are always set to a common state, and can thus be unified in one *switching module*, as depicted on the right side of the figure. The unified switches can be controlled by one common signal, which sets either a "straight" or a "swapping" connection pattern. We call an IBC network with unified switches an UIBC network. The $2^{n-1} = 2^7$ different connections patterns of the UIBC network realize exactly the elements of (a specific instance of) $\mathcal{H}_3$.

The control bits can be extracted from the Cartesian permutation a by simply selecting certain bits of a. Actually, the $n-1 = 7$ control bits can be seen as a special representation of the group elements of $\mathcal{H}_3$, which we call the *compact representation*. From the fact $|\mathcal{H}_s| = 2^{n-1}$ it is seen that the compact representation is non-redundant and hence optimal. Expanding the compact form to the Cartesian form is similarly simple, it can achieved by reproducing (copying) certain bits of the compact permutation. Note that the ease of the conversions is not a general feature but specific to the instance of H_s induced by the use of the UIBC network.

A great advantage of the compact representation is that it allows space-efficient storage of elements of $\mathcal{H}_s$. Note furthermore that since the Cartesian form of $b \in \mathcal{H}_s$ is redundant, more bits than actually necessary are transferred by the UIBC network while multiplying. By removing links and switching components from the network which convey redundant bits of b, the complexity of the scheme can be significantly reduced. The optimized scheme transmits merely the compact form of b. The resulting multiplier network, called MULAIB, is shown in Fig. 10 for $n = 8$.

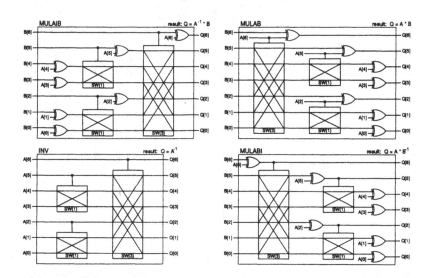

Fig. 10. Different multipliers and inverters working on compact permutations

Figure 10 illustrates further multiplier and inverter architectures deduced from the IBC network and respectively, from its inverse, the generalized cube network. All architectures work directly on the compact form of operands.

Above we followed an illustrative approach to introduce the arithmetic for binary groups. An accurate description of the construction of group $\mathcal{H}_s$ underlying the arithmetic, a formal definition of the compact representation, proofs of the multiplication algorithms, further multiplier and inverter schemes as well as a generalization of the theory to a large class of binary groups of arbitrary degree n have been omitted here in lack of space, but can be found in [14].

5 Conclusions

In the following, we give complexity and performance estimations for the presented multiplier architectures. The examined multipliers operate in the symmetric group S_{32} and in the binary group H_7 (degree $n = 128$), which have comparable orders: $|S_{32}| \approx 2^{117}$ and respectively $|H_7| = 2^{127}$. Estimations of complexity and delays have been made for the $0.7\,\mu m$ ES2 standard-cell CMOS technology of European Silicon Structures. The complexity of the typically extensive wiring of switching networks, indicated also by the measure *wiring width*, has been taken into account. The throughput of the networks has been calculated for a purely combinational, full-parallel, non-pipelined implementation. The estimation methodology as well as other implementation styles are detailed in [14]. Table 1 below summarizes the results.

Table 1. A comparison of multipliers for S_{32} and H_7

Multiplier Design	Topology		Complexity			Performance		
	depth	# sw. modules	gate count	wiring width	area mm²	gate delay	perf. MMPS	perf./ area
MUX-type crossbar	1	1024	8.68K	640	11.8	9	222	18.8
DMUX-type crossbar	1	1024	12.7K	640	15.8	12	167	10.5
3-operand crossbar	1	1024	25.0K	640	27.6	12	167	6.04
Linear oev. sorter	32	496	21.5K	0	16.3	72	27.8	1.70
Bitonic sorter	15	240	10.4K	820	14.4	38	52.6	3.65
Linear oev. sep.net	35	560	8.47K	526	10.2	71	28.2	2.76
Bitonic sep.net	25	400	5.95K	1030	10.7	51	39.2	3.67
Diamond sep.net	46	496	7.54K	0	5.73	93	21.5	3.75
Rotation sep.net	31	496	7.54K	526	9.26	63	31.7	3.43
MULAIB	7	756	1.61K	240	3.17	15	133	42.0
MULAB	7	756	1.61K	240	3.17	19	105	33.1
MULABI	7	756	1.61K	240	3.17	44	45.5	14.4
INV	7	756	1.33K	240	2.82	13	154	54.6

Among the multipliers for S_{32}, the crossbar architectures are very fast and cost-effective too. The bitonic separator network and the rotation separator network perform quite similarly, and are slightly smaller and slower than the well-known bitonic sorter. All multipliers for H_7 have extremely low gate-complexity, whereas about 60% of the total area is spent for global wiring in all designs. The reason that MULAB performs significantly worse than MULAIB is that control signals, that are to be distributed at a particular stage, are produced by the preceeding stage. Therefore, the delay of signal distribution adds to the total delay at each stage, a rather undesirable phenomenon.

To summarize, the multipliers for the binary groups outperform those for the symmetric group and because of their $O(n\log n)$ complexity, the gain becomes even more striking for larger groups. We stress here again that the very fundamental invention which allows both space-efficient storage and efficient computation in binary groups is that of the *compact representation*.

References

1. Alfred V. Aho, John E. Hopcroft, Jeffrey D. Ullmann:
   The Design and Analysis of Computer Algorithms
   Addison-Wesley, 1974.
2. Selim G. Akl: Parallel Sorting Algorithms
   Academic Press, 1985.
3. V. E. Beneš, Mathematical Theory of Connecting Networks and Telephone Traffic,
   Academic Press, 1965
4. Thomas H. Cormen, Charles E. Leiserson, Ronald L. Rivest:
   Introduction to Algorithms, MIT Press, 1990.
5. Tse-yun Feng, A Survey of Interconnection Networks, *IEEE Computer* , December
   1981, pp 12–27.
6. Alan Gibbons, Wojciech Rytter: Efficient Parallel Algorithms
   Cambridge University Press, 1988.
7. A.J. van de Goor: Computer Architecture and Design
   Addison-Wesley, 1989.
8. J.P. Hayes: Computer Architecture and Organization
   McGraw-Hill, 1988.
9. D. Knuth: The Art of Computer Programming
   Volume III: Sorting and Searching, Addison-Wesley, 1973.
10. C. P. Kruskal, M. Snir: "A Unified Theory of interconnection Network Structure",
    Theoretical Computer Science, Volume 48, 1986.
11. S. S. Magliveras, A cryptosystem from logarithmic signatures of finite groups, In
    Proceedings of the 29'th Midwest Symposium on Circuits and Systems, Elsevier
    Publishing Company (1986), pp 972–975.
12. S. S. Magliveras and N. D. Memon, Algebraic Properties of Cryptosystem PGM,
    in *Journal of Cryptology*, **5** (1992), pp 167–183.
13. T. Horváth, S. Magliveras, Tran van Trung, A Parallel Permutation Multiplier
    for a PGM Crypto-chip, *Advances in Cryptology - CRYPTO'94*, Springer-Verlag
    1994, pp 108–113.
14. T. Horváth, Secret-key Cryptosystem TST, *Ph.D. thesis, Institut for Experimental
    Mathematics, University of Essen, Germany, to be published in 1999.*

Fast Multiplication in Finite Fields GF(2^N)

Joseph H. Silverman

Mathematics Department, Brown University, Providence, RI 02912 USA
NTRU Cryptosystems, Inc., 3 Leicester Way, Pawtucket, RI 02860 USA
jhs@math.brown.edu, jhs@ntru.com, http://www.ntru.com

Abstract. A method is described for performing computations in a finite field GF(2^N) by embedding it in a larger ring R_p where the multiplication operation is a convolution product and the squaring operation is a rearrangement of bits. Multiplication in R_p has complexity $N + 1$, which is approximately twice as efficient as optimal normal basis multiplication (ONB) or Montgomery multiplication in GF(2^N), while squaring has approximately the same efficiency as ONB. Inversion and solution of quadratic equations can also be performed at least as fast as previous methods.

Introduction

The use of finite fields in public key cryptography has blossomed in recent years. Many methods of key exchange, encryption, signing and authentication use field operations in either prime fields GF(p) or in fields GF(2^N) whose order is a power of 2. The latter fields are especially pleasant for computer implementation because their internal structure mirrors the binary structure of a computer.

For this reason there has been considerable research devoted to making the basic field operations in GF(2^N) (especially squaring, multiplication, and inversion) efficient. Innovations include:

- use of optimal normal bases [15];
- use of standard bases with coefficients in a subfield GF(2^r) [26];
- construction of special elements $\alpha \in$ GF(2^N) such that powers α^e can be computed very rapidly [5, 6, 8];
- an analogue of Montgomery multiplication [14] for the fields GF(2^N) [13].

The discrete logarithm problem in finite fields can be used directly for cryptography, for example in Diffie–Hellman key exchange, ElGamal encryption, digital signatures, and pseudo-random number generation. (See [3, 9, 16, 17, 22] for a discussion of the difficulty of solving the discrete logarithm problem in GF(2^N).) An alternative application of finite fields to cryptography, as independently suggested by Koblitz and Miller, uses elliptic curves. In this situation the finite fields are much smaller (fields of order 2^{155} and 2^{185} are suggested in [10]), but the field operations are used much more extensively. Various methods have been suggested to efficiently implement elliptic curve cryptography over GF(2^N) in hardware [1] and in software [11, 23].

The standard way to work with $GF(2^N)$ is to write its elements as polynomials in $GF(2)[X]$ modulo some irreducible polynomial $\Phi(X)$ of degree N. Operations are performed modulo the polynomial $\Phi(X)$, that is, using division by $\Phi(X)$ with remainder. This division is time-consuming, and much work has been done to minimize its impact. Frequently one takes $\Phi(X)$ to be a trinomial, that is a polynomial $X^N + aX^M + b$ with only three terms, so as to simplify the division process. See, for example, [23] or [10, §6.3,6.4]. Montgomery multiplication replaces division by an extra multiplication [13], although this also exacts a cost.

A second way to work with $GF(2^N)$ is via normal bases, especially optimal normal bases [15], often abbreviated ONB. Using ONB, elements of $GF(2^N)$ are represented by exponential polynomials $a_0\beta + a_1\beta^2 + a_2\beta^4 + \cdots + a_{N-1}\beta^{2^{N-1}}$. Squaring is then simply a shift operation, so is very fast, and with an "optimal" choice of field, multiplication is computationally about the same as for a standard representation. More precisely, the computational complexity of multiplication is measured by the number of 1 bits in the multiplication transition matrix (λ_{ij}). The minimal complexity possible for a normal basis is $2N - 1$, and optimal normal bases are those for which the complexity is exactly $2N - 1$. (The ONB's described here are so-called Type I ONB's; the Type II ONB's are similar, but a little more complicated. Both types of ONB have complexity $2N - 1$.)

In this note we present a new way to represent certain finite fields $GF(2^N)$ that allows field operations, especially multiplication, to be done more simply and rapidly than either the standard representation or the normal basis representation. We call this method GBB, which is an abbreviation for *Ghost Bit Basis*, because as we will see, the method adds one extra bit to each field element. The fields for which GBB works are the same as those for which Type I ONB works, but the methods are quite different. Most importantly, the complexity of the multiplication transition matrix for GBB is $N + 1$, so multiplication using GBB is almost twice as fast (or, for hardware implementations, half as complex) as multiplication using ONB. Further, squaring in GBB is a rearrangement of bits that is different from the squaring rearrangement (cyclic shift) used by ONB. (We refer the reader to [24] for a description of all fields having a GBB-multiplication.)

[*Important Note.* The GBB construction is originally due to Ito and Tsujii [28]. See the note "Added in Proof" at the end of this article.]

1 Cyclotomic Rings Over GF(2)

We generate the field $GF(2^N)$ in the usual way as a quotient $GF(2)[X]/(\Phi(X))$, where we choose an irreducible cyclotomic polynomial of degree N,

$$\Phi(X) = X^N + X^{N-1} + X^{N-2} + \cdots + X^2 + X + 1.$$

As is well known, $\Phi(X)$ is irreducible in $GF(2)[X]$ if and only if

- $p = N + 1$ is prime.
- 2 is a primitive root modulo p.

The second condition simply means that the powers $1, 2, 2^2, 2^3, \ldots, 2^{p-1}$ are distinct modulo p, or equivalently, that

$$2^{N/\ell} \not\equiv 1 \pmod{p} \text{ for every prime } \ell \text{ dividing } N.$$

There are many N's that satisfy these properties, including for example the values $N = 148, 162, 180, 786,$ and 1018. (See Section 4 for a longer list.)

Remark 1. As noted above, the primes satisfying conditions (1) and (2) are exactly the primes for which there exists a Type I ONB. However, ONB's use these properties to find a basis for $\mathrm{GF}(2^N)$ of the form $\beta, \beta^2, \beta^4, \beta^8, \ldots, \beta^{2^N}$. For GBB, we will simply be using the fact that $\Phi(X)$ is irreducible.

We now observe that the field $\mathrm{GF}(2^N)$, when represented in the standard way as the set of polynomials modulo $\Phi(X)$, sits naturally as a subring of the ring of polynomials modulo $X^p - 1$. (Remember, $N = p - 1$.) In mathematical terms, there is an isomorphism

$$\frac{\mathrm{GF}(2)[X]}{(X^p - 1)} \cong \mathrm{GF}(2^N) \times \mathrm{GF}(2).$$

This is an isomorphism of rings, not fields, but as we will see, the distinction causes few problems.

For notational convenience, we let R_p denote the ring of polynomials modulo $X^p - 1$,

$$R_p = \frac{\mathrm{GF}(2)[X]}{(X^p - 1)}.$$

We interchangeably write polynomials as $a = a_N X^N + \cdots + a_1 X + a_0$ and as a list of coefficients $a = [a_N, a_{N-1}, \ldots, a_0]$.

Remark 2. Our method works more generally for fields $\mathrm{GF}(q^N)$ for any prime power q provided $p = N + 1$ is prime and q is a primitive root modulo p. In this setting, the ring $\mathrm{GF}(q)[X]/(X^N - 1)$ is isomorphic to $\mathrm{GF}(q^N) \times \mathrm{GF}(q)$. We leave to the reader the small adaptations necessary for $q \geq 3$. For most computer applications, $q = 2$ is the best choice, but depending on machine architecture other values could be useful, especially $q = 2^k$ for $k \geq 2$.

We now briefly discuss the complexity of operations in R_p. More generally, let R be any ring that is a $\mathrm{GF}(2)$-vector space of dimension n, so for example R could be R_p (with $n = p$), or R could be a field $\mathrm{GF}(2^N)$ (with $n = N$). Let $\mathcal{B} = \{\beta_0, \ldots, \beta_{n-1}\}$ be a basis for R as a $\mathrm{GF}(2)$-vector space. Then each product $\beta_i \beta_j$ can be written as a linear combination of basis elements,

$$\beta_i \beta_j = \sum_{k=0}^{n-1} \lambda_{ij}^{(k)} \beta_k.$$

The *complexity of multiplication relative to the basis* $\mathcal{B}$ is measured by the number of $\lambda_{ij}^{(k)}$'s that are equal to 1,

$$C(\mathcal{B}) = \frac{1}{n}\#\{(i,j,k) \ : \ \lambda_{ij}^{(k)} = 1\}.$$

It is easy to see that $C(\mathcal{B}) \geq 1$, and that if R is a field, then $C(\mathcal{B}) \geq n$. A more interesting example is given by a normal basis for $R = \text{GF}(2^N)$, in which case it is known [15] that $C(\mathcal{B}) \geq 2N - 1$. A normal basis for $\text{GF}(2^N)$ is called *optimal* if its complexity equals $2N - 1$. A complete description of all fields that possess an optimal normal basis is given in [7].

The complexity of the basis $\mathcal{B} = \{1, X, \ldots, X^{p-1}\}$ for the ring R_p is clearly $C(\mathcal{B}) = p$, since $\lambda_{ij}^{(k)} = 1$ if and only if $i + j \equiv k \mod p$. In other words, for each pair (i, j) there is exactly one k with $\lambda_{ij}^{(k)} = 1$. So taking $p = N + 1$ as usual, we see that an optimal normal basis for $\text{GF}(2^N)$ has complexity $2N - 1$, while the standard basis for R_p has complexity $N + 1$, making multiplication in R_p approximately twice as fast (or half as complicated) as in $\text{GF}(2^N)$. It is thus advantageous to perform $\text{GF}(2^N)$ multiplication by first moving to R_p and then doing the multiplication in R_p.

A second important property of a basis for finite field implementations, especially in hardware, is a sort of symmetry whereby the n^3 multipliers $\lambda_{ij}^{(k)}$ are determined by the n^2 multipliers $\lambda_{ij}^{(1)}$ by a simple transformation. We say that $\mathcal{B}$ is a *permutation basis* if there are permutations σ_k, τ_k such that

$$\lambda_{ij}^{(k)} = \lambda_{\sigma_k(i)\tau_k(j)}^{(1)} \quad \text{for all } 1 \leq i, j, k \leq n.$$

In practical terms, this means that the circuitry used to compute the first coordinate of a product ab can be used to compute all of the other coordinates simply by rearranging the order of the inputs.

To see why this is true, we write $a = \sum a_i \beta_i$ and $b = \sum b_j \beta_j$. Then (after a little algebra) the product ab is equal to

$$ab = \sum_{k=0}^{n-1} \left(\sum_{i,j=0}^{n-1} a_i b_j \lambda_{ij}^{(k)} \right) \beta_k.$$

If $\mathcal{B}$ is a permutation basis, we can rewrite this as

$$ab = \sum_{k=0}^{n-1} \left(\sum_{i,j=0}^{n-1} a_i b_j \lambda_{\sigma_k(i)\tau_k(j)}^{(1)} \right) \beta_k = \sum_{k=0}^{n-1} \left(\sum_{i,j=0}^{n-1} a_{\sigma_k^{-1}(i)} b_{\tau_k^{-1}(j)} \lambda_{ij}^{(1)} \right) \beta_k.$$

Thus the kth coordinate of ab is computed by first using the permutations σ_k and τ_k to rearrange the bits of a and b respectively, and then feeding the rearranged bit strings into the circuit that computes the first coordinate of ab.

It turns out that both ONB and GBB are permutation bases, but the corresponding permutations are slightly different. For ONB one has the relation

$\lambda_{ij}^{(k)} = \lambda_{i-k,j-k}^{(0)}$, while for GBB (i.e., for the standard basis in R_p) the relation is $\lambda_{ij}^{(k)} = \lambda_{i,j-k}^{(0)}$, where in both formulas we are taking the subscripts modulo n. Thus the permutation for GBB is a little easier to implement than ONB because only one of the inputs needs to be shifted.

2 Overview of Operations in R_p

In this section we briefly describe some of the advantages of working in the ring R_p. In Section 3 we will discuss these in more detail.

2.1 Moving Between GF(2^N) and R_p

An element of GF(2^N) is simply a list of N bits, and similarly an element of R_p is a list of $N + 1$ bits,

$$[a_{N-1}, \ldots, a_1, a_0] \in \text{GF}(2^N) \qquad \text{and} \qquad [a_N, a_{N-1}, \ldots, a_0] \in R_p.$$

We call the extra bit in R_p the "ghost bit". In order to do a computation in GF(2^N), we first move to R_p, next do all computations in R_p, and finally move the final answer back to GF(2^N). Movement between GF(2^N) and R_p is extremely fast, at most a single complement operation.

More precisely, the map from GF(2^N) to R_p is given by

$$\text{GF}(2^N) \longrightarrow R_p, \quad a = [a_{N-1}, \ldots, a_1, a_0] \longmapsto [0, a_{N-1}, \ldots, a_1, a_0].$$

That is, we simply pad a by setting the ghost bit equal to zero. Moving in the other direction is almost as easy. If the ghost bit is zero, we drop it, while if the ghost bit is one, we first take the complement:

$$R_p \to \text{GF}(2^N), \quad [a_N, a_{N-1}, \ldots, a_1, a_0] \mapsto \begin{cases} [a_{N-1}, \ldots, a_1, a_0] & \text{if } a_N = 0, \\ \sim [a_{N-1}, \ldots, a_1, a_0] & \text{if } a_N = 1. \end{cases}$$

Here $\sim$ means take the complement, that is, flip every bit. If this isn't available as a primitive operation, one can XOR with $1111\cdots111$.

2.2 Addition in R_p

Addition in R_p is the usual addition of vectors over GF(2). That is, the coordinates are added using the rules $0 + 0 = 1 + 1 = 0$ and $0 + 1 = 1 + 0 = 1$.

2.3 Squaring in R_p

The squaring operation in R_p is very fast. One simply interleaves the top order bits and the bottom order bits. Thus if $a = [a_N, a_{N-1}, \ldots, a_0] \in R_p$, then

$$a^2 = [a_{N/2}, a_N, a_{N/2-1}, a_{N-1}, \ldots a_2, a_{N/2+2}, a_1, a_{N/2+1}, a_0].$$

Fast squaring can be implemented using the operation that takes a w-bit word $[b_w, b_{w-1}, \ldots, b_1]$ and returns the two words $[b_w, 0, b_{w-1}, 0, \ldots, 0, b_{w/2+1}, 0]$ and $[b_{w/2}, 0, \ldots, b_2, 0, b_1, 0]$. This is trivial to implement in hardware, while in software it might be quickest to implement at the word level using a look-up table.

2.4 Multiplication in R_p

Multiplication in R_p is extremely fast, because the transition matrix for multiplication has complexity p (i.e., it is a p-by-p matrix with p entries equal to 1 and the rest 0). Multiplication in R_p is simply the convolution product of the coefficient vectors:

$$a(X)b(X) = \sum_{k=0}^{N} \left(\sum_{i+j=k} a_i b_j \right) X^k,$$

where we understand that the indices on a and b are taken modulo p. We will discuss in more detail below various ways in which to optimize the multiplication process.

Remark 3. Since multiplication $c = ab$ is simply the convolution product of the vectors a and b, it would be nice to use Fast Fourier Transforms to compute these convolutions. Unfortunately the vectors have dimension p, which is prime, so FFT does not help. On the other hand, some speed-up may be possible using the standard trick (Karatsuba multiplication) of splitting polynomials in half and replacing multiplications by additions, see for example [2, §3.1.2]).

2.5 Inversion in R_p

Inversion in R_p and in $GF(2^N)$ are extremely fast. Not all elements of R_p are invertible, but we are really interested in computing inverses in $GF(2^N)$. (Aside: $a \in R_p$ is invertible if and only if $a \neq \Phi$ and a has an odd number of 1 bits.) An especially efficient way to compute these inverses is the "Almost Inverse Algorithm" described in [23, §4.4]. Given a polynomial $a(X) \in GF(2)[X]$, the almost inverse algorithm efficiently finds a polynomial $A(X)$ so that

$$a(X)A(X) \equiv X^k \pmod{\Phi(X)}$$

for some exponent $0 \leq k < 2N$. Then $a(X)^{-1} = X^{-k}A(X)$, where the product $X^{-k}A(X)$ is easily computed as a cyclic right shift in R_p. Compare with [23], where the final step of dividing by X^k requires more work. (Use of the almost inverse algorithm is also efficient for computing inverses using ONB's, especially of Type I, see [21, §11.1].)

We also mention the well-known alternative method of inversion via multiplication using the relation

$$a(X)^{2^N-1} \equiv 1 \pmod{X^p - 1} \quad \text{for all invertible } a(X) \in R_p.$$

Thus we can compute the inverse of an invertible $a(X)$ by repeated squaring and multiplication

$$a(X)^{-1} \equiv a(X)^{2^N-2} \pmod{X^p - 1}.$$

2.6 Quadratic equations in R_p

For elliptic curve applications, it is important to be able to solve the equation $z^2 + z + c = 0$ in $GF(2^N)$, see [21, §6.5]. Not all such equations are solvable, the necessary and sufficient condition being $\mathrm{Tr}(c) = 0$, where Tr is the trace map $GF(2^N) \to GF(2)$. The analogous condition for R_p says that $z^2 + z + c = 0$ has a solution (actually 4 solutions) in R_p if and only if $c_0 + c_1 + \cdots + c_N = 0$ and $c_0 = 0$. If there is a solution, then a solution may be computed using a recursion coming from the formula

$$z = z^{1/2} + c^{1/2}.$$

The recursion is very simple because the squaring and square root operations in R_p are so simple. We also note that if $c_0 + c_1 + \cdots + c_N = c_0 = 1$, then there will still be a solution to $z^2 + z + c = 0$ in $GF(2^N)$. This solution may be found by first replacing c with its complement $\sim c$, next solving $z^2 + z + c = 0$ in R_p, and finally mapping the result back to $GF(2^N)$ in the usual way.

Remark 4. It is possible to use an (automatically "optimal") normal basis in the ring R_p. To do this, write each element of R_p in terms of the basis

$$X, X^2, X^4, \cdots, X^{2^{p-1}},$$

where it is understood that the exponents are reduced modulo p. All of the usual comments that apply to normal bases in $GF(2^N)$ apply to using a normal basis in the ring R_p. (See [15], [19], or [21, chapter 4] for information about optimal normal bases.) In particular, if a normal basis is used in R_p, then the complexity has the usual "optimal" value of $2p - 1$, and it is necessary to use a log table and an anti-log table to sort out the exponents when doing multiplications. Thus using a normal basis in R_p leads to slower multiplications than using the standard polynomial basis. On the other hand, squaring using a normal basis is simply a shift of bits, while squaring with the polynomial basis is interspersion of bits, so it is conceivable that situations or architectures might exist for which the normal basis is preferable.

Remark 5. For Diffie-Hellman key exchange, ElGamal encryption, and similar applications, there is no reason to move back and forth between $GF(2^N)$ and R_p. One could do all the work in R_p and move back to $GF(2^N)$ at the end. (Even in R_p, only one bit is exposed, namely evaluation at $X = 1$. Thus for the discrete logarithm problem $a(X)^k = b(X)$ in R_p, an attacker only deduces either $0^k = 0$ or $1^k = 1$ in $GF(2)$, so he gains no information about the exponent k.)

A similar comment applies when working with elliptic curves, keeping in mind that not all elements of R_p have inverses. Thus when computing the reciprocal of $a(X) \in R_p$, if a has an even number of 1 bits, then a must first be replaced by its complement.

Remark 6. For certain finite fields, essentially Type II ONB fields [8] and their generalizations [5, 6], it is possible to construct special elements called Gauss periods whose powers can be computed extremely rapidly. An interesting feature of these constructions is that the exponentiation process makes use of a "redundant representation" (see [6, page 345]), which is analogous to our "ghost bit". However, the bases used in [5, 6, 8] are normal bases and the fast exponentiation operation only applies to special elements, while the GBB construction in this paper gives a fast multiplication for arbitrary elements. Thus the two constructions are fundamentally different, as is also apparent from the fact that the fields to which they apply are different.

3 Bit-Level Description of Operations in $GF(2^N)$ and R_p

In this section we give bit-level descriptions of the basic operations described in Section 2. It is relatively straightforward to give analogous word-level descriptions, although for full efficiency it is important to use all of the usual programming tricks.

Remark 7. The algorithms in this section take as input an element of $GF(2^N)$ and return an element of $GF(2^N)$ using the standard basis for $GF(2)[X]/\Phi(X)$. As noted above, in practice one could do all computations in R_p and only move the answer back to $GF(2^N)$ as the very last step.

Remark 8. Polynomials in the following algorithms are written in the form

```
a[N]X^N+a[N-1]X^(N-1)+...+a[2]X^2+a[1]X+a[0].
```

Thus a[i] refers to the coefficient of X^i. We stress this point because when implementing these algorithms, it can be confusing (at least to the author) if the vector of coefficients is stored from high-to-low, instead of from low-to-high. We also note that $a(X) + \Phi(X)$ is the complement $\sim a(X)$ of $a(X)$. This is correct because $\Phi(X) = X^N + \cdots + X + 1 = [1, 1, \ldots, 1]$ has all of its bits set equal to 1. We also remind the reader again that $p = N + 1$.

Addition is simply addition of vectors with coordinates in $GF(2)$, so there is nothing further to say.

The squaring operation is an interleaving permutation of the coefficients.

Bit-Level Procedure for Squaring in $GF(2^N)$

```
Input:   a(X)
Output:  c(X)=a(X)^2 mod Phi(X)
Step 1:  b(X):=a[0]+a[1]X^2+a[2]X^4+...+a[N/2]X^N
Step 2:  c(X):=a[N/2+1]X+a[N/2+2]X^3+...+a[N]X^(N-1)
Step 3:  c(X):=b(X)+c(X)
Step 4:  if c[N]=1 then c(X):=c(X)+Phi(X)
```

The multiplication operation in R_p is what is commonly known as a convolution product. Here is how multiplication works at the bit level.

Bit-Level Procedure for Multiplication in GF(2^N)

```
Input:   a(X), b(X)
Output:  c(X)=a(X)b(X) mod Phi(X)
Step 1:  c(X):=0
Step 2:  for i=0 to p-1 do
Step 3:      if a[i]=1 then c(X):=c(X)+b(X)
Step 4:      cyclic shift c(X) right 1 bit
Step 5:  if c[p-1]=1 then c(X):=c(X)+Phi(X)
```

The most time-consuming part of multiplication is Step 3, since this step is inside the main loop and requires p additions (i.e., each of the p coefficients of b must be added or XOR'd to the corresponding coefficient of c). For comparison purposes, the analogous routine using ONB has the equivalent of two Step 3's, so it takes approximately twice as long (or alternatively requires twice as complicated a circuit). Similarly, Montgomery multiplication over GF(2^N) has the equivalent of two Step 3's, so also takes twice as long (cf. [13]).

Computation of inverses is relatively straightforward. We give below a slight adaptation of Schroeppel, Orman, O'Malley, and Spatscheck's "Almost Inverse Algorithm" [23] (with an improvement suggested by Schroeppel) that works quite well. The speed of the Inversion Procedure can be significantly enhanced by a number of implementation tricks, such as expanding the operations on b, c, f, g into inline loop-unrolled code. We refer the reader to [23] for a list of practical suggestions.

Bit-Level Inversion Procedure in GF(2^N)

```
Input:    a(X)
Output:   b(X)=a(X)^(-1) mod Phi(X)
Step 1:   Initialization:
              k:=0; b(X):=1; c(X):=0;
              f(X):=a(X); g(X):=Phi(X);
Step 2:   do while f[0]=0
Step 3:       f(X):=f(X)/X; k:=k+1;
Step 4:   do while f(X)!=1
Step 5:       if deg(f) < deg(g) then
Step 6:           exchange f and g; exchange b and c;
Step 7:       f(X):=f(X)+g(X)
Step 8:       b(X):=b(X)+c(X)
Step 9:       do while f[0]=0
Step 10:          f(X):=f(X)/X; c(X):=c(X)*X; k:=k+1;
Step 11:  b(X):=b(X)/X^k  modulo  X^p-1
Step 12:  if b[p-1]=1 then b(X):=b(X)+Phi(X)
```

Note that in Steps 3 and 10 of the inversion routine, $f(X)/X$ is f shifted right one bit and $c(X) * X$ is c shifted left one bit. Further, Step 11 in the Inversion Procedure is simply the cyclic shift

$$[b_{p-1}, b_{p-2}, \ldots, b_1, b_0] \longmapsto [b_{k-1}, b_{k-2}, \ldots, b_0, b_{p-1}, \ldots, b_{p-k+1}, b_{p-k}].$$

It is instructive to compare the simplicity of this step with the description [23, page 51] of how to compute $X^{-k}b(X)$ when $X^p - 1$ is replaced by a trinomial, even if the trinomial is selected to make this operation as simple as possible.

Finally, we describe how to solve a quadratic equation $z^2 + z + c = 0$ in $GF(2^N)$. The equivalent formula $z = z^{1/2} + c^{1/2}$ shows that the solution may be obtained recursively using the relation

$$z_i = z_{2i} + c_{2i}, \quad 0 \le i \le N,$$

where it is understood that the indices are always reduced modulo p into the range $[0, N]$. In particular, putting $i = 0$ shows that a necessary condition for a solution to exist in R_p is $c_0 = 0$.

Bit-Level Quadratic Formula in $GF(2^N)$

```
Input:    c(X)
Output:   z(X) satisfying z(X)^2+z(X)+c(X)=0 mod Phi(X)
Step 1:   if c[0]=1 then c(X):=c(X)+Phi(X)
Step 2:   z[0]:=0; z[1]:=1; j:=1;
Step 3:   do N-1 times
Step 4:      i:=j
Step 5:      j:=2*j mod p
Step 6:      z[j]:=z[i]+c[j]
Step 7:   if z[N/2+1]=z[1]+c[1] then
Step 8:      if z[N]=1 then z(X):=z(X)+Phi(X)
Step 9:      return z(X)
Step 10: else
Step 11:     return "Error: z^2+z+c=0 not solvable"
```

4 Selection of Good Fields $GF(2^N)$

Our first requirement in choosing $GF(2^N)$ is that $p = N + 1$ is prime and 2 is a primitive root modulo p, since this ensures that the cyclotomic polynomial

$$\Phi(X) = X^N + X^{N-1} + \cdots + X^2 + X + 1 = \frac{X^{N+1} - 1}{X - 1}$$

is irreducible in $GF(2)[X]$.

Table 1 lists all primes in the intervals $[100, 300]$, $[650, 850]$, and $[1000, 1200]$ for which $\Phi(X)$ is irreducible in $GF(2)[X]$. It is clear that there are lots of primes with this property. (Mathematical Aside: A conjecture of Emil Artin says that there are infinitely many primes p with this property. Artin's conjecture has not been proven unconditionally, but Hooley [12] has shown that Artin's conjecture would follow from the Riemann hypothesis.)

If one is merely interested in working in a field $GF(2^N)$ having a very fast multiplication method, then any of the primes in Table 1 will work (taking $N = p-1$). For example, this is the case for the many cryptographic applications

Table 1. Some primes p with $\Phi(X)$ irreducible in GF(2)[X]

$$
\boxed{
\begin{array}{c}
101,\ 107,\ 131,\ 139,\ 149,\ 163,\ 173,\ 179,\ 181,\ 197,\ 211,\ 227,\ 269,\ 293 \\
653, 659, 661, 677, 701, 709, 757, 773, 787, 797, 821, 827, 829 \\
1019,\ 1061,\ 1091,\ 1109,\ 1117,\ 1123,\ 1171,\ 1187
\end{array}
}
$$

that use elliptic curves over finite fields. For elliptic curve cryptography, one might take N to be one of the values 162, 172, 178, 180, or 196.

On the other hand, if one wishes to use the discrete logarithm problem (DLP) in $GF(2^N)$, for example with Diffie-Hellman key exchange or the ElGamal public key cryptosystem, then there is another very important issue to consider. The group of non-zero elements in the field $GF(2^N)$ is a cyclic group of order $2^N - 1$, and if $2^N - 1$ factors as a product of small primes, the Pohlig-Hellman algorithm [20] gives a reasonably efficient way to solve the DLP in $GF(2^N)$.

To investigate prime divisors of $2^N - 1$, we begin with the factorization of $X^N - 1$ as a product of cyclotomic polynomials,

$$
X^N - 1 = \prod_{d \mid N} \Phi_d(X).
$$

Here $\Phi_d(X)$ is the dth cyclotomic polynomial. That is, $\Phi_d(X)$ is the polynomial whose roots are the primitive d^{th} roots of unity,

$$
\Phi_d(X) = \prod_{1 \leq k \leq d,\ \gcd(k,d)=1} \left(X - e^{2\pi i k/d} \right).
$$

The polynomial $\Phi_d(X)$ has integer coefficients and is irreducible in $\mathbb{Q}[X]$. We will not need to use any special properties of the Φ_d's, but for further information on cyclotomic polynomials, see for example [25].

For cryptographic purposes, we want to choose a value for N so that $2^N - 1$ is divisible by a large prime. We always have the factorization

$$
2^N - 1 = \prod_{d \mid N} \Phi_d(2),
$$

so we look for cyclotomic polynomial values $\Phi_d(2)$ that have large prime divisors.

The problem of factoring numbers of the form $2^N - 1$ has a long history. Indeed, the Cunningham Project set itself the long-term task of factoring numbers of the form $b^N \pm 1$. Current results on the Cunningham Project are available on the web at [4]. The following two examples were devised using material from that site, but we include sufficient information here so that the interested reader can check that our examples have the stated properties.

Example 1. For our first example we take $p = 787$ and $N = 786$. Since 786 is divisible by 393, we see that $2^{786} - 1$ is divisible by $\Phi_{393}(2) = \frac{(2^{393}-1)}{(2^3-1)(2^{131}-1)}$. The Cunningham Project archive says that $\Phi_{393}(2)$ factors into primes as

$$
\Phi_{393}(2) = 36093121 \cdot 51118297 \cdot 58352641 \cdot q.
$$

Here $q \approx 2^{183} \approx 10^{55}$ is a prime. Hence $2^{786} - 1$ is divisible by the large prime q, so GF(2^{786}) is a suitable field for use with Diffie-Hellman and other schemes that depend on the intractability of the discrete logarithm problem.

Example 2. As a second example, consider $p = 1019$ and $N = 1018$. Since 1018 is divisible by 509, we see that $2^{1018} - 1$ is divisible by $\Phi_{509}(2) = 2^{509} - 1$. From the Cunningham Project archive, we find that $2^{509} - 1$ factors into primes as

$$2^{509} - 1 = 12619129 \cdot 19089479845124902223 \cdot 647125715643884876759057 \cdot q.$$

Here $q \approx 2^{242.26} \approx 10^{103.03}$ is prime. Hence $2^{1018} - 1$ is divisible by the large prime q, so GF(2^{1018}) is a suitable field for use with Diffie-Hellman and other schemes that depend on the intractability of the discrete logarithm problem.

Remark 9. There are many other p's listed in Table 1 with the property that $2^{p-1} - 1$ is divisible by a large prime. We have merely presented two examples for which GF(2^N) has approximately the same number of elements as the "First and Second Oakley Groups" described in [10]. However, we note that the discrete logarithm problem in GF(2^N) may be easier to solve than in GF(p) for $p \approx 2^N$, see for example [3, 9, 16, 17, 22].

Acknowledgments. I would like to thank John Platko for a serendipitous airborne conversation that piqued my interest in the problem of fast implementations of finite field arithmetic, and Andrew Odlyzko, Michael Rosing, Richard Schroeppel, Igor Shparlinski, and the referees for numerous helpful comments.

Added in Proof. Immediately prior to this article being sent to the publisher for printing, the author discovered a 1989 paper of Ito and Tsujii [28] containing the GBB construction, as well as the related papers [27] and [29]. Thus the GBB construction described here should be credited to Ito and Tsujii.

References

1. Agnew, G.B., Mullin, R.C., Vanstone, S.A.: An implementation of elliptic curve cryptosystems over $F_{2^{155}}$. IEEE Journal on Selected Areas in Communications. **11(5)** (June 1993) 804-813
2. Cohen, H. A course in computational algebraic number theory. Graduate Texts in Math., vol. 138. Springer Verlag, Berlin (1993)
3. Coppersmith, D.: Fast evaluation of discrete logarithms in fields of characteristic two. IEEE Transactions on Information Theory **30** (1984) 587-594
4. Cunningham Project. <ftp://sable.ox.ac.uk/pub/math/cunningham>
5. Gao, S., von zur Gathen, J., Panario, D.: Gauss periods and fast exponentiation in finite fields. In: Baeza-Yates, R., Goles, E., Poblete, P.V.(eds.): Latin American Symposium on Theoretical Informatics–LATIN '95. Lecture Notes in Computer Science, Vol. 911. Springer-Verlag, New York (1995) 311-322
6. Gao, S., von zur Gathen, J., Panario, D.: Gauss periods: orders and cryptographical applications. Mathematics of Computation **67** (1998) 343-352
7. Gao, S., Lenstra, H.W., Jr.: Optimal normal bases. Design, Codes, and Cryptography. **2** (1992) 315-323

8. Gao, S., Vanstone, S.A.: On orders of optimal normal basis generators. Mathematics of Computation **64** (1995) 1227–1233
9. Gordon, D.M., McCurley, K.S.: Massively parallel computation of discrete logarithms. In: Brickell, E.F. (ed.): Advances in cryptology–CRYPTO '92. Lecture Notes in Computer Science, Vol. 740. Springer-Verlag, New York (1993) 16–20
10. Harkins, D., Carrel, D.: The Internet Key Exchange, RFC 2409 Network Working Group (November 1998), <http://anreg.cpe.ku.ac.th/rfc/rfc2409.html>
11. Harper, G., Menezes, A., Vanstone, S.: Public-key cryptosystems with vey small key lengths. In: Rueppel, R.A. (ed.): Advances in Cryptology — EUROCRYPT 92. Lecture Notes in Computer Science, Vol. 658. Springer-Verlag, New York (1992) 163–173
12. Hooley, C.: On Artin's conjecture. J. Reine Angew. Math. **225** (1997) 209-220
13. Koç, Ç.K., Acar, T.: Montgomery multiplication in $GF(2^k)$. Design, Codes and Cryptography. **14** (1998) 57–69
14. Montgomery, P.L.: Modular multiplication without trial division. Math. Comp. **44** (1985) 519–521
15. Mullin, R., Onyszchuk, I., Vanstone, S., Wilson, R.: Optimal normal bases in $GF(p^n)$. Discrete Applied Mathematics. **22** (1988) 149–161
16. Odlyzko, A.M.: Discrete logarithms and smooth polynomials. In: Mullen, G.L., Shiue, P. (eds.): Finite Fields: Theory, Applications and Algorithms. Amer. Math. Soc., Contemporary Math. #168 (1994) 269–278
17. Odlyzko, A.M.: Discrete logarithms: The past and the future. Preprint, July 1999. <http://www.research.att.com/~amo/doc/crypto.html>
18. Omura, J., Massey, J.: Computational method and apparatus for finite field arithmetic. United States Patent 4587627 (May 6), 1986
19. Onyszchuk, I., Mullin, R., Vanstone, S.: Computational method and apparatus for finite field multiplication. United States Patent 4745568 (May 17), 1988
20. Pohlig, S.C., Hellman, M.E.: An improved algorithm for computing logarithms over $GF(p)$ and its cryptographic significance. IEEE Transactions on Information Theory. **24** (1978) 106–110
21. Rosing, M.: Implementing Elliptic Curve Cryptography. Manning Publications Greenwich, CT (1999)
22. Semaev, I. A.: An algorithm for evaluation of discrete logarithms in some nonprime finite fields. Math. Comp. **67** (1998) 1679–1689
23. Schroeppel, R., O'Malley, S., Orman, H., Spatscheck, O.: Fast key exchange with elliptic curve systems In: Coppersmith, D. (ed.): Advances in Cryptology — CRYPTO 95. Lecture Notes in Computer Science, Vol. 973. Springer-Verlag, New York (1995) 43–56
24. Silverman, J.H.: Low Complexity Multiplication in Rings. Preprint, April 14, 1999
25. Washington, L.: Introduction to Cyclotomic Fields. Springer-Verlag New York (1982)
26. De Win, E., Bosselaers, A., Vandenberghe, S., De Gersem, P., Vandewalle, J.: A fast software implementation for arithmetic operations in $GF(2^n)$. In: Advances in Cryptology — ASIACRYPT '96. Lecture Notes in Computer Science, Vol. 1163. Springer-Verlag New York (1996) 65–76
27. Drolet, G.: A new representation of elements of finite fields $GF(2^m)$ yielding small complexity arithmetic circuits. *IEEE Trans. Comput.* **47** (1998), no. 9, 938–946
28. Ito, B., Tsujii, S.: Structure of a parallel multipliers for a class of fields $GF(2^m)$. *Information and Computers* **83** (1989), 21–40
29. Wolf, J.K.: Efficient circuits for multiplying in $GF(2^m)$. *Topics in Discrete Mathematics* **106/107** (1992), 497–502

Efficient Finite Field Basis Conversion Involving Dual Bases

Burton S. Kaliski Jr. and Moses Liskov

RSA Laboratories
20 Crosby Drive, Bedford, MA 01730
{ burt, moses }@rsa.com

Abstract. Conversion of finite field elements from one basis representation to another representation in a storage-efficient manner is crucial if these techniques are to be carried out in hardware for cryptographic applications. We present algorithms for conversion to and from dual of polynomial and dual of normal bases, as well as algorithms to convert to a polynomial or normal basis which involve the dual of the basis. This builds on work by Kaliski and Yin presented at SAC '98.

1 Introduction

Conversion between different choices of basis for a finite field is an important problem in today's computer systems, particularly for cryptographic operations [1]. While it is possible to convert between two choices of basis by matrix multiplication, the matrix may be too large for some applications, hence the motivation for more storage-efficient techniques. The most likely such application would be in special-purpose hardware devices, but there are others as well.

The paper of Kaliski and Yin [2] introduced the shift-extract and technique of basis conversion, and also gave several storage-efficient algorithms based on those techniques for converting to a polynomial or normal basis. In this paper, we introduce techniques involving the dual of a polynomial or normal basis, including storage-efficient generation of a dual basis and storage-efficient shifting in such a basis. The new techniques result in several new storage-efficient algorithms for converting to and from the dual of a polynomial or normal basis, as well as additional algorithms for converting to a polynomial or normal basis.

2 Background

Elements of a finite field can be represented in a variety of ways, depending on the choice of basis for the representation [3]. Let $GF(q^m)$ be the finite field, and let $GF(q)$ be the ground field over which it is defined, where q is a prime or a prime power. We say that the *characteristic* of the field is p where $q = p^r$ for some $r \geq 1$. For even-characteristic fields, we have $p = 2$. The *degree* of the field is m; its *order* is q^m.

A *basis* for the finite field is a set of m elements $\omega_0, \ldots, \omega_{m-1} \in GF(q^m)$ such that every element of the finite field can be represented uniquely as a linear combination of basis elements. We write

$$\epsilon = B[0]\omega_0 + B[1]\omega_1 + \cdots + B[m-1]\omega_{m-1}$$

where $B[0], \ldots, B[m-1] \in GF(q)$ are the *coefficients*.

Two common types of basis are a *polynomial basis* and a *normal basis*. In a polynomial basis, the basis elements are successive powers of an element γ, called the *generator*:

$$\omega_i = \gamma^i.$$

In a normal basis, the basis elements are successive exponentiations of an element γ, again called the generator:

$$\omega_i = \gamma^{q^i}.$$

Another common type of basis is a *dual basis*. Let $\omega_0, \ldots, \omega_{m-1}$ be a basis and let h be a nonzero linear function from $GF(q^m)$ to $GF(q)$, i.e., a function such that for all $\epsilon, \phi \in GF(q^m)$ and $c \in GF(q)$, $h(\epsilon + \phi) = h(\epsilon) + h(\phi)$ and $h(c\epsilon) = ch(\epsilon)$. The dual basis of the basis $\omega_0, \ldots, \omega_{m-1}$ with respect to the function h is the basis $\eta_0, \ldots, \eta_{m-1}$ such that for $0 \le i, j \le m-1$,

$$h(\omega_i \eta_j) = 1 \text{ if } i = j, \ 0 \text{ otherwise.}$$

Duality is symmetric: the dual basis with respect to h of the basis $\eta_0, \ldots, \eta_{m-1}$ is the basis $\omega_0, \ldots, \omega_{m-1}$. A dual basis can be defined for a polynomial basis, a normal basis, or any other choice of basis, and with respect to a variety of functions.

The *basis conversion* or *change-of-basis* problem is to compute the representation of an element of a finite field in one basis, given its representation in another basis. The problem has two forms, where we distinguish between the internal basis in which finite field operations are performed, and the external basis to and from which we are converting:

- *Import problem.* Given an internal basis and an external basis for a finite field $GF(q^m)$ and the representation B of a field element in the external basis (the *external representation*), determine the corresponding representation A of the same field element in the internal basis (the *internal representation*).
- *Export problem.* Given an internal basis and an external basis for a finite field $GF(q^m)$ and the internal representation A of a field element, determine the corresponding external representation B of the same field element.

Normally, the import and export problem could be solved by using a change of basis matrix, which requires storage for $O(m)$ field elements. Since each field element consists of m base field coefficients, this is $O(m^2)$ coefficients. In constrained environments, this may be too large. What we want are algorithms which require storage for $O(1)$ field elements or $O(m)$ coefficients. The algorithms given in this paper for dual bases satisfy this requirement.

3 Overview of techniques

In the following, the dual of a polynomial basis is called a polynomial* basis, and the dual of a normal basis is called a normal* basis.

3.1 Import algorithms

Given an internal basis and an external basis for a finite field and the representation B of a field element in the external basis, an import algorithm determines the corresponding representation A of the same field element in the internal basis.

Two general methods for determining the internal representation A are described: the generate-accumulate method and the shift-insert method.

Generate-Accumulate method The generate-accumulate method computes the internal representation A by accumulating the products of coefficients $B[i]$ with successive elements of the external basis. The basic form of the algorithm for this method is as follows:

> **proc** IMPORTBYGENACCUM
> $A \leftarrow 0$
>   **for** i **from** 0 **to** $m - 1$ **do**
> $A \leftarrow A + B[i] \times W_i$
>   **endfor**
> **endproc**

As written, this algorithm requires storage for the m values $W_0, \ldots, W_{m-1}$, which are the internal representations of the elements of the external basis. To reduce the storage requirement, it is necessary to generate the values as part of the algorithm. This is straightforward when the external basis is a polynomial basis or a normal basis. For polynomial* and normal* bases, algorithms are given in this paper.

Shift-Insert method The shift-insert method computes the internal representation A by "shifting" an intermediate variable in the external basis and inserting successive coefficients between the shifts. This follows the same concept as the shift-extract method below. Let SHIFT be a function that shifts an element in the external basis, i.e., a function such as one which given the internal representation of an element with external representation

$$(B[0], B[1], \ldots, B[m - 2], B[m - 1])$$

computes the internal representation of the element with external representation

$$(B[m - 1], B[0], \ldots, B[m - 3], B[m - 2]).$$

(Other forms of shifting are possible, including shifting in the reverse direction, or shifting where the value 0 rather than $B[m - 1]$ is shifted in.)

The basic form of algorithm for this method is as follows ([2], Sec. 3.2, 3.3):

```
proc IMPORTBYSHIFTINSERT
    A ← 0
    for i from m − 1 downto 0 do
        SHIFT(A)
        A ← A + B[i] × W₀
    endfor
endproc
```

The direction of the **for** loop may vary depending on the direction of the shift. One advantage of the shift-insert method over the generate-accumulate method is that with a minor increase in storage, this algorithm can be parallelized. That is, if W_0 and $W_{m/2}$ are available, two elements can be inserted per shift. Since the shift is the most work-intensive part of the algorithm, this aids efficiency. This improvement is further discussed in [2].

3.2 Export algorithms

Given an internal basis and an external basis for a finite field and the representation A of a field element in the internal basis, an export algorithm determines the corresponding representation B of the same field element in the internal basis.

Two general methods for determining the external representation B are described: the generate*-evaluate method and the shift-extract method.

Generate*-Evaluate method The generate*-evaluate method computes the external representation B by evaluating products of A with successive elements of a dual of the external basis. For example, the following equation gives the ith coefficient of the external representation:

$$B[i] = h(AX_i)$$

where h is a linear function and $X_0, \ldots, X_{m-1}$ are the internal-basis representations of the elements of the dual of the external basis with respect to the function h. The basic form of algorithm for this method is as follows:

```
proc EXPORTBYGEN*EVAL
    for i from 0 to m − 1 do
        T ← A × Xᵢ
        B[i] ← h(T)
    endfor
endproc
```

This algorithm requires storage for the m values $X_0, \ldots, X_{m-1}$, which are the internal represenations of the dual of the external basis. As was the case for IMPORTBYGENACCUM, to reduce the storage requirement, it is necessary to generate the values as part of the algorithm.

Shift-Extract method The Shift-Extract method computes the external representation A by shifting an intermediate variable in the external basis and *extracting* successive coefficients between the shifts. This follows the same concept as the shift-insert method above, with a similar SHIFT function and an EXTRACT function that obtains a selected coefficient of the external representation. (The EXTRACT function is similar to the h function in the previous method.)

The basic form of algorithm for this method is as follows ([2], Sec. 3.4, 3.5):

> **proc** EXPORTBYSHIFTEXTRACT
>   **for** i **from** $m - 1$ **downto** 0 **do**
> $B[i] \leftarrow$ EXTRACT(A)
>   SHIFT(A)
>   **endfor**
> **endproc**

Again, the direction of the **for** loop may vary depending on the direction of the shift. As with the shift-insert method above, the shift-extract method can be parallelized to extract multiple coefficients per iteration.

3.3 Summary

For these methods to accomplish our goal of being storage efficient, we depend on the efficiency of some additional functions. For the generate*-evaluate and generate-accumulate methods, we need an efficient dual basis generator. For the shift-insert and shift-extract methods, we need an efficient SHIFT function that works when the external basis is a normal* or polynomial* basis. An efficient *Extract* function (and hence an efficient method of evaluating a linear function h) is given in [2] (cf. Lemma 3).

4 Polynomial* basis techniques

This section discusses the structure of a polynomial* basis, and presents an efficient basis generation function and an efficient external shift function.

Theorem 1. *Let* $1, \gamma, \ldots, \gamma^{m-1}$ *be a polynomial basis for* $GF(q^m)$, *and let* $h(\epsilon)$ *be a linear function from* $GF(q^m)$ *to* $GF(q)$. *Let* $h_0(\epsilon)$ *be the 1-coefficient of the representation of the element* ϵ *in the polynomial basis. Let* ζ *be the element of* $GF(q^m)$ *such that* $h_0(\zeta\epsilon) = h(\epsilon)$. *A formula for the dual basis* $\eta_0, \ldots, \eta_{m-1}$ *of this polynomial basis with respect to* h *is*

$$\eta_i = \zeta^{-1}\xi_i,$$

where $\xi_0 = 1$ *and* $\xi_i = \gamma^{-1}\xi_{i-1} - h_0(\gamma^{-1}\xi_{i-1})$.

Proof. We first observe that the value ζ exists since there is a one-to-one correspondence between linear functions and field elements (cf. Lemma 3 of [2]). To prove the correctness of the formula, we use the definition of the dual basis and induction. First, we consider η_0 and observe that $h(\gamma^i \eta_0) = h_0(\gamma^i)$, which is 1 if $i = 0$ and 0 if $1 \le i \le m - 1$, meeting the definition. Now suppose we know that for $j > 0$ the first $j - 1$ elements are correct elements of the dual basis. Then we get the following for the jth element:

$$h(\gamma^i \eta_j) = h_0(\gamma^i \xi_j)$$
$$= h_0(\gamma^{i-1} \xi_{j-1}) - h_0(\gamma^i h_0(\gamma^{-1} \xi_{j-1})).$$

For $i = 0$, this reduces to $h_0(\gamma^{-1} \xi_{j-1}) - h_0(\gamma^{-1} \xi_{j-1}) = 0$. For $1 \le i \le m - 1$, the equation becomes $h_0(\gamma^{i-1} \xi_{j-1})$, which by induction is 1 if $i = j$ and 0 if $i \ne j$. In both cases the definition is met.

In the following algorithms, Z will be the internal representation of ζ, G will be the internal representation of γ, and I will be the internal representation of the identity element. The value V_0 corresponds to the element such that $(A \times V_0)[0] = h_0(A)$. The value Z corresponds to the function $h(\epsilon) = h_0(\zeta \epsilon)$, and contains the information specific to the choice of dual basis in the following algorithms. Note that if Z is 0, $h(\epsilon) = h_0(0) = 0$, and therefore, h would not be a nonzero linear function. Thus, we can assume Z is nonzero.

4.1 GENPOLY*

The algorithm GENPOLY* generates the internal representation of the dual basis elements. GENPOLY* is an *iterator*; it is meant to be called many times in succession. The first time an iterator is called, it starts from the **iter** line. When a **yield** statement is reached, the iterator returns the value specified by the yield. The next time the iterator is called, it starts immediately after the last **yield** executed; all temporary variables are assumed to retain their values from one call to the next. An iterator ends when the **enditer** line is reached.

```
iter GENPOLY*
     W ← I
     yield Z⁻¹
     for i from 1 to m − 1 do
          W ← W × G⁻¹
          T ← W × V₀
          W ← W − T[0] × I
          yield W × Z⁻¹
     endfor
enditer
```

4.2 SHIFTPOLY*

With our knowledge of the formula for a polynomial* basis, we can also devise a method for shifting an element's representation in the polynomial* basis. The algorithm simply uses the recursive formula for generating ξ_i from ξ_{i-1}, namely

$$s(\epsilon) = \gamma^{-1} \epsilon - h_0(\gamma^{-1} \epsilon).$$

Theorem 2. *s performs an external shift in the polynomial* basis with respect to h_0, such that $s(\xi_i) = \xi_{i+1}$ for $0 \le i \le m-2$ and $s(\xi_{m-1}) = 0$.*

Proof. First we observe that s is linear, i.e., that for all $\epsilon, \phi \in GF(q^m)$, $c \in GF(q)$, $s(\epsilon + \phi) = s(\epsilon) + s(\phi)$ and $s(c\epsilon) = cs(\epsilon)$. Since s is linear, we only have to show that applying it to basis elements is correct. Since s is merely the recursive formula for generating ξ_i from ξ_{i-1}, we know it is correct for all basis elements except ξ_{m-1}. Thus, it remains to show that $s(\xi_{m-1}) = 0$. To see this, define ξ_m as $s(\xi_{m-1})$ and apply the equation from the proof above:

$$h_0(\gamma^i \xi_m) = h_0(\gamma^{i-1}\xi_{m-1}) - h_0(\gamma^i h_0(\gamma^{-1}\xi_{m-1})).$$

For $i = 0$, this cancels to 0. For $1 \le i \le m-1$, the equation becomes $h_0(\gamma^{i-1}\xi_{m-1})$, which is 0. Since the values $h_0(\gamma^i \xi_m)$ as i varies correspond to coefficients of the representation of ξ_m in the basis $\xi_0, \ldots, \xi_{m-1}$ and they are all zero, it follows that $\xi_m = 0$.

Since the dual basis $\eta_0, \ldots, \eta_{m-1}$ is just the basis $\xi_0, \ldots, \xi_{m-1}$ scaled by ζ^{-1}, shifting in the dual basis $\eta_0, \ldots, \eta_{m-1}$ is accomplished by computing the function $\zeta^{-1}s(\zeta\epsilon)$. The following is an algorithm for shifting in the polynomial* basis based on this technique.

> **proc** SHIFTPOLY* (A)
> $\quad A \leftarrow A \times ZG^{-1}$
> $\quad T \leftarrow A \times V_0$
> $\quad A \leftarrow A - T[0] \times I$
> $\quad A \leftarrow A \times Z^{-1}$
> **endproc**

Also note that we can use SHIFTPOLY* to make a new version of GENPOLY* that generates by repeated shifting.

5 Techniques involving the dual of a normal basis

This section discusses the structure of a normal* basis, and presents an efficient basis generation function and an efficient external shift function.

Theorem 3. *Let $\gamma, \ldots, \gamma^{q^{m-1}}$ be a normal basis for $GF(q^m)$, and let $h(\epsilon)$ be a linear function from $GF(q^m)$ to $GF(q)$. Let $h_0(\epsilon)$ be the γ-coefficient of the representation of the element ϵ in the normal basis. Let ζ be the element of $GF(q^m)$ such that $h_0(\zeta\epsilon) = h(\epsilon)$. A formula for the dual basis $\eta_0, \ldots, \eta_{m-1}$ of this normal basis with respect to h is*

$$\eta_i = \zeta^{-1}\xi_i,$$

where $\xi_0 = 1$ and $\xi_i = \sigma\xi_{i-1}^q$, and where σ is the element such that $h_0(\gamma^{q^i}\sigma)$ is 1 for $i = 1$ and 0 for $i = 0$ and $2 \le i \le m-1$.

Proof. First, we observe that ζ and σ exist, the latter being an element of the dual basis with respect to h_0. We also observe that $h_0(\sigma\epsilon^q) = h_0(\epsilon)$ for all $\epsilon \in GF(q)$. To prove that the formula is correct, we use the definition of the dual basis and induction. First, we consider η_0 and observe that $h(\gamma^{q^i}\eta_0) = h_0(\gamma^{q^i})$, which is 1 if $i = 0$ and 0 if $1 \le i \le m - 1$, meeting the definition. For the induction step, we get the following for the jth element, where $j > 0$:

$$h(\gamma^{q^i}\eta_j) = h_0(\gamma^{q^i}\xi_j)$$
$$= h_0(\gamma^{q^i}\sigma\xi_{j-1}^q).$$

For $i = 0$, the equation becomes $h_0(\gamma^{q^{m-1}}\xi_{j-1})$, which by induction is 0. (Note that $\gamma = \gamma^{q^m}$.) For $1 \le i \le m - 1$, it becomes $h_0(\gamma^{q^{i-1}}\xi_{j-1})$, which by induction is 1 if $i = j$ and 0 if $i \ne j$. In both cases the definition is met.

In the following algorithms, G will be the internal representation of γ, S will be the internal representation of σ, and Z will be the internal representation of ζ. As before, we can assume Z is nonzero.

5.1 GenNormal*

Now that we know the general formula for the dual of a normal basis, we can demonstrate a technique for efficiently generating the dual of a normal basis. Like GenPoly*, GenNormal* is written as an iterator.

```
iter GenNormal*
    T ← Z⁻¹
    W ← S
    yield T
    for i from 1 to m − 1 do
        T ← T × W
        W ← Wq
        yield T
    endfor
enditer
```

Theorem 4. *The iterator* GenNormal* *generates the elements of the normal* basis.

Proof. After the first iteration, GenNormal* outputs the internal representation of ζ^{-1}. At each successive step, the basis is multiplied by successively higher powers of q of σ, so we get $\zeta^{-1}, \zeta^{-1}\sigma, \zeta^{-1}\sigma^{q+1}, \zeta^{-1}\sigma^{q^2+q+1}$, and so on. By our formula, this is the correct list of normal* basis elements.

5.2 ShiftNormal*

There is also an efficient method for doing a rotation of an element in the normal* basis. The algorithm simply uses the recursive formula for generating ξ_i from ξ_{i-1}, namely

$$s(\epsilon) = \sigma\epsilon^q.$$

Theorem 5. *s performs an external shift (actually, a rotation) in the normal* basis with respect to h_0, such that $s(\xi_i) = \xi_{i+1}$ for $0 \le i \le m - 2$ and $s(\xi_{m-1}) = \xi_0$.*

Proof. As before, we first observe that s is linear. We again only have to show that that the formula is correct for ξ_{m-1}. To see this, define ξ_m as $s(\xi_{m-1})$ and apply the equation from the proof above:

$$h_0(\gamma^{q^i} \xi_m) = h_0(\gamma^{q^i} \sigma \xi_{m-1}^q).$$

For $i = 0$, the equation becomes $h_0(\gamma^{q^{m-1}} \xi_{m-1})$, which is 1. For $1 \le i \le m - 1$, it becomes $h_0(\gamma^{q^{i-1}} \xi_{m-1})$, which is 0. Since the values $h_0(\gamma^{q^i} \xi_m)$ as i varies correspond to coefficients of the representation of ξ_m in the basis $\xi_0, \ldots, \xi_{m-1}$ and they are all zero except for the ξ_0-coefficient, which is 1, it follows that $\xi_m = \xi_0$.

Shifting in the dual basis $\eta_0, \ldots, \eta_{m-1}$ is accomplished by computing the function $\zeta^{-1} s(\zeta \epsilon)$, as before. Based on this, we have the algorithm SHIFTNOR-MAL*.

> **proc** SHIFTNORMAL* (A)
> $\quad A \leftarrow A^q$
> $\quad A \leftarrow A \times SZ^{q-1}$
> **endproc**

Note that this only requires storage for one value, as SZ^{q-1} can be precomputed. Also note that we can use SHIFTNORMAL* to make a new version of GENNORMAL*.

6 Conclusion

We have demonstrated efficient algorithms for external shifting and efficient basis generation in the polynomial* and normal* bases. Using these algorithms in the storage-efficient basis conversion methods described above, we can implement the following basis conversion methods: IMPORTBYSHIFTINSERT, IMPORT-BYGENACCUM, and EXPORTBYSHIFTEXTRACT for a polynomial* or normal* external basis, and EXPORTBYGEN*EVAL for a polynomial or normal basis.

References

1. *IEEE P1363: Standard Specifications for Public-Key Cryptography*, draft 11, July 1999. http://grouper.ieee.org/groups/1363/draft.html.
2. B.S. Kaliski Jr. and Y.L. Yin. Storage-efficient finite field basis conversion. In S. Tavares and H. Meijer, editors, *Selected Areas in Cryptography '98 Proceedings*, volume 1556 of *Lecture Notes in Computer Science*, pages 81–93. Springer, 1999.
3. R. Lidl and H. Niederreiter. *Finite Fields*, volume 20 of *Encyclopedia of Mathematics and Its Applications*. Addison-Wesley, 1983.

Power Analysis Attacks of Modular Exponentiation in Smartcards

Thomas S. Messerges[1], Ezzy A. Dabbish[1], Robert H. Sloan[2,3]

[1]Motorola Labs, Motorola
1301 E. Algonquin Road, Room 2712, Schaumburg, IL 60193
{tomas, dabbish}@ccrl.mot.com

[2]Dept. of EE and Computer Science, University of Illinois at Chicago
851 S. Morgan Street, Room 1120, Chicago, IL 60607
sloan@eecs.uic.edu

Abstract. Three new types of power analysis attacks against smartcard implementations of modular exponentiation algorithms are described. The first attack requires an adversary to exponentiate many random messages with a known and a secret exponent. The second attack assumes that the adversary can make the smartcard exponentiate using exponents of his own choosing. The last attack assumes the adversary knows the modulus and the exponentiation algorithm being used in the hardware. Experiments show that these attacks are successful. Potential countermeasures are suggested.

1 Introduction

Cryptographers have been very successful at designing algorithms that defy traditional mathematical attacks, but sometimes, when these algorithms are actually implemented, problems can occur. The implementation of a cryptographic algorithm can have weaknesses that were unanticipated by the designers of the algorithm. Adversaries can exploit these weaknesses to circumvent the security of the underlying cryptographic algorithm. Attacks on the implementations of cryptographic systems are a great concern to operators and users of secure systems. Implementation attacks include power analysis attacks [1,2], timing attacks [3,4], fault insertion attacks [5,6], and electromagnetic emission attacks [7]. Kelsey et al. [8] review some of these attacks and refer to them as "side-channel" attacks. The term "side-channel" is used to describe the leakage of unintended information from a supposedly tamper-resistant device, such as a smartcard.

In a power analysis attack the side-channel is the device's power consumption. An adversary can monitor the power consumption of a vulnerable device, such as a smartcard, to defeat the tamper-resistance properties and learn the secrets contained inside the device [1]. Although it is preferable to design secure systems that do not rely on secrets contained in the smartcard, there are applications where this may not be possible or is undesirable. In these systems, if the secret, for instance a private key, is compromised, then the entire system's security may be broken.

---

[3] Partially supported by NSF Grant CCR-9800070

In this paper we examine the vulnerabilities of public-key cryptographic algorithms to power analysis attacks. Specifically, attacks on the modular exponentiation process are described. These attacks are aimed at extracting the secret exponent from tamper-resistant hardware by observing the instantaneous power consumption signals into the device while the exponent is being used for the exponentiation. Experimental results on a smartcard containing a modular exponentiation circuit are provided to confirm the threats posed by these attacks.

Three types of attacks are described that can be mounted by adversaries possessing various degrees of capabilities and sophistication. The first attack requires that, in addition to exponentiating with the secret exponent, the smartcard will also exponentiate with at least one exponent known to the attacker. This attack, referred to as a "Single-Exponent, Multiple-Data" (SEMD) attack, requires the attacker to exponentiate many random messages with both the known and the secret exponent. The SEMD attack is demonstrated to be successful on exponentiations using a small modulus (i.e., 64 bits) with 20,000 trial exponentiations, but with a large modulus might require 20,000 exponentiations per exponent bit. The second attack we introduce requires that the attacker can get the smartcard to exponentiate using exponents of his own choosing. Our experiments showed that this attack, referred to as a "Multiple-Exponent, Single-Data" (MESD) attack, requires the attacker to run about 200 trial exponentiations for each exponent bit of the secret exponent. The last attack that we discovered does not require the adversary to know any exponents, but does assume the attacker can obtain basic knowledge of the exponentiation algorithm being used by the smartcard. With this attack, referred to as a "Zero-Exponent, Multiple-Data" (ZEMD) attack, we can successfully extract a secret exponent with about 200 trial exponentiations for each secret exponent bit.

The organization of this paper is as follows; first, the related work is reviewed and the motivation for research into power analysis attacks is given. Next, implementations of modular exponentiation and the basic principles of power analysis attacks are reviewed. The equipment and software needed for these attacks is described and detailed descriptions of the MESD, SEMD and ZEMD attacks are given. Finally, potential countermeasures are suggested.

1.1 Related Work

Previous papers that describe power analysis attacks mainly examine the security of symmetric key cryptographic algorithms. Kocher, et al. [1] review a Simple Power Analysis (SPA) attack and introduce a Differential Power Analysis (DPA) attack, which uses powerful statistical-based techniques. They describe specific attacks against the Digital Encryption Standard (DES)[9], and their techniques can also be modified for other ciphers. Kelsey, et al. [8] show how even a small amount of side-channel information can be used to break a cryptosystem such as the DES. An alternate approach is taken in [2], where techniques to strengthen the power consumption attack by maximizing the side-channel information are described. The Advanced Encryption Standard (AES) candidate algorithms are analyzed in [10-12] for their vulnerabilities to power analysis attacks. These papers advise that the vulnerabilities of the AES algorithms to power analysis attacks should be considered when choosing the next encryption standard.

1.2 Research Motivation

Tamper-resistant devices, such as smartcards, can be used to store secret data such as a person's private key in a two-key, public-key cryptosystem. Familiar examples of such systems are an RSA cryptosystem [13] and an elliptic-curve cryptosystem [14,15]. In a typical scenario, the owner of a smartcard needs to present the card in order to make a payment, log onto a computer account, or gain access to a secured facility. In order to complete a transaction, the smartcard is tested for authenticity by a hardware device called a reader. The reader is provided by a merchant or some other third party and, in general, may or may not be trusted by the smartcard owner. Thus, it is important that when a user relinquishes control of her smartcard, she is confident that the secrecy of the private key in the card can be maintained. In an RSA cryptosystem, the authenticity of the card is tested by asking the card to use its internally stored private key to modularly exponentiate a random challenge. Since it is possible that the card is being accessed by a malicious reader, the power consumption of the card during the exponentiation process should not reveal the secret key.

2 Review of Modular Exponentiation Implementations

Modular exponentiation is at the root of many two-key, public-key cryptographic implementations. The technique used to implement modular exponentiation is commonly known as the "square-and-multiply" algorithm. Elliptic curve cryptosystems use an analogous routine called the "double-and-add" algorithm. Two versions of the square-and-multiply algorithm are given in Fig. 1. The first routine in Fig. 1, $exp1$, starts at the exponent's most significant nonzero bit and works downward. The second routine, $exp2$, starts at the least significant bit of the exponent e and works upward. Both routines are vulnerable to attack and both return the same result, $M^e \bmod N$. Common techniques to implement modular exponentiation (i.e., particular implementations of the modular square and modular multiply operations) can be found in [16-21]. One pop-

```
exp1(M, e, N)                    exp2(M, e, N)
{   R = M                        {   R = 1
    for (i = n-2 down to 0)          S = M
    {   R = R² mod N                 for (i = 0 to n-1)
        if (ith bit of e is a 1)     {   if (ith bit of e is a 1)
            R = R·M mod N }                  R = R·S mod N
    return R }                          S = S² mod N }
                                     return R }
```

Fig. 1. Exponentiation Routines Using the Square-and-Multiply Algorithm
Two versions of the square-and-multiply algorithm used for smartcard authentication are given above. The routine $exp1$ starts at the most significant bit and works down and the routine $exp2$ does the opposite. The routine $exp2$ requires extra memory to store the S variable. The exponent, e, has n bits, where the least significant bit is numbered 0 and the most significant nonzero bit is numbered $n-1$

ular method to speed up the exponentiation is to use Montgomery's modular multiplication algorithm [22] for all the multiplies and squares.

The attacks in this paper are a potential threat to all of these implementations. The MESD and SEMD attacks are against the square-and-multiply method. Every implementation executes the square-and-multiply method in some manner, so all are potentially vulnerable. The ZEMD attack works on intermediate data results and is only possible if an attacker possesses basic knowledge of the implementation. The attacker needs to know which of the types of square-and-multiply algorithms is being used and the technique used for the modular multiplications. Even if the attacker does not know the implementation, there is a fairly small number of likely possibilities. In our attack, all that was necessary to know was that the exponentiation was done using the *exp1* algorithm and Montgomery's method was used for the modular multiplies.

3 Review of Power Analysis Attacks

Power analysis attacks work by exploiting the differences in power consumption between when a tamper-resistant device processes a logical zero and when it processes a logical one. For example, when the secret data on a smartcard is accessed, the power consumption may be different depending on the Hamming weight of the data. If an attacker knows the Hamming weight of the secret key, the brute force search space is reduced and given enough Hamming weights of independent functions of the secret key, the attacker could potentially learn the entire secret key. This type of attack, where the adversary directly uses a power consumption signal to obtain information about the secret key is referred to as a Simple Power Analysis (SPA) attack and is described in [1].

Differential Power Analysis (DPA) is based on the same underlying principle of an SPA attack, but uses statistical analysis techniques to extract very tiny differences in power consumption signals. DPA was first introduced in [1] and a strengthened version was reported in [2].

3.1 Simple Power Analysis (SPA)

SPA[1] on a single-key cryptographic algorithm, such as DES, could be used to learn the Hamming weight of the key bytes. DES uses only a 56-bit key so learning the Hamming weight information alone makes DES vulnerable to a brute-force attack. In fact, depending on the implementation, there are even stronger SPA attacks. A two-key, public-key cryptosystem, such as an RSA or elliptic curve cryptosystem, might also be vulnerable to an SPA attack on the Hamming weight of the individual key bytes, however it is possible an even stronger attack can be made directly against the square-and-multiply algorithm.

If exponentiation were performed in software using one of the square-and-multiply algorithms of Fig. 1, there could be a number of potential vulnerabilities. The main problem with both algorithms is that the outcome of the "if statement" might be observable in the power signal. This would directly enable the attacker to learn every bit of the secret exponent. A simple fix is to always perform a multiply and to only save the result if the exponent bit is a one. This solution is very costly for performance and still may be vulnerable if the act of saving the result can be observed in the power signal.

3.2 Differential Power Analysis (DPA)

The problem with an SPA attack is that the information about the secret key is difficult to directly observe. In our experiments, the information about the key was often obscured with noise and modulated by the device's clock signal. DPA can be used to reduce the noise and also to "demodulate" the data. Multiple-bit DPA [2] can be used to attack the DES algorithm by defining a function, say D, based on the guessed key bits entering an S-box lookup table. If the D-function predicts high power consumption for a particular S-box lookup, the power signal is placed into set S_{high}. If low power consumption is predicted, then the signal is placed into an alternate set, S_{low}. If the predicted power consumption is neither high or low, then the power signal is discarded. The result of this partitioning is that when the average signal in set S_{high} is subtracted from the average signal in set S_{low}, the resulting signal is demodulated. Any power biases at the time corresponding to the S-box lookup operation are visible as an obvious spike in the difference signal and much of the noise is eliminated because averaging reduces the noise variance. Correct guesses of the secret key bits into an S-box are verified by trying all 2^6 possibilities and checking which one produces the strongest difference signal.

All of the attacks described in this paper use averaging and subtracting and so are similar to a DPA attack. The averaging reduces the noise and the subtracting demodulates the secret information and enhances the power biases.

4 Power Analysis Equipment

A smartcard with a built-in modular exponentiation circuit was used to evaluate the attacks described in this paper. The exponentiation circuit on this smartcard is a typical implementation of the square-and-multiply algorithm using a Montgomery multiplication circuit to speed up the modular reductions. The exponentiation circuit was accessed via a software program residing in the card's memory. This software executed a simple ISO7816 smartcard protocol [23] which supports a command similar to the standard "internal authenticate" command.

5 Attacking a Secret Exponent

The objective of the attacks described in this paper is to find the value of e, the secret exponent stored in the smartcard's internal memory. The attacker is assumed to have complete control of the smartcard. He can ask the card to exponentiate using e and can monitor all input and output signals. The card will obey all commands of the attacker, except a command to output the secret key. The main command that is needed is the "internal authenticate" command which causes the card to receive an input value, M, and output $M^e \bmod N$. Some smartcard systems require the user to enter a Personal Identification Number (PIN) prior to allowing access to the card. This feature is not considered in our attacks. Also, the number of times the attacker can query the card is assumed unlimited. All of these assumptions are reasonable since smartcard systems have been implemented that allow such access. Other assumptions used for particular attacks are stated in the sections that describe the specific attack details.

5.1 A Simple Correlation Experiment

We performed a correlation experiment to determine if e could be revealed by simply cross-correlating the power signal from a single multiply operation with the entire exponentiation's power signal. This attack was designed to see how easy it is to distinguish the multiplies from the squares, thus revealing the bits of e. Let the multiply's power signal be $S_m[j]$ and the exponentiation's power signal be $S_e[j]$. The cross-correlation signal, $S_c[j]$ is calculated as

$$S_c[j] = \sum_{\tau = 0}^{W} S_m[\tau] S_e[j + \tau]$$

where W is the number of samples in the multiply's power signal. That is, $W = T_m/T$, where T_m is the time needed for a multiply operation and T is the sampling rate. An attacker can learn the approximate value of W through experimentation or from the smartcard's documentation.

The power signals and cross-correlation signal obtained from running this experiment are shown in Fig. 2. The exponentiation and multiply power signals were obtained by running the smartcard with constant input data and averaging 5,000 power signals to reduce the measurement noise. This experiment was first tested on a known exponent, so the locations of the squares and multiplies are known and are labeled in the Fig. 2. The resulting cross-correlation signal shows peaks at the locations of the individual squares and multiplies, but the height of the peaks are uncorrelated with the type of operation. Thus, this cross-correlation technique is not useful to differentiate between squares and multiplies. However, it is interesting to point out that the time needed for each operation in the square-and-multiply algorithm can be determined from the cross

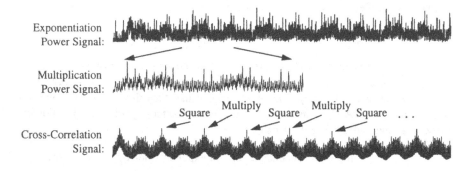

Fig. 2. Cross-Correlation of Multiplication and Exponentiation Power Signals
The above signals were obtained using the power analysis equipment described in Section 4. The signals were averaged for 5,000 exponentiations using a constant input value. The results show an ability to determine the time between the square-and-multiply operations, but cannot be used to distinguish multiply operations from squaring operations

correlation signal. This information could lead to a combined power analysis and timing attack in implementations where the time to multiply is slightly different than the time to square. Such an attack would be more powerful than previously documented timing attacks because the cross-correlation signal would yield the timing of all intermediate operations. Fortunately, the smartcards we examined do not have this problem.

5.2 Single-Exponent, Multiple-Data (SEMD) Attack

The SEMD attack assumes that the smartcard is willing to exponentiate an arbitrary number of random values with two exponents; the secret exponent and a public exponent. Such a situation could occur in a smartcard system that supports the ISO7816 [23] standard "external authenticate" command. Whereas the "internal authenticate" command causes the smartcard to use its secret key, the "external authenticate" command can be used to make the smartcard use the public key associated with a particular smartcard reader. It is assumed that the exponent bits of this public key would be known to the attacker.

The basic premise of this attack is that by comparing the power signal of an exponentiation using a known exponent to a power signal using an unknown exponent, the adversary can learn where the two exponents differ, thus learn the secret exponent. In reality, the comparison is nontrivial because the intermediate data results of the square-and-multiply algorithm cause widely varying changes in the power signals, thereby making direct comparisons unreliable. The solution to this problem is to use averaging and subtraction. This simple DPA technique begins by using the secret exponent to exponentiate L random values and collects their associated power signals, $S_i[j]$. Likewise, L power signals, $P_i[j]$, are collected using the known exponent. The average signals are then calculated and subtracted to form $D[j]$, the DPA bias signal,

$$D[j] = \frac{1}{L}\sum_{i=1}^{L} S_i[j] - \frac{1}{L}\sum_{i=1}^{L} P_i[j] = \bar{S}[j] - \bar{P}[j]$$

The portions of the signals $\bar{S}[j]$ and $\bar{P}[j]$ that are dependent on the intermediate data will average out to the same constant mean μ, thus:

$$\bar{S}[j] \approx \bar{P}[j] \approx \mu \qquad \text{if} \qquad j = \text{a data dependent sample point}$$

The portion of the signals $\bar{S}[j]$ and $\bar{P}[j]$ that are dependent on the exponent bits will average out to different values, μ_s or μ_m, depending on whether a square or multiply operation is performed. Thus, if μ_s and μ_m are not equal, then their difference will be nonzero and the DPA bias signal, $D[j]$, can be used to determine the exact location of the squares and multiplies in the secret exponent:

$$D[j] \approx \begin{cases} 0 & \text{if } j = \text{data dependent point or exponentiation operations agree} \\ \text{nonzero} & \text{if } j = \text{point where the exponentiation operations differ} \end{cases}$$

The SEMD attack was performed on a smartcard and the result is shown in Fig. 3. For this experiment the exponentiation was simplified by using a modulus and data with only 64 bits. This simplification was done only for illustrative purposes. Using smaller

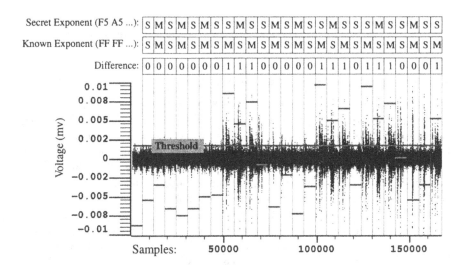

Fig. 3. Single-Exponent, Multiple-Data (SEMD) Attack Results
The above plot is the DPA signal comparing the exponentiation power signal produced with a known exponent and an unknown exponent. The energy in the DPA signal is greater when the two exponent operations are different. The shaded horizontal bars show the output of an integrate-and-dump filter indicating the energy associated with each interval of time. The above signal was obtained using 20,000 trial exponentiations

data made it possible to store more of the exponentiation's power signal in the digital oscilloscope, so many more exponent bits could be attacked with each test. In an actual attack against full-sized data, one would likely be able to attack only a small portion of the exponentiation at a time. The number of bits attacked at one time depends on the size of the memory in the attacker's digital oscilloscope. The DPA signal in Fig. 3 shows an attack on about 16 exponent bits and was obtained with $L=10,000$; thus 20,000 trial exponentiations were needed. In a real attack, using full-sized data, this attack might need 20,000 exponentiations for each exponent bit. In this case a sliding window approach is needed, where only a windowed portion of the power trace is attacked at one time.

The DPA bias signal in Fig. 3 is labeled to show the squares (S) and multiplies (M) associated with the secret and known exponents. The regions where these operations differ exhibit a corresponding increase in the amplitude of the DPA bias signal. An integrate-and-dump filter was used to compute the signal energy associated with each region and the output of the filter is graphed as shaded horizontal line segments in Fig. 3. The output of the integrate-and-dump filter is given as:

$$I_{out}[j] = \sum_{i = jT_m}^{(j+1)T_m} F_{clip}(D[i]) \qquad \text{where, } F_{clip}(x) = \max(x^2, V_{clip})$$

In this equation, V_{clip} is chosen to eliminate the overwhelming influence of spurious spikes in the bias signal. The final result shows that the output of the integrate-and-dump filter is good at indicating where the secret and known exponent operations differ. Thus, the SEMD attack is an important attack that implementors of smartcard systems need to consider when designing a secure system.

5.3 Multiple-Exponent, Single-Data (MESD) Attack

The MESD attack is more powerful than the SEMD attack, but requires a few more assumptions about the smartcard. The previously described SEMD attack is a very simple attack requiring little sophistication on the part of the adversary, but the resulting DPA bias signal is sometimes difficult to interpret. The Signal-to-Noise Ratio (SNR) can be improved using the MESD attack. The assumption for the MESD attack is that the smartcard will exponentiate a constant value[1] using exponents chosen by the attacker. Again, such an assumption is not unreasonable since some situations might allow the smartcard to accept new exponents that can be supplied by an untrusted entity. Also, the smartcard does not have unlimited memory, so it is impossible for it to keep a history of previous values it has exponentiated. Thus, the card cannot know if it is being repeatedly asked to exponentiate a constant value.

The algorithm for the MESD attack is given in Fig. 4. The first steps of the algorithm are to choose an arbitrary value, M, exponentiate M using the secret exponent e, and then collect the corresponding average power signal $S_M[j]$. Next, the algorithm progresses by successively attacking each secret exponent bit starting with the first bit used in the square-and-multiply algorithm and moving towards the last. To attack the ith secret exponent bit, the adversary guesses the ith bit is a 0 and then a 1 and asks the card to exponentiate using both guesses. It is assumed that the adversary already knows the first through $(i-1)$st exponent bits so the intermediate results of the exponentiation up to the $(i-1)$st exponent bit will be the same for the guessed exponent and the secret exponent. If the adversary guesses the ith bit correctly, then the intermediate results will also agree at the ith position. If the guess is wrong, then the results will differ. This difference can be seen in the corresponding power traces. Let e_g be the current guess for the exponent. The average power signals for exponentiating M using an e_g with the ith

```
M = arbitrary value and e_g = 0
Collect S_M[j]
for (i = n-1 to 0)
{  guess (ith bit of e_g is a 1) and collect S_1[j]
   guess (ith bit of e_g is a 0) and collect S_0[j]
   Calculate two DPA bias signal:
       D_1[j] = S_M[j] - S_1[j] and D_0[j] = S_M[j] - S_0[j]
   Decide which guess was correct using DPA result
   update e_g }
e_g is now equal to e (the secret exponent)
```

Fig. 4. Algorithm for the Multiple-Exponent, Single Data (MESD) Attack
This algorithm gradually makes e_g equal to the secret exponent bit, by using the DPA signal to decide which guess is correct at the ith iteration

---

[1] This value may or may not be known to the attacker.

bit equal to 1 is $S_1[j]$ and an e_g with the ith bit equal to 0 is $S_0[j]$. Two DPA bias signals can be calculated:

$$D_1[j] = S_M[j] - S_1[j] \quad \text{and} \quad D_0[j] = S_M[j] - S_0[j]$$

Whichever exponent bit was correct produces a power signal that agrees with the secret exponent's power signal for the larger amount of time. Thus, whichever bias signal is zero for a longer time corresponds to the correct guess.

The resulting bias signal for a correct and incorrect guess are shown in Fig. 5. It is clear that the SNR in Fig. 5 is much improved over the SEMD attack. The higher SNR of the MESD attack means that fewer trial exponentiations are needed for a successful attack. Also, an experienced attacker really needs to calculate only one DPA bias signal. For example, the attacker could always guess the exponent bit is a 1. If the guess is correct, the DPA bias signal will remain zero for the duration of a multiply and a square operation. If the guess is wrong, then the bias signal will only remain zero for the duration of the square operation. This technique effectively cuts the running time of the algorithm of Fig. 4 in half. Our experiments showed that as few as 100 exponentiations were needed per exponent bit. Memory limitations in a digital oscilloscope also might require a moving window approach to collect the secret exponent's power signal, thus resulting in 200 exponentiations per exponent bit. The circumstances allowing an MESD attack definitely need to be addressed by implementors concerned with power analysis attacks.

5.4 Zero-Exponent, Multiple-Data (ZEMD) Attack

The ZEMD attack is similar to the MESD attack, but has a different set of assumptions. One assumption for the ZEMD attack is that the smartcard will exponentiate many random messages using the secret exponent. This attack does not require the adversary know any exponents, hence the zero-exponent nomenclature. Instead, the adversary needs to be able to predict the intermediate results of the square-and-multiply algorithm using an off-line simulation. This usually requires that the adversary know the algorithm being used by the exponentiation hardware and the modulus used for the exponentiation. There are only a few common approaches to implementing modular exponentiation algorithms, so it is likely an adversary can determine this information. It is also likely that the adversary can learn the modulus because this information is usually public.

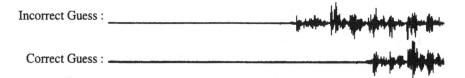

Incorrect Guess :

Correct Guess :

Fig. 5. Multiple-Exponent, Single-Data (MESD) Attack Results
The above plot is the DPA signal obtained when the next bit is guessed correctly compared to when the next bit guess is wrong. The correct guess is clearly seen to be the signal that remains zero the longest. This signal was obtained using 1,000 trial exponentiations

The algorithm for the ZEMD attack is given in Fig. 6. The ZEMD attack starts by attacking the first bit used during the exponentiation and proceeds by attacking each successive bit. In this algorithm, the variable e_g gradually becomes equal to the secret exponent. After each iteration of the attack, another exponent bit is learned and e_g is subsequently updated. At the ith iteration of the algorithm, it is assumed that the correct exponent, e_g, is correct up to the $(i-1)$st bit. The algorithm then guesses that the ith bit of the secret exponent is a 1 and a DPA bias signal is created to verify the guess. The DPA bias signal is created by choosing a random input, M, and running a simulation to determine the power consumption after the multiply in the ith step of the square-and-multiply algorithm. This simulation is possible because the exponent e_g is known up to the ith bit and the power consumption can be estimated using the Hamming weight of a particular byte in the multiplication result. Previous power analysis experiments showed that a higher number of ones correspond to higher power consumption.

If the multiply at the ith step actually occurred, then the power analysis signals can be accurately partitioned into two sets, thereby creating biases (or spikes) in the DPA bias signal when the average signals in each partition are subtracted. If the guess is incorrect, then the partitioning will not be accurate and the power biases will not occur. A natural error-correcting feature of this algorithm is that if there is ever a mistake, all subsequent steps will fail to show any power biases.

The ZEMD attack was implemented and an example of the DPA bias signals for a correct and an incorrect guess are given in Fig. 7. The DPA bias signals in Fig. 7 were generated using an 8-bit partitioning function based on the Hamming weight of the multiplication result. Power signals corresponding to results with Hamming weight eight were subtracted from power signals corresponding to results with Hamming weight zero. This partitioning technique creates a larger SNR and is further described in [2].

```
e_g = 0
for (i = n-1 to 0)
{   guess (ith bit of e_g is a 1)
    for (k = 1 to L)
    {   choose a random value: M
        simulate to the ith set the calculation of M^(e_g) mod N
        if (multiplication result has high Hamming weight)
            run smartcard and collect power signal: S[j]
            add S[j] to set S_high
        if (multiplication result has low Hamming weight)
            run smartcard and collect power signal: S[j]
            add S[j] to set S_low }
    Average the power signals and get DPA bias signal:
        D[j] = S̄_low[j] - S̄_high[j]
    if DPA bias signal has spikes
        the guess was correct: make ith bit of e_g equal to 1
    else
        the guess was wrong: make ith bit of e_g equal to 0 }
e_g is now equal to e (the secret exponent)
```

Fig. 6. Algorithm for the Zero-Exponent, Multiple Data (ZEMD) Attack
This algorithm gradually makes e_g equal to the secret exponent bit, by using the DPA signal to decide if the guess of the ith bit being a one is correct

Guess was
incorrect ($e_i = 0$):

Guess was
correct ($e_i = 1$):

Fig. 7. Zero-Exponent, Multiple-Data (ZEMD) Attack Results
The above plot is the DPA signal comparing the DPA bias signal produced when the guess of the ith exponent bit is correct compared to when it is incorrect. The spikes in the correct signal can be used to confirm the correct guess. This signal was obtained using 500 trial exponentiations

The signals in Fig. 7 were obtained by averaging 500 random power signals, but we have also been able to mount this attack with only 100 power signals per exponent bit.

In the attack we implemented it was necessary to collect power signals using a windowing approach. This meant it was necessary to collect new power signals for each exponent bit being attacked. With optimizations to the equipment and algorithm, more exponent bits could be attacked simultaneously requiring even fewer trial exponentiations. The exact number of trial exponentiations necessary is dependent on the equipment of the adversary, the size of the power biases, and the noise in the signals. Implementors need to keep the ZEMD attack in mind when designing modular exponentiation hardware and software.

6 Countermeasures

Potential countermeasures to the attacks described in this paper include many of the same techniques described to prevent timing attacks on exponentiation. Kocher's [3] suggestion for adapting the techniques used for blinding signatures [24] can also be applied to prevent power analysis attacks. Prior to exponentiation, the message could be blinded with a random value, v_i and unblinded after exponentiation with $v_f = (v_i^{-1})^e \bmod N$. Kocher suggests an efficient way to calculate and maintain (v_i, v_f) pairs.

Message blinding would prevent the MESD and ZESD attacks, but since the same exponent is being used, the SEMD attack would still be effective. To prevent the SEMD attack, exponent blinding, also described in [3], would be necessary. In an RSA cryptosystem, the exponent can be blinded by adding a random multiple of $\phi(N)$, where $\phi(N) = (p-1)(q-1)$ and $N=pq$. In summary, the exponentiation process would go as follows:

1. Blind the message M: $\hat{M} = (v_i M) \bmod N$
2. Blind the exponent e: $\hat{e} = e + r\phi(N)$
3. exponentiate: $\hat{S} = (\hat{M}^{\hat{e}}) \bmod N$
4. unblind the result: $S = (v_f \hat{S}) \bmod N$

Another way to protect against power analysis attack is to randomize the exponentiation algorithm. One way this can be accomplished is to combine the two square-and-multiply algorithms of Fig. 1. A randomized exponentiation algorithm could begin by selecting a random starting point in the exponent. Exponentiation would proceed from this random starting point towards the most significant bit using *exp2* of Fig. 1. Then, the algorithm would return to the starting point and finish the exponentiation using *exp1* and moving towards the least significant bit. It would be difficult for an attacker to determine the random starting point from just one power trace (an SPA attack), so this algorithm would effectively randomize the exponentiation. The amount of randomization that is possible depends on the number of bits in the exponent. For large exponents this randomization might be enough to make power analysis attacks impractical to all but the most sophisticated adversaries. All the attacks presented in this paper would be significantly diminished by randomizing the exponentiation.

7 Conclusions

The potential threat of monitoring power consumption signals to learn the private key in a two-key, public-key cryptosystem has been investigated. A variety of vulnerabilities have been documented and three new attacks were developed. The practicality of all three attacks was confirmed by testing on actual smartcard hardware. Table 1 summarizes the attacks and some of the assumptions and possible solutions.

The goal of this research is to point out the potential vulnerabilities and to provide guidance towards the design of more secure tamper-resistant devices. Hopefully the results of this paper will encourage the design and development of solutions to the problems posed by power analysis attacks.

TABLE 1: Summary of Power Analysis Attacks on Exponentiation

Attack Name	Number of trial exponentiations	Assumptions	Possible Solution
SEMD	20,000	attacker knows one exponent	exponent blinding
MESD	200	attacker can choose exponent	message blinding
ZEMD	200	attacker knows algorithm and modulus	message blinding

References

1. P. Kocher, J. Jaffe, and B. Jun, "Introduction to Differential Power Analysis and Related Attacks," http://www.cryptography.com/dpa/technical, 1998.
2. T. S. Messerges, E. A. Dabbish and R. H. Sloan, "Investigations of Power Analysis Attacks on Smartcards," *Proceedings of USENIX Workshop on Smartcard Technology*, May 1999, pp. 151-61.
3. P. Kocher, "Timing Attacks on Implementations of Diffie-Hellman, RSA, DSS, and Other Systems," in *Proceedings of Advances in Cryptology–CRYPTO '96*, Springer-Verlag, 1996, pp. 104-13.

4. J. F. Dhem, F. Koeune, P. A. Leroux, P. Mestré, J. J. Quisquater and J. L. Willems, "A Practical Implementation of the Timing Attack," in *Proceedings of CARDIS 1998*, Sept. 1998.

5. D. Boneh and R. A. Demillo and R. J. Lipton, "On the Importance of Checking Cryptographic Protocols for Faults," in *Proceedings of Advances in Cryptology–Eurocrypt '97*, Springer-Verlag, 1997, pp. 37-51.

6. E. Biham and A. Shamir, "Differential Fault Analysis of Secret Key Cryptosystems," in *Proceedings of Advances in Cryptology–CRYPTO '97*, Springer-Verlag, 1997, pp. 513-25.

7. W. van Eck, "Electromagnetic Radiation from Video Display Units: An Eavesdropping Risk," *Computers and Security*, v. 4, 1985, pp. 269-86.

8. J. Kelsey, B. Schneier, D. Wagner, and C. Hall, "Side Channel Cryptanalysis of Product Ciphers," in *Proceedings of ESORICS '98*, Springer-Verlag, September 1998, pp. 97-110.

9. ANSI X.392, "American National Standard for Data Encryption Algorithm (DEA)," American Standards Institute, 1981.

10. J. Daemen, V. Rijmen, "Resistance Against Implementation Attacks: A Comparative Study of the AES Proposals," *Second Advanced Encryption Standard (AES) Candidate Conference*, http://csrc.nist.gov/encryption/aes/round1/conf2/aes2conf.htm, March 1999.

11. E. Biham, A. Shamir, "Power Analysis of the Key Scheduling of the AES Candidates," *Second Advanced Encryption Standard (AES) Candidate Conference*, http://csrc.nist.gov/encryption/aes/round1/conf2/aes2conf.htm, March 1999.

12. S. Chari, C. Jutla, J.R. Rao, P. Rohatgi, "A Cautionary Note Regarding Evaluation of AES Candidates on Smart-Cards," *Second Advanced Encryption Standard (AES) Candidate Conference*, http://csrc.nist.gov/encryption/aes/round1/conf2/aes2conf.htm, March 1999.

13. R. L. Rivest, A. Shamir, and L. Adleman, "A Method for Obtaining Digital Signatures and Public-Key Cryptosystems," *Comm. ACM*, vol. 21, 1978, pp. 120-126.

14. N. Koblitz, "Elliptic Curve Cryptosystems," *Mathematics of Computation*, vol. 48, 1987, pp. 203-9.

15. V. S. Miller, "Uses of Elliptic Curves in Cryptography," in *Proceedings of Advances in Cryptology–CRYPTO '85*, Springer-Verlag, 1986, pp. 417-26.

16. E. F. Brickel, "A Survey of Hardware Implementations of RSA," in *Proceedings of Advances in Cryptology–CRYPTO '89*, Springer-Verlag, 1990, pp. 368-70.

17. A. Selby and C. Mitchel, "Algorithms for Software Implementations of RSA," *IEE Proceedings*, vol. 136E, 1989, pp. 166-70.

18. S. E. Eldridge and C. D. Walter, "Hardware Implementations of Montgomery's Modular Multiplication Algorithm," *IEEE Transactions on Computers*, vol. 42, No. 6, June 1993, pp. 693-9.

19. S. R. Dussé and B. S. Kaliski Jr., "A Cryptographic Library for the Motorola 56000," in *Proceedings of Advances in Cryptology–Eurocrypt '90*, Springer-Verlag, 1991, pp. 230-44.

20. G. Monier, "Method for the Implementation of Modular Multiplication According to the Montgomery Method," *United States Patent*, No. 5,745, 398, April 28, 1998.

21. C. D. Gressel, D. Hendel, I. Dror, I. Hadad and B. Arazi, "Compact Microelectronic Device for Performing Modular Multiplication and Exponentiation over Large Numbers," *United States Patent*, No. 5,742,530, April 21, 1998.

22. P. L. Montgomery, "Modular Multiplication Without Trial Division," *Mathematics of Computation*, vol. 44, 1985, pp. 519-21.

23. ISO7816, "Identification Cards-Integrated Circuit(s) Cards with Contacts," International Organization for Standardization.

24. D. Chaum, "Blind Signatures for Untraceable Payments," in *Proceedings of Advances in Cryptology–CRYPTO '82*, Plenum Press, 1983, pp. 199-203.

DES and Differential Power Analysis
The "Duplication" Method*

Louis Goubin, Jacques Patarin
Bull SmartCards and Terminals
68, route de Versailles - BP45
78431 Louveciennes Cedex - France
{L.Goubin, J.Patarin}@frlv.bull.fr

Abstract. Paul Kocher recently developped attacks based on the electric consumption of chips that perform cryptographic computations. Among those attacks, the "Differential Power Analysis" (DPA) is probably one of the most impressive and most difficult to avoid.
In this paper, we present several ideas to resist this type of attack, and in particular we develop one of them which leads, interestingly, to rather precise mathematical analysis. Thus we show that it is possible to build an implementation that is provably DPA-resistant, in a "local" and restricted way (*i.e.* when – given a chip with a fixed key – the attacker only tries to detect predictable local deviations in the differentials of mean curves). We also briefly discuss some more general attacks, that are sometimes efficient whereas the "original" DPA fails. Many measures of consumption have been done on real chips to test the ideas presented in this paper, and some of the obtained curves are printed here.

Note: An extended version of this paper can be obtained from the authors.

1 Introduction

This paper is about a way of securing a cryptographic algorithm that makes use of a secret key. More precisely, the goal consists in building an implementation of the algorithm that is not vulnerable to a certain type of physical attacks – so-called "Differential Power Analysis".

These DPA attacks belong to a general family of attacks that look for information about the secret key by studying the electric consumption of the electronic device during the execution of the computation. In this family, we usually distinguish between SPA attacks ("Simple Power Analysis") and DPA attacks.

In SPA attacks, the aim is essentially to guess – from the values of the consumption – which particular instruction is being computed at a certain time and with which input or output, and then to use this information to deduce some part of the secret. Figure 1 shows the electric consumption of a chip, measured during a DES computation on a real smartcard. The fact that the 16 rounds of the DES algorithm are clearly visible is a good sign that power analysis attacks may indeed provide information about what the chip is doing.

---

* Patents Pending

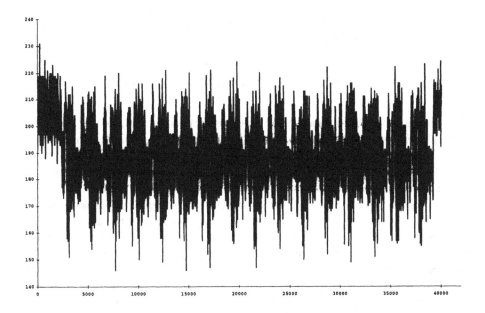

Fig. 1. Electric consumption measured on the 16 rounds of a DES computation

In DPA attacks, some differentials on two sets of average consumption are computed, and the attacks succeed if an unusual phenomenon appears – on these differentials of consumption – for a good choice of some of the key bits (we give details below), so that we are able to find out those key bits. What makes DPA attacks so impressive, when they work, is the fact that they can find out the secret key of a public algorithm (for example DES, but also many other algorithms) without knowing anything (nor trying to find anything) about the particular implementation of that algorithm. Implementations exist that are DPA-resistant (differentials do not show anything special) but not SPA-resistant (some critical information can be deduced from the consumption curves). On the contrary, other implementations exist that are SPA-resistant but not DPA-resistant (some critical information can be found by studying differentials of two mean curves of consumption). Finally, some implementations can be found that resist both types of attack (at least at the present), or none of them.

Throughout this paper, we study more particularly DPA and we will not deal any longer with SPA. Indeed, as we see below, DPA can easily be analyzed in a mathematical way (and not only in an empirical way). There exist many attacks based on the electric consumption. We do not claim to give here solutions to all the problems that may result from these attacks.

The cryptographic algorithms we consider here make use of a secret key in order to compute an output information from an input information. It may be a ciphering, a deciphering or a signature operation. In particular, all the material described in this paper applies to "secret key algorithms" and also to the so-called "public key algorithms".

2 The "Differential Power Analysis" attacks

The "Differential Power Analysis" attacks, developped by Paul Kocher and Cryptographic Research (see [1]), start from the fact that the attacker can get many more information (than the knowledge of the inputs and the outputs) during the execution of the computation, such as for instance the electric consumption of the microcontroller or the electromagnetic radiations of the circuit. The "Differential Power Analysis" (DPA to be brief) is an attack that allows to obtain information about the secret key (contained in a smartcard for example), by performing a statistical analysis of the electric consumption records measured for a large number of computations with the same key. Let us consider for instance the case of the DES algorithm (Data Encryption Standard). It executes in 16 steps, called "rounds". In each of these steps, a transformation F is performed on 32 bits. This F function uses eight non-linear transformations from 6 bits to 4 bits, each of which is coded by a table called "S-box". The DPA attack on the DES can be performed as follows (the number 1000 used below is just an example):

Step 1: We measure the consumption on the first round, for 1000 DES computations. We denote by E_1, ..., E_{1000} the input values of those 1000 computations. We denote by C_1, ..., C_{1000} the 1000 electric consumption curves measured during the computations. We also compute the "mean curve" MC of those 1000 consumption curves.

Step 2: We focus for instance on the first output bit of the first S-box during the first round. Let b be the value of that bit. It is easy to see that b depends on only 6 bits of the secret key. The attacker makes an hypothesis on the involved 6 bits. He computes – from those 6 bits and from the E_i – the expected (theoretical) values for b. This enables to separate the 1000 inputs E_1, ..., E_{1000} into two categories: those giving $b = 0$ and those giving $b = 1$.

Step 3: We now compute the mean MC' of the curves corresponding to inputs of the first category (i.e. the one for which $b = 0$). If MC and MC' show an appreciable difference (in a statistical meaning, i.e. a difference much greater than the standard deviation of the measured noise), we consider that the chosen values for the 6 key bits were correct. If MC and MC' do not show any sensible difference, we repeat step 2 with another choice for the 6 bits.

Note: In practice, for each choice of the 6 key bits, we draw the curve representing the difference between MC and MC'. As a result, we obtain 64 curves, among which one is supposed to be very special, i.e. to show an appreciable difference, compared to all the others.

Step 4: We repeat steps 2 and 3 with a "target" bit b in the second S-box, then in the third S-box, ..., until the eighth S-box. As a result, we finally obtain 48 bits of the secret key.

Step 5: The remaining 8 bits can be found by exhaustive search.

Note: It is also possible to focus (in steps 2, 3 and 4) on the set of the four output bits for the considered S-boxes, instead of only one output bit. This is what we actually did for real smartcards. In that case, the inputs are separated into 16 categories: those giving 0000 as output, those giving 0001, ..., those giving 1111. In step 3, we may compute for example the mean MC' of the curves corresponding to the last category (*i.e.* the one which gives 1111 as output). As a result, the mean MC' is computed on approximately $\frac{1}{16}$ of the curves (instead of approximately half of the curves with step 3 above): this may compel us to use a number of DES computations greater than 1000, but it generally leads to a more appreciable difference between MC and MC'.

We presented in figures 2 and 3 two mean curves, resulting from steps 2 and 3, for a classical implementation of DES on a real smartcard (with '1111' as target output of the first S-box and with 2048 different inputs, even if we noted that 512 inputs are sufficient). A detailed analysis of the 64 obtained curves (that we cannot all print here, due to the lack of place) shows that the one corresponding to the correct choice of the 6 key-bits can easily be detected (it contains much greater peaks than all the others).

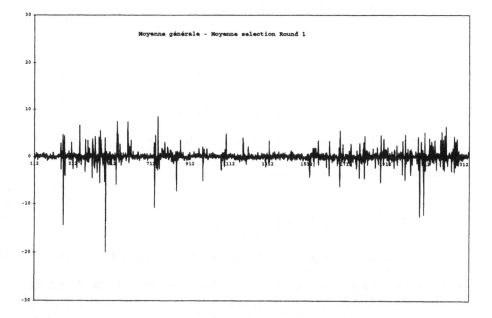

Fig. 2. An example of difference of the curves MC and MC' when the 6 bits are false

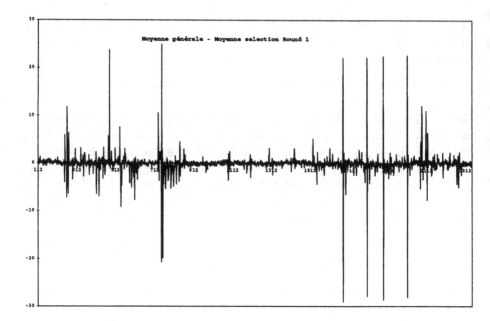

Fig. 3. Difference of the curves MC and MC' when the 6 bits are correct

This attack does not require any knowledge about the individual electric consumption of each instruction, nor about the position in time of each of these instructions. It applies exactly the same way as soon as the attacker knows the outputs of the algorithm and the corresponding consumption curves. It only relies on the following fundamental hypothesis:

Fundamental hypothesis: *There exists an intermediate variable, that appears during the computation of the algorithm, such that knowing a few key bits (in practice less than 32 bits) allows us to decide whether two inputs (respectively two outputs) give or not the same value for this variable.*

All the algorithms that use S-boxes, such as DES, are potentially vulnerable to the DPA attack, because the "natural" implementations generally remain within the hypothesis mentioned above.

3   Securing the algorithm

Several countermeasures against DPA attacks can be conceived. For instance:

1. Introducing random timing shifts, so that the computed means do not correspond any longer to the consumption of the same instruction. The crucial point consists here in performing those shifts so that they cannot be easily eliminated by a statistical treatment of the consumption curves.

2. Replacing some of the critical instructions (in particular the basic assembler instructions involving writings in the carry, readings of data from an array, etc) by assembler instructions whose "consumption signature" is difficult to analyze.

3. For a given algorithm, giving an explicit way of computing it, so that DPA is provably unefficient on the obtained implementation. For instance, for a DES-like algorithm, we detail in section 4 how to compute the non-linear transformations of the S-boxes in order to avoid some DPA attacks.

In the present paper, we essentially study the third idea because it leads to a quite precise mathematical analysis. We give in this section a general method to implement an algorithm with a secret key so as to avoid the DPA attacks described above. The basic principle consists in programming the algorithm so that the fundamental hypothesis above is not true any longer (*i.e.* an intermediate variable never depends on the knowledge of an easily accessible subset of the secret key).

The main idea

In this paper, we mainly study how this can be done by using the following main idea: replacing each intermediate variable V, occuring during the computation and depending on the inputs (or the outputs), by k variables V_1, ..., V_k, such that $V_1, V_2, ..., V_k$ allows us – if we want – to retrieve V. More precisely, to guarantee the security of the algorithm in its new form, it is sufficient to choose a function f satisfying the identity $V = f(V_1, ..., V_k)$, together with the two following conditions:

Condition 1: *From the knowledge of a value v and for any fixed value i, $1 \leq i \leq k$, it is not feasible to deduce information about the set of the values v_i such that there exist a $(k-1)$-uple $(v_1, ..., v_{i-1}, v_{i+1}, ..., v_k)$ satisfying the equation $f(v_1, ..., v_k) = v$.*

Condition 2: *The function f is such that the transformations to be performed on V_1, V_2, ..., or V_k during the computation (instead of the transformations usually performed on V) can be implemented without calculating V.*

First example for condition 1: If we choose $f(v_1, ..., v_k) = v_1 \oplus v_2 \oplus ... \oplus v_k$, where $\oplus$ denotes the bit-by-bit "exclusive-or" function, condition 1 is obviously satisfied, because – for any fixed index i between 1 and k – the considered set of the values v_i contains all the possible values and thus does not depend on v.

Second example for condition 1: If we consider some variable V whose values lie in the multiplicative group of $\mathbf{Z}/n\mathbf{Z}$, we can choose the function $f(v_1, ..., v_k) = v_1 \cdot v_2 \cdot ... \cdot v_k \bmod n$, where the new variables v_1, v_2, ..., v_k also have values in the multiplicative group of $\mathbf{Z}/n\mathbf{Z}$. Condition 1 is also obviously true because – for any fixed index i between 1 and k – the considered set of the values v_i contains all the possible values and thus does not depend on v.

We then "translate" the algorithm by replacing each intermediate variable V depending on the inputs (or the outputs) by the k variables V_1, ..., V_k. In the following sections, we study how conditions 1 and 2 can be achieved in the case of the DES or RSA algorithms.

4 The DES algorithm: First example of implementation for DPA resistance

In this section, we consider the particular case of the DES algorithm. We choose here to separate each intermediate variable V, occuring during the computation and depending on the inputs (or the outputs), into two variables V_1 and V_2 (i.e. we take $k = 2$). Let us choose the function $f(v_1, v_2) = v = v_1 \oplus v_2$ (see the first example of section 3), which satisfies condition 1. From the construction of the algorithm, it is easy to see that the transformations performed on v always belong to one of the five following categories:

1. permutation of the bits of v;
2. expansion of the bits of v;
3. "exclusive-or" between v and another variable v' of the same type;
4. "exclusive-or" between v and a variable depending only on the key;
5. transformation of v using a S-box.

The first two cases correspond to linear transformations on the bits of the variable v. For these ones, condition 2 is thus very easy to satisfy: we just have – instead of the transformation usually performed on v – to perform the permutation or the expansion on v_1, then on v_2, and the identity $f(v_1, v_2) = v$, which was true before the transformation, is also true afterwards.

In the same way, in the third case, we just have to replace the computation of $v'' = v \oplus v'$ by the computation of $v_1'' = v_1 \oplus v_1'$ and $v_2'' = v_2 \oplus v_2'$. The identities $f(v_1, v_2) = v$ and $f(v_1', v_2') = v'$ give indeed $f(v_1'', v_2'') = v''$, so that condition 2 is true again.

As concerns the exclusive-or between v and a variable c depending only on the key, condition 2 is also very easy to satisfy: we just have to replace the computation of $v \oplus c$ by $v_1 \oplus c$ (or $v_2 \oplus c$) and that gives condition 2.

Finally, instead of the non-linear transformation $v' = S(v)$, given under the form of a S-box (which in that example has 6-bits inputs and 4-bits outputs), we implement the transformation $(v_1', v_2') = S'(v_1, v_2)$ by using two new S-boxes (each of which sending 12 bits onto 4 bits). In order to keep the identity $f(v_1', v_2') = v'$, we may choose:

$$(v_1', v_2') = S'(v_1, v_2) = (A(v_1, v_2), S(v_1 \oplus v_2) \oplus A(v_1, v_2)).$$

where A denotes a *randomly* chosen *secret* transformation from 12 bits to 4 bits (see figure 4). The first of the new S-boxes corresponds to the table of the transformation $(v_1, v_2) \mapsto A(v_1, v_2)$, and the second one corresponds to the table of the transformation $(v_1, v_2) \mapsto S(v_1 \oplus v_2) \oplus A(v_1, v_2)$. Thanks to the randomly

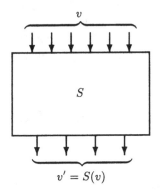

**Initial implementation: the predictable values
v and v' appear in RAM at some time**

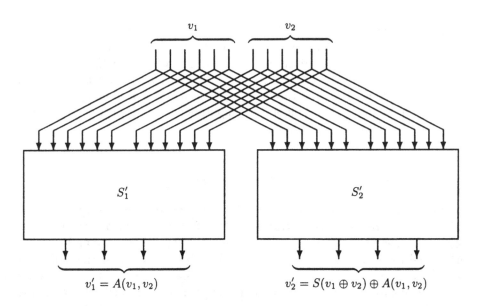

**Modified implementation: the values $v = v_1 \oplus v_2$ and
$v' = v_1' \oplus v_2'$ never explicitly appear in RAM**

Fig. 4. Standard transformation of a S-box

chosen function A, condition 1 is satisfied. Moreover, the use of tables allows us to avoid the computation of $v_1 \oplus v_2$, so that condition 2 is also true.

The solution presented in this section is quite realistic for chips that compute DES in hardware (and are not embedded in a card), or for PCs, because – in those cases – enough memory is available. More precisely, the size of the memory required to store the S-boxes is 32 Kbytes for the method described in this section. It is too much for smartcards, for which specific variations using less memories are described in section 5 below.

5   Smartcard implementations of DES

First variation

In order to reduce the ROM used by the algorithm, it is quite possible to use the same random function A for the eight S-boxes (of the initial description of the DES), so that we have only nine (new) S-boxes (*i.e.* 18 Kbytes) to store in ROM, instead of sixteen S-boxes.

Second variation

In order to reduce the size of the ROM needed to store the S-boxes, we can also use the following method: instead of each non-linear transformation $v' = S(v)$ of the initial implementation, given under the form of a S-box (with 6-bits inputs and 4-bits outputs in the case of the DES), we implement the transformation $(v_1', v_2') = S'(v_1, v_2)$ by using two S-boxes, each of which sending 6 bits onto 4 bits. The initial implementation of the computation $v' = S(v)$ is replaced by the two following successive computations:

$$- v_0 = \varphi(v_1 \oplus v_2)$$
$$- (v_1', v_2') = S'(v_1, v_2) = (A(v_0), S(\varphi^{-1}(v_0)) \oplus A(v_0))$$

where φ is a *bijective* and *secret* function from 6 bits to 6 bits and where A denotes a *random* and *secret* transformation from 6 bits to 4 bits. The first of the two new S-boxes corresponds to the table of the transformation $v_0 \mapsto A(v_0)$ and the second one corresponds to the table of the transformation $v_0 \mapsto S(\varphi^{-1}(v_0)) \oplus A(v_0)$. From this construction, the identity $f(v_1', v_2') = v'$ is always true. Thanks to the random function A, condition 1 is satisfied. Moreover, the use of tables allows us to avoid the computation of $\varphi^{-1}(v_0) = v_1 \oplus v_2$, so that condition 2 is also true. This solution (shown in figure 5) requires 512 bytes to store the S-boxes.

In order to satisfy condition 2, it remains to choose the bijective transformation φ such that the computation of $v_0 = \varphi(v_1 \oplus v_2)$ is feasible without computing $v_1 \oplus v_2$. We give below two examples of possible choice for the function φ.

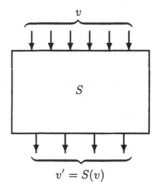

Initial implementation: the predictable values
v **and** v' **appear in RAM at some time**

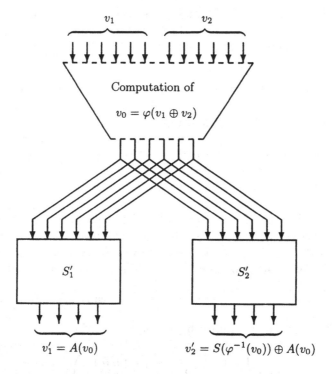

Modified implementation: the values $v = v_1 \oplus v_2$ **and**
$v' = v_1' \oplus v_2'$ **never explicitly appear in RAM**

Fig. 5. Transformation of a S-box (second variation)

Exemple 1: a linear bijection

We choose φ as a *linear secret* and *bijective* function from 6 bits to 6 bits (we consider the set of the 6-bits values as a vectorial space of dimension 6 on the finite field $\mathbf{F}_2$ with 2 elements). In practice, choosing φ is equivalent to choosing a *random* and *invertible* 6×6 *matrix* whose coefficients are 0 or 1. With this choice of φ, it is easy to see that condition 2 is satisfied. Indeed, to compute $\varphi(v_1 \oplus v_2)$, we just have to compute $\varphi(v_1)$, then $\varphi(v_2)$ and finally to compute the "exclusive-or" of the two obtained results.

For instance, the matrix
$$\begin{pmatrix} 1 & 1 & 0 & 1 & 0 & 0 \\ 1 & 1 & 0 & 1 & 0 & 1 \\ 0 & 1 & 1 & 0 & 1 & 0 \\ 1 & 1 & 1 & 0 & 1 & 0 \\ 0 & 1 & 1 & 1 & 1 & 0 \\ 0 & 0 & 1 & 1 & 0 & 1 \end{pmatrix}$$
is invertible. It corresponds to the

linear bijection φ from 6 bits to 6 bits defined by $\varphi(u_1, u_2, u_3, u_4, u_5, u_6) = (u_1 \oplus u_2 \oplus u_4, u_1 \oplus u_2 \oplus u_4 \oplus u_6, u_2 \oplus u_3 \oplus u_5, u_1 \oplus u_2 \oplus u_3 \oplus u_5, u_2 \oplus u_3 \oplus u_4 \oplus u_5, u_3 \oplus u_4 \oplus u_6)$.

Let $v_1 = (v_{1,1}, v_{1,2}, v_{1,3}, v_{1,4}, v_{1,5}, v_{1,6})$ and $v_2 = (v_{2,1}, v_{2,2}, v_{2,3}, v_{2,4}, v_{2,5}, v_{2,6})$. To compute $\varphi(v_1 \oplus v_2)$, we successively compute:

- $\varphi(v_1) = (v_{1,1} \oplus v_{1,2} \oplus v_{1,4}, v_{1,1} \oplus v_{1,2} \oplus v_{1,4} \oplus v_{1,6}, v_{1,2} \oplus v_{1,3} \oplus v_{1,5}, v_{1,1} \oplus v_{1,2} \oplus v_{1,3} \oplus v_{1,5}, v_{1,2} \oplus v_{1,3} \oplus v_{1,4} \oplus v_{1,5}, v_{1,3} \oplus v_{1,4} \oplus v_{1,6})$
- $\varphi(v_2) = (v_{2,1} \oplus v_{2,2} \oplus v_{2,4}, v_{2,1} \oplus v_{2,2} \oplus v_{2,4} \oplus v_{2,6}, v_{2,2} \oplus v_{2,3} \oplus v_{2,5}, v_{2,1} \oplus v_{2,2} \oplus v_{2,3} \oplus v_{2,5}, v_{2,2} \oplus v_{2,3} \oplus v_{2,4} \oplus v_{2,5}, v_{2,3} \oplus v_{2,4} \oplus v_{2,6})$

Then we compute the "exclusive-or" of the two obtained results.

Exemple 2: a quadratic bijection

We choose φ as a *quadratic secret* and *bijective* function from 6 bits to 6 bits. Here, "quadratic" means that each bit of the output is given by a polynomial function of *total degree two* of the 6 bits of the input (which are identified to 6 elements of the finite field $\mathbf{F}_2$). In practice, we may choose the function φ defined by $\varphi(x) = t(s(x)^5)$, where s is a secret linear bijection from $(\mathbf{F}_2)^6$ to $\mathcal{L}$, t is a secret linear bijection from $\mathcal{L}$ to $(\mathbf{F}_2)^6$ and $\mathcal{L}$ denotes an algebraic extension of degree 6 over the finite field $\mathbf{F}_2$. The bijectivity of this function φ follows from the fact that $a \mapsto a^5$ is a bijection on the extension $\mathcal{L}$ (whose inverse is $b \mapsto b^{38}$). To be convinced that condition 2 is still satisfied, just notice that we can write:

$$\varphi(v_1 \oplus v_2) = \psi(v_1, v_1) \oplus \psi(v_1, v_2) \oplus \psi(v_2, v_1) \oplus \psi(v_2, v_2),$$

where the function ψ is defined by $\psi(x, y) = t(s(x)^4 \cdot s(y))$.

For instance, if we identify $\mathcal{L}$ to $\mathbf{F}_2[X]/(X^6 + X + 1)$ and if we choose s and t

whose matrices are
$$\begin{pmatrix} 1 & 1 & 0 & 1 & 0 & 0 \\ 1 & 1 & 0 & 1 & 0 & 1 \\ 0 & 1 & 1 & 0 & 1 & 0 \\ 1 & 1 & 1 & 0 & 1 & 0 \\ 0 & 1 & 1 & 1 & 1 & 0 \\ 0 & 0 & 1 & 1 & 0 & 1 \end{pmatrix} \text{ and } \begin{pmatrix} 0 & 1 & 0 & 0 & 1 & 1 \\ 1 & 1 & 0 & 1 & 0 & 0 \\ 1 & 0 & 1 & 0 & 1 & 1 \\ 0 & 1 & 1 & 1 & 0 & 0 \\ 1 & 0 & 1 & 0 & 1 & 0 \\ 0 & 0 & 1 & 0 & 1 & 1 \end{pmatrix}$$
with respect to the basis

$(1, X, X^2, X^3, X^4, X^5)$ of $\mathcal{L}$ over $\mathbf{F}_2$ and to the canonical basis of $(\mathbf{F}_2)^6$ over $\mathbf{F}_2$, we obtain the following quadratic bijection φ from 6 bits to 6 bits:

$$\varphi(u_1, u_2, u_3, u_4, u_5, u_6) = (u_2u_5 \oplus u_1u_4 \oplus u_4 \oplus u_6 \oplus u_6u_2 \oplus u_4u_6 \oplus u_2 \oplus u_5 \oplus u_3 \oplus$$
$$u_4u_3, u_2u_5 \oplus u_5u_1 \oplus u_1u_4 \oplus u_4 \oplus u_6 \oplus u_4u_5 \oplus u_2 \oplus u_3 \oplus u_3u_1, u_2u_5 \oplus u_5u_1 \oplus u_6u_5 \oplus$$
$$u_1u_4 \oplus u_3u_5 \oplus u_1 \oplus u_4u_6 \oplus u_6u_3 \oplus u_4u_3 \oplus u_3u_1, u_1u_4 \oplus u_2u_3 \oplus u_6u_1 \oplus u_4u_6 \oplus u_5 \oplus$$
$$u_6u_3 \oplus u_4u_3, u_5u_1 \oplus u_1u_4 \oplus u_6 \oplus u_3u_5 \oplus u_4u_5 \oplus u_1 \oplus u_6u_1 \oplus u_4u_6 \oplus u_3 \oplus u_6u_3 \oplus$$
$$u_4u_2, u_4 \oplus u_6 \oplus u_3u_5 \oplus u_1 \oplus u_4u_6 \oplus u_6u_3).$$

To compute $\varphi(v_1 \oplus v_2)$, we use the function $\psi(x, y) = t(s(x)^4 \cdot s(y))$ from 12 bits to 6 bits, which gives the 6 output bits from the 12 input bits as follows:

$$\psi(x_1, x_2, x_3, x_4, x_5, x_6, y_1, y_2, y_3, y_4, y_5, y_6) = (x_3y_5 \oplus x_6y_2 \oplus x_6y_3 \oplus x_6y_4 \oplus x_3y_1 \oplus$$
$$x_6y_1 \oplus x_1y_3 \oplus x_1y_5 \oplus x_5y_2 \oplus x_5y_5 \oplus x_5y_1 \oplus x_6y_6 \oplus x_1y_6 \oplus x_1y_2 \oplus x_1y_4 \oplus x_2y_1 \oplus$$
$$x_2y_2 \oplus x_4y_4 \oplus x_3y_3 \oplus x_3y_6 \oplus x_4y_3 \oplus x_5y_3, x_4y_5 \oplus x_3y_1 \oplus x_6y_1 \oplus x_2y_5 \oplus x_5y_1 \oplus x_6y_6 \oplus$$
$$x_1y_6 \oplus x_1y_2 \oplus x_2y_1 \oplus x_2y_2 \oplus x_4y_1 \oplus x_4y_4 \oplus x_3y_3, x_6y_2 \oplus x_6y_3 \oplus x_6y_4 \oplus x_6y_5 \oplus x_3y_1 \oplus$$
$$x_6y_1 \oplus x_2y_5 \oplus x_5y_1 \oplus x_1y_6 \oplus x_1y_1 \oplus x_1y_2 \oplus x_1y_4 \oplus x_2y_1 \oplus x_2y_4 \oplus x_4y_2 \oplus x_2y_6 \oplus$$
$$x_3y_4 \oplus x_5y_3, x_3y_1 \oplus x_6y_2 \oplus x_2y_6 \oplus x_5y_3 \oplus x_5y_4 \oplus x_5y_6 \oplus x_6y_3 \oplus x_2y_3 \oplus x_4y_6 \oplus x_6y_5 \oplus$$
$$x_1y_3 \oplus x_5y_5 \oplus x_2y_4 \oplus x_4y_2 \oplus x_4y_5 \oplus x_3y_5 \oplus x_4y_3 \oplus x_6y_1 \oplus x_4y_1, x_3y_1 \oplus x_6y_6 \oplus x_5y_3 \oplus$$
$$x_5y_6 \oplus x_5y_2 \oplus x_1y_5 \oplus x_1y_1 \oplus x_1y_2 \oplus x_2y_1 \oplus x_2y_3 \oplus x_3y_6 \oplus x_6y_5 \oplus x_1y_3 \oplus x_2y_4 \oplus$$
$$x_3y_3 \oplus x_4y_5 \oplus x_2y_5 \oplus x_6y_1 \oplus x_4y_1 \oplus x_6y_4 \oplus x_3y_2, x_6y_6 \oplus x_4y_4 \oplus x_5y_4 \oplus x_5y_6 \oplus x_6y_3 \oplus$$
$$x_1y_6 \oplus x_1y_1 \oplus x_1y_2 \oplus x_2y_1 \oplus x_6y_5 \oplus x_2y_4 \oplus x_4y_2 \oplus x_4y_5 \oplus x_3y_5 \oplus x_6y_1 \oplus x_6y_4).$$

By using these formulas, we successively compute $\psi(v_1, v_1)$, $\psi(v_1, v_2)$, $\psi(v_2, v_1)$ and $\psi(v_2, v_2)$. Finally, we compute the "exclusive-or" of the four obtained results.

Third variation

To further reduce the size of the ROM needed to store the S-boxes, we can apply simultaneously the ideas of both variations 1 and 2: we use the second variation, with the same secret bijection φ (from 6 bits to 6 bits) and the same secret random function A (from 6 bits to 6 bits) in the new implementation of each non-linear transformation given by a S-box. This variation thus requires only 288 bytes to store the S-boxes. We have applied the Differential Power Analysis on real smartcard implementations of this third variation. Two examples of differential mean curves (with 2048 inputs and with '1111' as target output of the first S-box) are presented in figures 6 and 7. A precise analysis of the 64 curves given by the DPA (see note after step 3, in section 2) shows that none of them appears to be "very special", compared to the others, so that we can say that this implementation resists the DPA attack (at least in its basic form, see appendix 2 for a possible generalization that could still be dangerous).

Fourth variation

In this last variation, instead of implementing the transformation $(v'_1, v'_2) = S'(v_1, v_2)$ (which replaces the non-linear transformation $v' = S(v)$ of the initial implementation, given by a S-box) by using two S-boxes, we perform the computation of v'_1 (respectively v'_2) by using a simple algebraic function (i.e. the bits

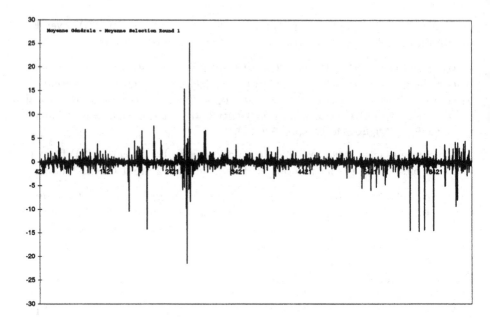

Fig. 6. An example of difference of the curves MC and MC' when the 6 bits are false

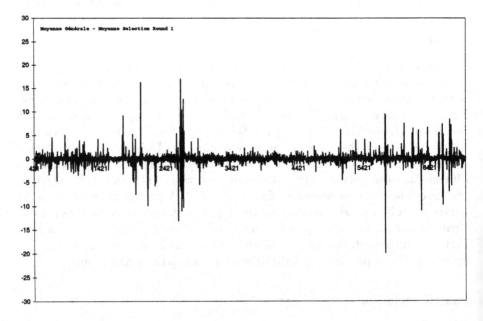

Fig. 7. Difference of the curves MC and MC' when the 6 bits are correct

of v_1' (respectively v_2') are given by a polynomial function of total degree 1 or 2 of the bits of v_1 and v_2), then we compute v_2' (respectively v_1') by using a table. This enables to reduce again the needed ROM for the implementation. This last variation requires only 256 bytes to store the S-boxes.

6   The RSA algorithm

The "Power Analysis" attacks also threaten the classical implementations of the RSA algorithm. Indeed, these implementations often use the so-called "square-and-multiply" principle to perform the computation of $x^d \bmod n$. It consists in writing the binary decomposition $d = d_{m-1}2^{m-1} + d_{m-2}2^{m-2} + ... + d_1 2^1 + d_0 2^0$ of the secret exponent d, and then in performing the computation as follows:

1. $z \leftarrow 1$;
   For i going backwards from $m - 1$ to 0 do:
2. $z \leftarrow z^2 \bmod n$;
3. if $d_i = 1$ then $z \leftarrow z \times x \bmod n$.

In this computation, we see that – among the successive values taken by the z variable – the first ones depend on only a few bits of the secret key d. The fundamental hypothesis that enables the DPA attack is thus satisfied. As a result, we can guess for instance the 10 most significant bits of d by studying the consumption measures on the part of the algorithm corresponding to i going from $m - 1$ to $m - 10$. We can then continue the attack by using comsumption measures on the part of the algorithm corresponding to i going from $m - 11$ to $m - 20$, which gives the 10 next bits of d, and so on. We finally find all the bits of the secret exponent d.

The method described in section 3 also applies to securing the RSA algorithm. We use here a separation of each intermediate variable V (whose values lie in the multiplicative group of $\mathbf{Z}/n\mathbf{Z}$), occuring during the computation and depending on the inputs (or the outputs), into two variables V_1 and V_2 (i.e. we take $k = 2$), and we choose the function $f(v_1, v_2) = v = v_1 \cdot v_2 \bmod n$. We already saw in section 3 (cf "second example") that this function f satisfies condition 1.

We thus replace x by (x_1, x_2) such that $x = x_1 \cdot x_2 \bmod n$ and z by (z_1, z_2) such that $z = z_1 \cdot z_2 \bmod n$ (in practice, we can for instance choose x_1 randomly and deduce x_2). Considering again the three steps of the "square-and-multiply" method, we perform the following transformations:

1. $z \leftarrow 1$ is replaced by $z_1 \leftarrow 1$ and $z_2 \leftarrow 1$;
2. $z \leftarrow z^2 \bmod n$ is replaced by $z_1 \leftarrow z_1^2 \bmod n$ and $z_2 \leftarrow z_2^2 \bmod n$;
3. $z \leftarrow z \times x \bmod n$ is replaced by $z_1 \leftarrow z_1 \times x_1 \bmod n$ and $z_2 \leftarrow z_2 \times x_2 \bmod n$.

It is easy to check that the identity $z = f(z_1, z_2)$ remains true all along the computation, which shows that condition 2 is satisfied.

Let us notice that the computations performed respectively on the z_1 variable and on the z_2 variable are completely independent. We thus can imagine to perform the two computations either in a sequential way, or in an overlapped way, or simultaneously in the case of multiprogrammation, or simultaneously in different processors working concurrently.

7 Generalized Attacks

Recently, more general attacks were introduced, where the attacker tries to correlate different points of a power consumption curve. We have no place here to analyze in detail the effect of this idea on the "Duplication Method". However, it is possible to show that if each variable is splitted in, say, k variables, then the complexity of the implementation increases in $\mathcal{O}(k)$, while the complexity of the attack increases exponentially in k.

As concerns DES implementations, we also recommend, when it is possible, to use different S-Boxes for each smartcard (stored in EEPROM). In particular, this avoids some attacks which use a smartcard with a known key to help finding the key in another smartcard whose key is unknown.

8 Conclusion

In this paper, we investigate how the study of the electric consumption measures of an electronic device can be used by an attacker to get information about the secret key of the cryptographic algorithm computed by the chip. More precisely, we focus on the so-called Differential Power Attacks, which were recently introduced by Paul Kocher, and which use a statistical analysis of a set of consumption curves measured for many different inputs of the cryptographic algorithm.

We study more precisely how DPA attacks work, and what precise hypotheses they rely on. We then present several ways of securing cryptosystems. In particular, concrete examples of such countermeasures are described in the cases of DES and RSA, which are the most used cryptographic algorithms at the present.

To secure those algorithms, we essentially study the main idea that consists in splitting each intermediate variable, occuring in the computation, into two (or more) variables, such that the values of these new variables cannot be easily predicted. The obtained implementations can be proved to resist the "local" version of Differential Power Analysis (where the attacker only tries to detect local deviations in the differentials of mean curves). Nevertheless other attacks can be conceived, still using the analysis of electric consumption. We do not pretend to solve all security problems linked to these threats. These latter attacks are not only theoretical, since we found real products that are defeated by them, but it also shows that theoretical investigations have to be continued in that sensitive subject.

References

1. Paul Kocher, Joshua Jaffe, Benjamin Jun, *Introduction to Differential Power Analysis and Related Attacks*, 1998. This paper is available at http://www.cryptography.com/dpa/technical/index.html

IPA: A New Class of Power Attacks

Paul N. Fahn*
and Peter K. Pearson**

Certicom Corp.
25801 Industrial Blvd.
Hayward, CA 94545, USA

Abstract. We present Inferential Power Analysis (IPA), a new class of attacks based on power analysis. An IPA attack has two stages: a profiling stage and a key extraction stage. In the profiling stage, intratrace differencing, averaging, and other statistical operations are performed on a large number of power traces to learn details of the implementation, leading to the location and identification of key bits. In the key extraction stage, the key is obtained from a very few power traces; we have successfully extracted keys from a single trace. Compared to differential power analysis, IPA has the advantages that the attacker does not need either plaintext or ciphertext, and that, in the key extraction stage, a key can be obtained from a small number of traces.

1 Introduction

Recent years have seen significant progress in what are called "power attacks" on cryptographic modules, attacks in which one monitors the power drawn by the module and from these measurements extracts some secret quantity that the module manipulates during some cryptographic operation. In 1998, Kocher et al. [5] described Differential Power Analysis (DPA), in which power measurements from many repeated cryptographic operations are cleverly combined. More recently, Biham and Shamir [1] showed how to derive key information by combining power measurements on different cryptographic modules.

This paper describes a class of attacks called Inferential Power Analysis (IPA) attacks. An IPA attack is characterized by two stages, the first a lengthy profiling stage, and the second a simpler key extraction stage. The profiling step is typically based on comparisons of repeated parts of a selected cryptographic operation, such as the different rounds in a DES encryption. These comparisons can be performed on a single cryptographic module, requiring many measured operations, and result in a profile that can subsequently be used to extract keys from other modules using as little as a single cryptographic operation. Unlike DPA, these attacks do not require knowledge of the operation's inputs or outputs.

---

* pfahn@certicom.com
** ppearson@certicom.com

More generally, these attacks illustrate a class of attacks in which a one-time effort requiring just one module produces information with which keys can be easily extracted from other modules of the same design. Such attacks can be applied not only by a cardholder against a smartcard in his possession, but also by a terminal owner against smartcards that use his terminal.

Due to the rapidly advancing state of knowledge about power analysis, we cannot make conclusive statements about the effectiveness of specific counter-measures. Nevertheless we suggest several possible defenses in §5 that may make power attacks more difficult and raise the level of effort and expertise required of the attacker.

2 Background

More and more cryptographic systems are embedding keys in portable electronic modules such as smartcards and PC cards. These modules usually provide both storage for the key and processor power sufficient to allow the key to be used *in situ*, so that the key is never exposed to the outside world. When the holder of the module (which we will henceforth assume is a smartcard) has a stake in keeping the key secret, such modules provide strong, convenient, and inexpensive security.

On the other hand, when the cardholder has an incentive to violate the secrecy of the key, protecting the key is a difficult challenge to the system's designer. For example, in the case of stored-value cards, learning the card's key may enable the cardholder to defraud a bank. Since the cardholder has physical possession of the card, many avenues of attack are available:

- The cardholder can subject the card to unusual conditions like out-of-range supply voltage, out-of-range clock frequency, extreme temperatures, radia-tion, or unusual commands, in order to induce errors. Some errors may di-rectly expose keys, while others may produce incorrect cryptographic results from which keys can be computed [2].
- The cardholder can physically dissect the card and reset protection bits, or directly read electrical charges in memory cells, or measure voltages on bus traces while sensitive data are passing between memory and processor [6].
- While the card is performing cryptographic calculations, the cardholder can measure currents, voltages, electric fields, or execution times [4], any of which might exhibit correlation with the key being used.

The current drawn through the card's power connector (V_{cc}) is easy to mea-sure with a digital oscilloscope, and provides much revealing information. For example, if the smartcard uses a hardware multiplier for modular exponenti-ations of large integers, each multiplication is visible as a distinct period of increased current consumption. The fastest implementations of modular expo-nentiation handle ordinary multiplications differently from squarings; but if this technique is used in a smartcard, squarings and nonsquare multiplications can be distinguished in an oscilloscope trace of current consumption. From the order in which these operations occur, one can deduce the exponent, which is often a vital secret.

Almost all processes display a similarly rich variability in current consumption patterns. In a plot of current consumption versus time during the execution of a single high-level command, the eye easily discerns several separate computational phases, distinguished by mean current consumption and by "fuzziness" (short-term variability in current consumption). Because of such variations, the 16 rounds of a DES encryption are generally easy to recognize as a train of 16 identical boxcars. (On closer inspection, one finds that boxcars 1, 2, 9, and 16 are slightly shorter than the rest, due to key schedule idiosyncrasies.)

Since the current drawn by the smartcard is, at constant voltage, proportional to the power consumed by the card, attacks based on current measurements are usually referred to as *power attacks*. We will henceforth refer to power instead of current.

Differential Power Analysis (DPA), developed by Kocher et al. [5], is a powerful extension of these techniques. In a DPA attack on a DES key, the smartcard is repeatedly induced to encrypt various plaintexts with the key to be found, while digitized "traces" of power consumption are recorded along with the plaintexts of the encryptions.[1] When a large number (often on the order of 1000) of traces and plaintexts have been accumulated, averages of subsets of the traces are computed and compared in order to test guesses of various key bits.

Specifically, one sorts the traces into two classes according to conjectured values of a particular bit B computed during the encryption, the conjectured values being computed from the plaintext and a guess at some subset of key bits. If the guessed key bits are correct, the conjectured value of B will match the true value of B in all traces, and the mean $B = 1$ trace will differ significantly from the mean $B = 0$ trace wherever bit B is handled. On the other hand, if the guessed key bits are wrong, the sorting of traces into subsets is (one expects) uniformly random, and no significant difference will be observed.

Two beautiful virtues of DPA are the following: (1) although the attacker makes the assumption that the DES code computes the value B, it is not necessary to know where that computation occurs; and (2) if chip designers add random noise to mask power consumption, the attacker can compensate for the lower signal-to-noise ratio by increasing the number of traces. On the other hand, practical problems in mounting a DPA attack include: (1) protocols may be designed to keep the attacker from seeing the plaintext or ciphertext; and (2) sometimes it is not possible to get enough traces.

More recently, Biham and Shamir [1] described a power attack in which many traces from each of several different cards are compared to identify when key bits are being handled. Instants with large same-card power variations are assumed to be data-handling, not key-handling, instants. Non-data-handling instants with large between-card power variations are assumed to be key-handling instants. Once the attacker has located the instructions handling parts of the key, their

---

[1] Decryptions may be used instead of encryptions; and independently, either plaintexts or ciphertexts can be used.

power consumptions can be measured to attack the key. Since its "profiling" stage can be separated from its key extraction stage, this attack falls into the class described in the present paper.

3 IPA Attacks

Here we describe an IPA attack. Stage 1, the profiling stage, contains almost all of the effort. Stage 2, the key extraction stage, is then quick and simple. One can think of the profiling stage as a long precomputation, after which one can obtain each subsequent key with only a small incremental effort.

The first steps in the profiling stage are familiar from other sophisticated power attacks, such as DPA: collecting a large number of power traces, and then aligning and averaging the traces; the details of these steps need to be specifically tailored to IPA. Next come the two tasks necessary to finding the key: locating and identifying the key bits. These steps are described in detail below.

As our primary example we will consider an IPA attack on DES, since DES is not only well-known but is perhaps the most widely implemented of all cryptographic algorithms. We have successfully performed IPA attacks on DES smart card implementations and have extracted DES keys from a single power trace in Stage 2 of our IPA attacks.

3.1 Context and Assumptions

We assume the following: we have a smart card containing a known cryptographic algorithm but we do not know the details of the implementation, i.e., we do not possess the source code (knowledge of the source code would enable a far simpler attack). Furthermore, we can cause hundreds of executions of the algorithm with different plaintexts (not necessarily uniformly distributed), and we can record the amount of power (current) used at each step within each such execution.

For the purposes of this exposition, we will assume that the algorithm has been implemented in a straightforward manner, without introducing elements designed to thwart power attacks; in particular, the algorithm execution is a deterministic function of the plaintext and key. If the card did employ defenses, modified versions of IPA might still work, depending on which defenses were used; space does not permit discussion of these modified versions. In §5, we discuss some defensive measures that can be used by system implementers.

These assumptions appear to represent the standard context facing someone wishing to attack a smart card. In practice we have successfully performed IPA attacks in this context using only moderate resources, e.g., only a few hundred power traces and a low sample rate (3.57 million samples per second, or one per clock cycle) on an oscilloscope.

3.2 Stage 1: Profiling

The goal of the profiling stage is to locate and identify the key bits as they are used during the computation. To do this, we often need to learn about other aspects of the implementation we are facing: the order of operations, how key bits are handled, and other details that were decided by the programmer.

The attacker starts by facing an unknown implementation and then gradually learns more and more during the course of the profiling stage, until he finds when, and often how and where, the program engages the key bits. At the end of the profiling stage, the attacker knows many of the decisions that were made by the implementer.

The next three sections describe the important steps in the profiling stage.

3.3 Profiling: First Steps

As the attacker, we first cause the smart card (or other device) to execute its cryptographic algorithm a large number of times to obtain a large number of traces. In practice, we have found that a few hundred traces usually suffices; as lower and upper bounds, we estimate that in general the number of traces needed will be between 100 and 1000.

For simplicity of exposition, we will suppose that the executions are all with the same key and with varying plaintexts. This is not essential, as the attack will work even if the keys vary. Of course, if both the key and the plaintext are constant, then we are merely resampling the same data point and the multiple traces are practically useless. The plaintext needs not be either random or uniformly distributed, but merely non-constant.

The traces are then averaged together, in order to remove the effects of the varying data bits while keeping the effects of the constant key bits. Before averaging, we must first align the traces so that the power consumptions of every operation are matched across all the traces. In practice, we have found that each implementation of each algorithm requires a slightly different alignment technique, and the alignment effort ranges from quite simple to quite cumbersome.

We now have a single "average" trace containing the average power consumed throughout the execution. This represents the average, over all plaintexts, of the power consumed using the constant, unknown key.[2]

3.4 Profiling: Key Location

The next major task is to locate the key bits within the average trace. We first describe the basic procedure, followed by a more mathematical description, and then some mention some implementation notes.

Basic Description
Almost all cryptographic algorithms contain repetitive structures, used with

---

[2] Even if the plaintext is not uniformly distributed, the "data" bits soon become uniform, due to the randomizing effect of the rounds; for example, even if the plaintext into DES has its top 50 bits set to one fixed value, the bits in the L and R registers become uniform after the first couple of rounds.

changing pieces of the overall key. For example, in most symmetric ciphers, repeating rounds use differing subkeys. Public key algorithms such as RSA use different key bits while repeatedly performing modular multiplications. We call these repeating structures "rounds", and the corresponding key bits "subkeys", with the understanding that in an algorithm such as RSA, the modular multiplications play the role of rounds.

Suppose there are n rounds, and let K_i denote the subkey used in round i.

Due to code space limitations on smart cards and other devices, the repeating structures are generated by the same source code being executed over and over; therefore, each round's subkey is handled identically.

The key location proceeds as follows: we chop the average trace into rounds to obtain traces $R_1, R_2, \ldots, R_n$, representing "average round 1", "average round 2", and so on. Then we average these together to obtain a single "super-average round" $\mathcal{R}$, i.e., the average of the average rounds.

Next, we take the difference, for each round i, between R_i and $\mathcal{R}$ to obtain the "round i difference trace" Δ_i. Finally, we square and then average together the Δ_i's. These last few steps are equivalent to computing the variances of the instruction offsets in the R_i's. The final trace, the mean square of the Δ_i's, contains peaks that reveal the key bit locations.

Why does this work? At an intuitive level, note that the first averaging (of different traces) removed the effects of the data bits but left the effects of the key bits: the only differences between the R_i's are due to the differences between the subkeys (see Figure 1 for an example). The second averaging (of different rounds) removed the effects of the key bits as well, leaving only "code" features. When we then take the difference between average round R_i and the super-average round $\mathcal{R}$, the code features cancel out, leaving only the effects of the specific subkey K_i. The subsequent squaring and averaging produces clean peaks at all subkey bit locations.

Since we know the algorithm, we know the number of key bits comprising each subkey, and therefore we know how many peaks to look for. In DES, we look for 48 peaks in the average of the squared Δ_i's. An example is shown in Figure 2.

It is a good idea at this point to observe the distribution of power levels at the detected peaks, in order to verify that the peaks represent key bits and to determine the power threshold that separates a 0 from a 1 bit value. Since the actual key bits are presumably 0 or 1 with probability $\frac{1}{2}$, an instruction that handles a single key bit should exhibit a bimodal distribution, with half the probability in each mode. An instruction that handles more than one key bit should exhibit the appropriate binomial distribution. Thus the shape of the power distribution can reveal the way the key bits are being handled.

In practice, the straightforward cryptographic implementations we have seen most commonly on smart cards handle key bits individually, and we have therefore seen bimodal distributions at the peaks. If key bits are not handled individually, one can use the resulting binomial distributions to learn the Hamming

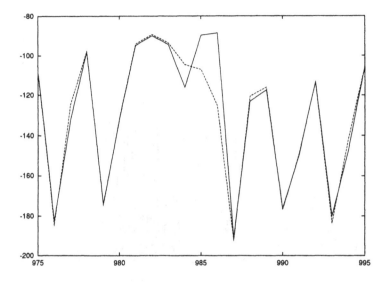

Fig. 1. A section of two round traces R_i and R_j. They differ only where a key bit is being handled, at positions 984 through 986.

weights of key bit groupings; one way to proceed in this case is discussed by Biham and Shamir [1].

Mathematical Description

Let $T_{i,j}(t)$ denote the power consumed at time t within the i-th round in power trace j. In general, the power consumed at any time t is a function f_t of some key bits $k = \langle k_1, \ldots, k_r \rangle$, and some data bits $d = \langle d_1, \ldots, d_s \rangle$. If we write $d_{i,j}$ and k_i for the actual bit values of d and k in round i in the j-th trace (k depends only on the round, not the trace), we have[3]

$$T_{i,j}(t) = f_t\left(d_{i,j}, k_i\right). \tag{1}$$

Assuming that the number of traces m is large enough and the numbers of dependent key bits r and data bits s are small enough,[4] we will find that the power $R_i(t)$ in the average round trace is approximately the same as if we had

---

[3] In all equations we ignore the random power fluctuations due to internal and external noise. In practice, these effects disappear once we have done the first averaging.

[4] We would like to see $m > \max\{2^r, 2^s\}$. This is reasonable since typical values might be $m \approx 300$, and $r, s \leq 4$.

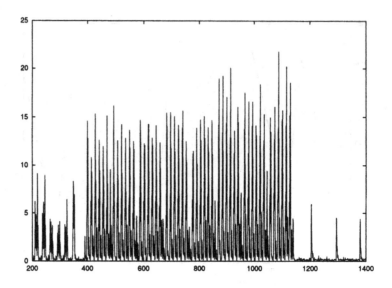

Fig. 2. Peaks in Δ_i^2 in a DES implementation. In this implementation, there are 48 peaks, one for each subkey bit, and they are clustered in 8 groups of 6, for the 8 S-boxes.

averaged over all 2^s possible values of d:

$$R_i(t) = \frac{1}{m} \sum_{j=1}^{m} T_{i,j}(t)$$

$$= \frac{1}{m} \sum_{j=1}^{m} f_t(d_{i,j}, k_i)$$

$$\approx \frac{1}{2^s} \sum_{d} f_t(d, k_i). \tag{2}$$

Furthermore, taking the average of the R_i's to get the super-average trace $\mathcal{R}$ will have the effect of averaging over all 2^r values of k, so that $\mathcal{R}(t)$ is:

$$\mathcal{R}(t) = \frac{1}{n} \sum_{i=1}^{n} R_i(t)$$

$$\approx \frac{1}{n} \sum_{i=1}^{n} \frac{1}{2^s} \sum_{d} f_t(d, k_i) \quad \text{from Equation 2}$$

$$\approx \frac{1}{2^{r+s}} \sum_{k,d} f_t(d, k). \tag{3}$$

Consider an instruction, at, say, time t_1, that does not involve any key bits; then the power consumption function f_1 depends only on the data bits d, and the value of $R_i(t_1)$ from Equation 2 becomes

$$R_i(t_1) \approx \frac{1}{2^s} \sum_d f_1\left(\boldsymbol{d}\right). \tag{4}$$

The important point about Equation 4 is that $R_i(t_1)$ does not depend on i at all, and is therefore constant among all the rounds and in the super-average round $\mathcal{R}$, i.e., $\mathcal{R}(t_1) = R_i(t_1)$. So when we take the difference between the average round i trace, R_i, and the super-average round trace, $\mathcal{R}$, we find, at position t_1,

$$\Delta_i(t_1) = \mathcal{R}(t_1) - R_i(t_1) = 0. \tag{5}$$

Now consider another instruction, at time t_2, which handles some key bits $\boldsymbol{k}$ as well as data bits $\boldsymbol{d}$; its power consumption function f_2 then looks like that in Equation 1, and the values of $R_i(t_2)$ and $\mathcal{R}(t_2)$ look like Equations 2 and 3, respectively. In the round i difference trace Δ_i we then find, at position t_2,

$$\begin{aligned} \Delta_i(t_2) &= \mathcal{R}(t_2) - R_i(t_2) \\ &= \frac{1}{2^{r+s}} \sum_{\boldsymbol{k},\boldsymbol{d}} f_2\left(\boldsymbol{d},\boldsymbol{k}\right) - \frac{1}{2^s} \sum_{\boldsymbol{d}} f_2\left(\boldsymbol{d},\boldsymbol{k}_i\right) \end{aligned} \tag{6}$$

which in general is *not* 0, but some function that depends on the specific values of the subkey bits $\boldsymbol{k}_i \subseteq K_i$.

The end result is that the difference traces Δ_i will be close to zero whenever key bits are *not* being handled. Therefore, when we square and average the Δ_i's, we find peaks exactly at those times (like t_2 in our example) when key bits *are* being handled.

Implementation Notes

An actual implementation of an IPA attack may encounter difficulties; here we mention a couple of the most common.

The super-average $\mathcal{R}$ will work as planned only if the average rounds R_i are aligned at use of the key bits. Therefore, in practice, one may need to do some alignment of the R_i's before averaging. For example, in DES, some rounds contain more shifts of the key registers than others; therefore the offsets of instructions at which the key bits are used may differ from round to round, and this difference must be accounted for before averaging. Depending on the algorithm, this may mean identifying other features within the round traces; this is one instance in which the "profiling" can involve learning more than just the key usage patterns.

Another important point is that the number of spikes in Δ_i may differ from the size of the subkey, depending on exactly how the key bits are being handled in the (unknown) implementation. However, the number of spikes and the number of subkey bits should have a simple mathematical relation.

At the end of the key location step, we know the times when the power consumption depends on the key bits, i.e., we know where to find the key.

3.5 Profiling: Key Bit Identification

Given the list of key bit locations, we still can't read off the key, because we don't know which location corresponds to which key bit. This process of determining the identity of the key bits is the final step in the profiling stage of an IPA attack.

The actual method of key bit identification varies greatly with the algorithm and the implementation. So rather than give overall rules, we restrict ourselves to some general comments on algorithmic features, followed by specific remarks about some of the more common algorithms.

An algorithm's specification may (or may not) restrict the order in which the key bits are used. In RSA, for example, the secret key is used as a modular exponent, and the exponentiation uses the bits of the key sequentially. However, the programmer's decisions still affect the order in which the key bits are used: one can start from either the most significant bit or the least significant bit of the exponent, and the bits can be used one at a time or two at a time. But these are a relatively small number of choices, and therefore the key identification step for RSA is usually fairly simple.

In DES, on the other hand, key identification can be much more difficult, since there are fewer restrictions on key bit order. A DES subkey consists of 48 bits used as inputs to 8 S-boxes. The S-box operations within a round do not depend on one another, and thus all 8! orderings of S-boxes are possible. Furthermore, there is no restriction on the order in which the 6 key bits used in an S-box are loaded from the key registers. A DES programmer may in fact choose a key bit order not based on any S-box order, but based on, say, the key bit locations inside the key registers.

Because of the large subkey size in DES, we have also run into difficulties when we have found less than 48 key locations. Suppose we find only 32 key locations. Then the key location step must determine, first, which of the possible $\binom{48}{32}$ subsets, and, second, which of the 32! permutations of that subset, we are seeing. Of course, in a straightforward implementation, some key bit orderings are much more common than others. The S-box order is more likely to be 12345678 or 87654321 than 53821467, for example. Still, we caution against assuming any particular ordering, and after the most obvious guesses have failed, it may be unclear where to turn next.

When the key identification becomes non-trivial, one can turn to the key scheduling section of the algorithm's specification, and select patterns that can be sought for in the empirical key location data. In DES, for example, the key schedule specifies the patterns in which individual key bits move from one S-box position to another in consecutive rounds. This allows a key identification hypothesis to be tested by comparing the movement from round to round of 1's and 0's in our observed key locations against the movement of fixed key bits in the DES key schedule.

At the end of a successful key identification step, and thus at the end of the profiling stage of the attack, we have a table of locations (inside the round traces) and the corresponding key bit identity. If we are attacking DES, for example, our table might look something like that in Table 1, where the numbers in the location column are offsets into the aligned, averaged round traces (R_i) and the key indices refer to bits within the round subkeys (K_i).

Location	Subkey Bit
380	k_4
672	k_1
1022	k_9
⋮	⋮

Table 1. The final result of the profiling stage of an IPA attack: the key table.

3.6 Stage 2: Fast Key Extraction

Armed with the key location table, we can easily find the subkey bits and then the master key bits from the traces we have. But the profiling data depend only on the software implementation that we are attacking, and not at all on the key that was used in the traces we processed. Therefore the information in the table will be equally valid for all other instances of the same software running on identical hardware, and so we can easily find the key in any such instance, not just the instance whose power traces we have already recorded.

In short, the profiling stage needs only be done once, and then key extraction can be done quickly and efficiently from new instances with unknown keys. One can think of the profiling stage as a long precomputation of the key location table; after the precomputation we can then quickly solve any similar instance of the same problem. For example, given a second smart card, identical to the first except for a different key, the data in our key location table immediately point us to the new key, without taking hundreds of new traces involving the new key.

To extract the key from a new instance of the same implementation, we take a single power trace, chop it into rounds, and measure the power consumed at the locations specified in the key location table. Using our knowledge of the key bit power distribution, which we obtained during the profiling stage, we can tell whether the key bit is a 0 or a 1.

Due to the particularities of the algorithm and the implementation, a single power trace may not suffice, in which case we would take, say, 5 traces, average them together, and then measure the power levels at the key locations. The issue of whether one trace will suffice depends, among other things, on whether there are instructions that handle key bits separately from data bits. In most of the implementations that we have seen, each key bit is handled by itself in at least one instruction (for example, the bit is loaded into a register) without the interference of data bits; therefore we have been able to extract keys using a single trace.

In any case, the number of traces needed in the key extraction stage is far less than the number needed during profiling, where we needed enough traces to average away any effects of the data bits and to ensure that any peaks we found were due to key bits only. For key extraction, on the other hand, we already know where the key bits are, and only care about whether data bits may affect readings at the key bit locations, and can safely ignore the effects of data bits elsewhere in the trace.

4 Strengths of IPA

There are several aspects of IPA attacks that make them effective in situations where DPA attacks would be difficult to mount.

First we note that in order to mount a DPA attack an attacker needs the plaintext (or ciphertext) associated with every trace; but this is not required for an IPA attack. Thus one of the important defenses against DPA — protocol designs that hide plaintext and ciphertext when master keys are used — is useless against IPA.

Also, a DPA attack is restricted to points in the algorithm where the plaintext (or ciphertext) interacts directly with the key; this is because the differential traces are based on a "selection function" that predicts a bit value based on a small number of plaintext bits and a small number of key bits. In practice this usually means that a DPA attack is restricted to the beginning (if plaintext) or ending (if ciphertext) of a cryptographic algorithm; for example, DPA attacks on DES generally concentrate on either the first or last couple of rounds.

In contrast, an IPA attack is as capable of looking at the middle of an algorithm as at the beginning or end. This can be an important advantage in cases where some intervening processing is applied to the plaintext before the key is directly applied. For example, in a recent DPA attack on the AES candidate cipher Twofish [3], the attackers could only extract a certain "whitening key," after which significantly more analysis (including an exhaustive search of 9^8 possibilities) was needed to derive the master key. An IPA attack, on the other hand, can focus its attention on any (or all) of the intervening rounds, and thus extract the round keys without any further analysis.

Another set of advantages of IPA derives from its ability to do fast key extraction after a single lengthy profiling stage. This means that the cost of the profiling stage can be amortized over many key extractions, thus making an IPA attack economically feasible even if the cost of obtaining hundreds of power traces is large. In a DPA attack, by contrast, the attacker must collect a large number of traces for every key to be broken. If the cost of obtaining those traces is greater than the benefit of the key itself, the DPA attack is rendered impractical, whereas an IPA attack remains viable.

Fast key extraction also overcomes another defense against DPA: a protocol may disable a card after only a small number of operations, if the operator does not know the secret. Such a protocol can block DPA, but does not block IPA, where the many profiling traces can be obtained using a "friendly" card (one where we are the legitimate owner and know the secret), and only a small number of traces are needed for each "unfriendly" card being attacked.

5 Defenses

Here are some suggestions to system implementers trying to protect against this class of attacks.

First, avoid handling key bits one at a time. Some ciphers are more amenable to this approach than others. Still, even a bit-oriented cipher like DES can sometimes be effectively protected: when a DES master key is inserted, its key schedule can be computed once for all time, and stored as six bits in each of 16×8 bytes, ready to be used with no further bit manipulation. For further protection, unused bits in any byte may be filled with irrelevant values instead of being set to zero.

Randomize the execution of the code. Where the order of operations is unimportant, such as S-box evaluation in DES, vary the order instead of using a fixed order. Insert random delays, even if only one instruction-time in duration.

Randomize the representation of data. Sometimes a quantity can be "blinded" by combining it with a randomly chosen constant; for example, value A may be maintained as $A \oplus K_1$, value B as $B \oplus K_2$, and $A \oplus B$ computed as $(A \oplus K_1) \oplus (B \oplus K_2) \oplus (K_1 \oplus K_2)$. When a single bit must be handled, consider representing the bits 0 and 1 as "01" and "10".

Limit the number of times a key can be used without confirmation of legitimacy, while simultaneously reducing the attacker's signal-to-noise ratio with filters or generators of random noise. The addition of noise will prevent key extraction from a captured trace of a legitimate transaction, and the limit on key probes will discourage key extraction from a stolen card.

Although no single defense makes a system impervious to IPA, and new attacks can be expected in the future, adding a variety of these countermeasures will likely increase the difficulty of IPA attacks, reducing, one would hope, both the number of potential attackers and the probability of any given attacker's succeeding.

References

1. Eli Biham and Adi Shamir. "Power Analysis of the Key Scheduling of the AES Candidates," *Second Advanced Encryption Standard Candidate Conference*, Rome, March 1999.
2. Eli Biham and Adi Shamir. "Differential Fault Analysis of Secret Key Cryptosystems," in *Advances in Cryptology — Crypto '97*, Lecture Notes in Computer Science Vol. 1294, p. 513–525. 1997.
3. Suresh Chari, Charanjit Jutla, Josyula R. Rao, and Pankaj Rohatgi. "A Cautionary Note Regarding Evaluation of AES Candidates on Smart-Cards," *Second Advanced Encryption Standard Candidate Conference*, Rome, March 1999.

4. Paul Kocher. "Timing Attacks on Implementations of Diffie-Hellman, RSA, DSS and Other Systems," in *Advances in Cryptology — Crypto '96*, Lecture Notes in Computer Science Vol. 1109, p. 104–113. 1996.
5. Paul Kocher, Joshua Jaffe, and Benjamin Jun. "Introduction to Differential Power Analysis and Related Attacks".
   http://www.cryptography.com/dpa/technical/index.html. 1998.
6. Oliver Kömmerling and Markus G. Kuhn. "Design Principles for Tamper-Resistant Smartcard Processors," in *Proceedings of the USENIX Workshop on Smartcard Technology (Smartcard '99)*, USENIX Association, p. 9–20. 1999.

Security Evaluation Schemas for the Public and Private Market with a Focus on Smart Card Systems

Eberhard von Faber

debis IT Security Services, Rabinstraße 8, D-53111 Bonn, Germany
e-vonfaber@itsec-debis.de

Abstract. Even users must have some understanding of the different evaluation schemas. They must be able to rate the outcomes they rely on and use the opportunities to steer the processes. Some evaluation schemas are designed for general purposes others for specific application contexts. The elements of evaluation schemas are introduced first. Then observations about smart card evaluations are discussed demonstrating that the evaluation or approval process itself effects the evidence of the assurance and the value of evaluation verdicts. Especially trade-off situations typical of smart card evaluations are discussed.

Table of Contents

1 Introduction

Modern companies use information technology more and more to support their traditional business activities, to offer them in a better way or to more customers. The commercial goals of a company can only be reached if the information technology operates perfectly. Nowadays information is a critical resource that enables companies to succeed in their business. Therefore, many products and systems provide security functions exercising proper control of the information. Companies and the individuals using such products expect that the sensitive information remains private and that unauthorised modifications are detected.

It is key to know (i) whether the products and systems actually and properly respond to the security needs in a specific application context and (ii) whether they provide a sufficient level of protection. Customers usually do not wish to rely solely on the promises given by the product vendor. Therefore, IT-products or systems which already exist or which are still in the development stage are to be evaluated by *independent specialists* to proof if and to what extend the security objectives are met. In addition, the developer himself often likes to have some proof and indication how to improve his solution.

Security assessments (evaluations) are sub-contracted to independent (third) parties (Evaluation Facilities or labs). They have to have the knowledge, expertise and resources necessary to judge whether the product or system is "secure". The Evaluation Facility is eventually expected to pass a verdict. For this it is required to formulate the *question* the lab has to answer as precise as possible. If the question or the definition of the Security Target is hazy the evaluation result will not be very helpful. For that reason international evaluation criteria specify requirements for content and presentation of the Security Target.

Too little attention is given to the fact that the evaluation or approval process itself effects the evidence of the assurance and the value of evaluation verdicts. The evaluation schema, its structure and rules, the criteria and their evaluation methodology actually effects the outcome of an evaluation and the information given to the user. There are different evaluation schemas being designed for general purposes or for specific application contexts. After having introduced the elements of an evaluation schema trade-off situations typical of smart card evaluations are discussed. Studying these examples give valuable information to individuals, companies and organizations using evaluated products.

1 Evaluation Schemas

According to almost all security evaluation schemas three parties are involved when an evaluation is being carried out:

⊔ developer and manufacturer of the product or system,
⊔ Evaluation Facility, and
⊔ Overseer (Certification Body or Approval Authority).

The developer and manufacturer applies for a security certificate (to be used in the public market) or an approval (for a closed private market). He has to provide information about all the construction details allowing the Evaluation Facility to assess the security provided. The Overseer (Certification Body or Approval Authority) monitors the evaluation activities, reviews and analyses the evaluation report(s) to assess conformance to the evaluation criteria, the evaluation methodology, and the evaluation schema. Finally, the Overseer issues the certificate or the approval. The decision made by the overseer is based upon the evaluation report prepared by the Evaluation Facility.

In the public market the developer/manufacturer wants to demonstrate to customers that they can have confidence in the security provided by the product or system. The certificate issued by the Overseer (Certification Body in this context) confirms that the evaluation has successfully been performed according to the criteria. The users' decision whether to use the product is additionally based on information given in the Certification Report which contains the non-confidential evaluation results (major findings) and perhaps extra guidelines for operational use.

In the private market the developer/manufacturer wants to get the approval that the product or system can be used in the specific application context given. For instance in banking applications (like the POS debit electronic cash system which uses the eurocheque card in Europe and many smart card based electronic purse systems like GeldKarte in Germany) the card issuers require a successful evaluation before the component can be sold to the banks or payment processors and used in the payment system.

In fact approvals are often used for marketing in other markets since they demonstrate that the developer/manufacturer is able to provide high-quality products. But unlike in public markets the evaluation results are to be accepted by specific institutions. To some extend this changes the security evaluations and the co-operation of the parties being involved in many ways.

Regardless of the market or evaluation schema, the Evaluation Facility uses the same set of information to perform the assessment. This is visualized in Figure 1.

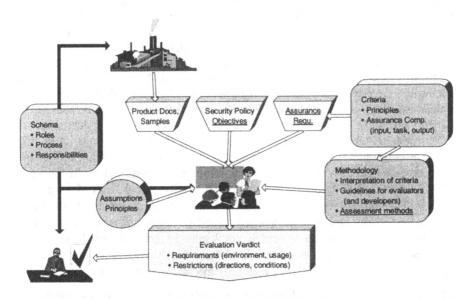

Fig. 1. Evaluation Process as seen by the Evaluation Facility

The basic input for the Evaluation Facility is as follows: (i) All the information about the product's construction is provided by the developer. He also provides samples of the product for penetration and testing. (ii) The Evaluation Facility needs to have a list of the security objectives since the product is checked whether to meet the security objectives defined for the context given. (iii) The assurance requirements define on what grounds evidence can be given that the product meets its security objectives.

The assurance requirements are defined in the evaluation criteria. One can find them in the "Information Technology Security Evaluation Criteria (ITSEC)" [4], in the "Common Criteria" (Part 3: Security Assurance Requirements) [3] and in the "Trusted Computer System Evaluation Criteria (TCSEC)" [7] as well. When working on the different assurance aspects the Evaluation Facility uses guidelines describing the evaluation methodology (for instance [5] and [6]).

The relations between the three parties are described in a document called Evaluation Schema. The Evaluation Schema defines the process together with the roles and responsibilities of the three parties. Additional regulations may be given in form of assumptions or principles if needed in a specific application context.

1.1 Public Market (Business to Customer)

The developer/manufacturer wants to demonstrate to his customers that they can have confidence in the security provided by the smart card integrated circuit or another product he develops. For that reason he prefers to have an "official" certificate issued by an officially recognised Certification Body. The evaluation carried out by specialised laboratories (as third parties) gives evidence of the product's assurance. Assurance in turn gives the confidence needed by the customers and users.

Independent from the application contexts the products are used, evaluation schemas were set-up by governmental organisations and assurance requirements were defined. Such national evaluation schemas shall meet general market needs. Here criteria such as the "Information Technology Security Evaluation Criteria (ITSEC)" [4] and the "Common Criteria" ([1], [2], [3]) are used and international recognition agreements were signed. In practice recognition of certificates (evaluation results) is a problem especially if the Evaluation Facility is not known and the methods of the laboratory are not clearly documented in the Certification Report being published.

The evaluation shall give assurance (or trust) that the product meets its security target. The corresponding requirements checked during the evaluation are defined in a document called Security Target. The result of the evaluation is given in form of a verdict (pass or fail). If needed, directions for the developer or the user of the product are given. The criteria set out in the ITSEC or Common Criteria permit the developer to define the security functions without any restriction and to choose one out of seven evaluation levels representing increasing confidence in the ability of the product to meet its Security Target. The evaluation level determines the assurance on the verdict. In addition, a minimum strength of the security mechanisms shall be claimed.

There are clear advantages of having an evaluation schema set-up and maintained by governments. Perhaps the most important ones are:

⊔ independence from influence of manufacturers and service providers,
⊔ broad recognition of the evaluation results and the Certificates, and
⊔ possibility to incorporate many institutions, organizations, and laboratories especially for definition and maintenance of the evaluation methodology.

Therefore, a high quality is expected. Again the quality of the assessment depends on the knowledge, expertise and market position of the Evaluation Facilities. The national bodies being responsible for the accreditation of laboratories and the Certification Bodies monitoring the evaluations have to have detailed know-how and enjoy a high reputation. Otherwise the schema will not be valuable but a burden for the laboratories (and the vendors) trying to maintain a high standard.

1.2 Private Market (Banking Applications)

Especially in banking applications controlled by specific providers of payment systems, card issuers or banking associations the approach is little different. For example, the use of a smart card in the German GeldKarte is permitted by Zentraler Kreditausschuß (ZKA, the common organisation of the German credit sector associations) only after successful evaluation of its hardware and software. This evaluation must be carried out by a laboratory (as trusted third party) being accredited by ZKA. The evaluation must show that the ZKA-Criteria ([9] or [10]) are fulfilled.

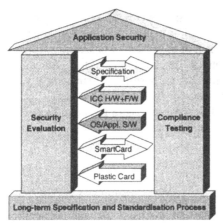

Fig. 2. Foundation of Application Security in Smart Card based Payment Systems

In the late 80's the banks in Germany began to develop their own Evaluation Schema. In the same time governmental organizations developed security evaluation criteria [8]. But the banks decided not to participate in the schema to have the freedom to control the evaluation process independently according to their specific needs.

In banking applications the products (for example smart cards) are designed to provide specific services. The cards are purchased by experts of the banks or their service companies. The specifications are not developed by the card manufacturer but by security experts commissioned by the banks, their associations or working groups. Note that there are approximately 40 million cards in Germany each equipped with the same functionality. So, there are a lot of differences compared to other public markets.

As shown in Figure 2 the security in banking applications like the electronic cash system which uses the eurocheque card or the GeldKarte in Germany is supported by a long-term specification process, security evaluations as well as compliance testing. The specification of the application must be subject to an evaluation. It is used again when performing the exhaustive compliance testing.

There are some advantages of having an independent evaluation schema. Perhaps the most important ones are:

⊔ flexible definition of assessment rulers,
⊔ possibility to require the analysis of specific attack scenarios and
⊓ possibility to require improvements of the products.

The approval is given if all successful attacks considered are so expensive or difficult that the value of gathered information is less than the expenditure. There is some opportunity for interpretation which in turn introduces flexibility since the aspects (i) attacker's skill, (ii) attacker's knowledge, (iii) money and equipment, (iv) time, and (v) availability of samples (components) must somehow be combined to yield an overall verdict.

Since Zentraler Kreditausschuß (ZKA), as the Approval Authority for the banking applications just mentioned, is held responsible for the security it is in the position to demand the improvement of the system. For instance, some years ago a plan has been presented how to attack the Data Encryption Standard (DES) using hardware especially designed for that purpose [12]. Our company worked out all details and presented this information to ZKA [13]. As a result, the German banks decided to move to the Triple-DES. Note that there are more than 250,000 terminals and about 40 million smart cards in the field.

2 Observations

In the following examples are discussed showing difficulties in practical evaluations. First they help to understand the peculiarities of different evaluation schemas. Then the examples shall give indications on how to develop such schemas and the ability of the parties being involved to treat with them.

2.1 Case Study #1: Who defines the Security Target?

The Security Target shall define the user's requirements because the product or system is checked against it. The evaluations results in turn (verdict, answer to the question defined in the Security Target) guides users whether or not to purchase and use the product. Therefore users (knowing the application context), developers (knowing the product) and third parties (being familiar with the assessment schemas and methodologies) should co-operate to define the Security Target.

The criteria set out in the ITSEC or Common Criteria permit the sponsor to define the security functions without any restriction and to choose the evaluation level. But often the Security Target does not exactly meet the user's requirements. Then the evaluation will fail to give the evidence exactly needed by the user.

The user finds himself in a bad position: He has either to demonstrate to the developer that the Security Target does not meet exactly his needs or he has to live without having a certified product. Note that only a small percentage of the products have been evaluated.

The Common Criteria [1] allow to define such requirements for a set of products all intended to respond to the same security needs identified for similar environments. Such a set of requirements is called "Protection Profile". A Protection Profile holds for a group of products (implementations). Before using such a Protection Profile in an evaluation process, it must be evaluated and then filled with all the information identifying a special implementation (product).

As a consequence, a Protection Profile like [14] is a helpful tool for manufacturers (when developing their products etc.) and for users (to articulate their security needs). Evaluations based on Security Targets which in turn are based on the same Protection Profile are expected to yield comparable results.

Unfortunately, Protection Profiles are often written by the manufacturers and focus on specific aspects only [14]. It is therefore up to the users to clearly express interests. This can be done by writing Protection Profiles or by defining similar sets of requirements. An outstanding example for the second way are the regulations of the German "Digital Signature Act" [15] and its "Digital Signature Ordinance" [16]. Here standard assurance requirements are used [4]. But functional requirements are also defined for services and components to be provided for a public key infrastructure planned to partly replace the hand written signatures. Users must form consortiums to define functional requirements not only assurance requirements.

2.2 Case Study #2: Logical versus Physical Security

Software often being evaluated provides security against hostile access on a well-defined interface or on an external channel. The software itself is not subject to an attack. Security function and threat agent are well separated. Software provides logical security. But in many cases one can not guarantee nor even assume that the attacker is not able to attack the security functions themselves. If the module is in a hostile environment and not protected by other means it can be subject to tampering or other types of influencing its behavior. Physical security is required.

In the nineteenth century Kerckhoff stated that secrecy must reside entirely on the key. So, it is assumed that an attacker may have complete knowledge of the cryptographic algorithm and its implementation. For smart cards this assumption does not hold. Especially for hardware the secrecy of design information is important. All details about the design and layout relieve attacks since modern equipment such as a focused ion beam (FIB) can be used to re-wire the chip. If design data are available prior reverse engineering is not required. But Kerckhoff's assumption does not always hold for software too. For instance the concrete software implementation of a cryptographic function may provide measures to avert the Differential Power Analysis (DPA, refer to chapter 2.6). Knowing the method of data coding or data processing the attack can be refined and perhaps successfully be carried out. So, smart card security is also based upon concealing information. If the attacker is able to manipulate or intentionally disturb the card's operation, the construction of the countermeasures in hardware (and sometimes in software) must be kept secret. Then the attacker must carry out a costly reverse-engineering. But hiding information has a disadvantage. The more experts had analyzed a solution the higher is to estimate the assurance of its security. Restricting the availability of information reduces the number of experts having the possibility to approve, disapprove or improve the mechanisms considered.

It seems that the differences between hardware and software measures have not been thoroughly taken into account. The ITSEC and the Common Criteria and their evaluation guidelines do not consider the study time needed to prepare the attack. The ZKA approach distinguishes between first and following attacks.

2.3   Case Study #3: Information versus Protection

In fact all security mechanisms realized in the hardware can be disabled or bypassed by direct manipulation based on previous reverse-engineering. The hardware developer may describe his measures built in to counter direct manipulation or reverse-engineering in the Security Target. This would inform the user about such important security measures but discloses the details to the public. If such kind of information is kept secret it is up to the Evaluation Facility to assess them. But for the same reason it is difficult to inform the user about the effectiveness of counter-measures against direct manipulation and reverse-engineering, for instance. But especially card issuers taking the risk for fraud in payment systems have legitimate interests to be informed about the existence and the effectiveness of the security measures built in.

All kind of information about such details implicitly may yield information about things not thoroughly being addressed and therefore give hints to possible attacks (possible weaknesses). And it will not only inform attackers but competitors. Of course, security shall not be founded on confidentiality of design details, but not publishing details often supports security. This can be understood by evaluators working for chip manufacturers when they read papers on chip card security being published by experts not having the same level of detailed information.

2.4   Case Study #4: Smart Cards as Composite Components

Smart card hardware and software is developed by different companies. These components are assembled in a particular way. Then security relevant data are injected into the card. The process is depicted in Figure 3.

Different entities are involved when producing a smart card: software develop-ment, chip and mask manufacturing, module manufacturing, card manufacturing (plastic #1 and #2), and personalization in two steps. Even for the smart card chip itself there are at least two companies providing components.

Hence for practical reasons and to prevent disclosure of details of the design to the other party, hardware and software are evaluated separately.[1] Apart from such confidentiality aspects each company must take the responsibility for his own developments or services.

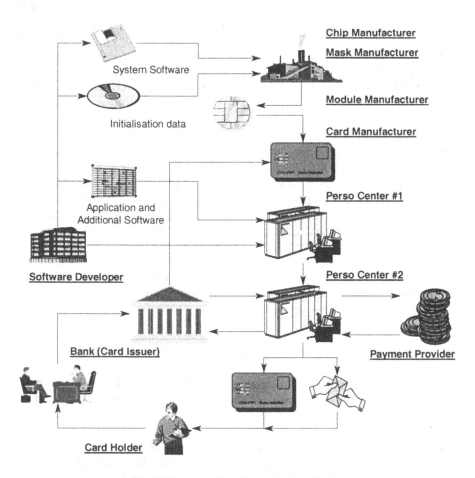

Fig. 3. Process of Producing Smart Cards

The security objectives for a smart card are twofold:

[] Ensure "security" for the card when being in the field.
[] Maintain "security" throughout the development and production process.

Although many specialists concentrate on the security in the field since the smart card is delivered into a hostile environment without any security regulations and may be subject to tampering, the security in the development, production and personalization process is also important. More precise, the organization of the

---

[1] Of course, for instance for DPA both companies have to co-operate to provide samples for analysis soon.

personalization process turns out to affect the security functionality to be provided by the smart card. So, the security objectives, a smart card component is being assessed against, depend very much on the application context which in this case includes the production and personalization process.

Therefore, one has to start with a concept of the smart card supply chain (refer to Figure 3). It addresses security and should be a subject to a separate assessment. From this a list of requirements for the individual components (and processes using other equipment) is gained including information about the number, extend, rigor, and depth of the evaluations to be carried out.

In principle, the evaluation criteria used in public markets (ITSEC and Common Criteria) cover all the aspects of a product's life-cycle (especially: development, system generation, delivery, configuration, and effectiveness in the field). Nevertheless, a central authority is required being responsible to ensure that all the measures fit together. Note that there are several components and processes. The published result of an ITSEC or Common Criteria evaluation (certification report) usually do not contain enough information to decide whether continual security is guaranteed.

In case of complicate composite products evaluation schemas like that of Zentraler Kreditausschuß (ZKA), as the Approval Authority for the banking applications mentioned above, are very efficient. Because of nowadays rapid changes in technology, evaluation overhead should be avoided. Simultaneously, requirements must be defined soon. ZKA is in the position both to approve components and processes (based on evaluation results provided by the laboratories) and to represent and improve the "user's requirements" since the payment systems are operated for the banking community the ZKA belongs to.

2.5 Case Study #5: Security Target Definition for Hardware

The hardware's countermeasures are often characteristics of the device which can not easily be described. Sometimes there are countermeasures not being designed as such. Nevertheless, the evaluator has to check whether the device has vulnerabilities. For example, an attacker may try to cause faulty operations of a smart card processors to compromise the modules security [17]. It is well-known that RSA-keys (Bellcore attack) or DES-keys (Differential Fault Analysis, DFA) can be read out if the attacker succeeds in causing specific faults by exposing the device to radiation or changing the environmental conditions in another way.

The principle of the Differential Fault Analysis (DFA) is shown in Figure 4 (last round of the DES, one S-Box i and the associated lines are considered). The attacker looks for the key component $K(i)$ but he does not know $C(i)$. Due to a single bit fault in R_{15} one has one or two values i with $I(i)' \neq I(i)$. (The faulty values are marked with a prime.) Comparing error-free and faulty values one has

$$S_i[I(i) \oplus K(i)] \oplus S_i[I(i)' \oplus K(i)] = O(i) \oplus O(i)' \tag{1}$$

The unknown constant value $C(i)$ disappeared.

There are lots of environmental conditions which may cause erratic operation. For instance our laboratory read out keys by superimposing glitches on the power supply. Of course, there are rather simple measures to be implemented in the DES calculation to prevent this kind of attack. But what are the issues to be checked by the evaluator when he considers the bare hardware only. According to the security evaluation criteria mentioned above (ITSEC and Common Criteria) the developer must explicitly define security functions or mechanisms designed to avert threats.

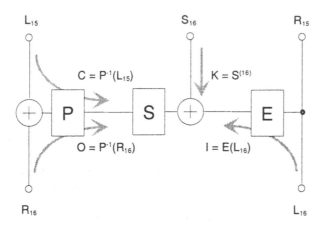

Fig. 4. Differential Fault Analysis: Last Round of the DES

Hardware and software may be evaluated separately to avoid disclosure of design details. In addition, each company must take the responsibility for his own developments. But it is often difficult to define a threat for the hardware since the cryptographic algorithm is realized in the software (outside the Target of Evaluation). General statements like robustness against failures are hard to check since there are many ways to affect the chip and the effect of a malfunction caused by an attacker is difficult to rate without knowing the application context.

But listing the security measures to be implemented in the hardware does not solely solve the problem. The smart card must withstand attacks. The analysis of such attack scenarios require to assess the suitability, binding and strength of a set of many security measures and characteristics of the hardware (layout for example). Even the latter are often not claimed to be a security measure. A smart card hardware offering many state of the art security measures may have fundamental vulnerabilities. If defining such detailed security requirements then the user (not the hardware developer) unexpectedly will design smart card security.[2]

User groups like banking consortiums shall carefully use lists of security measures. They are helpful as a first guidance. But its again the Evaluation Facility performing the detailed investigations. The evaluation schema operated for the users shall ensure that skill and knowledge of the laboratories is developed. Lists of attack scenarios and methods are mandatory.

In the case of logical security it can rather easily be decided whether a solution is "secure". For example the effort for an exhaustive key search can be calculated and

---

[2] In addition, if a solution has been disapproved the developer may lead this back to the requirements he is responsible for.

then the probability for a successful attack can turn out to be almost zero. The identification of individual mechanisms often supports this analysis. For hardware the rating is more difficult. The effort for a successful attack is often significantly smaller. When evaluating a complex of many mechanisms, characteristics and properties the verdict can be quite clear. But if one considers individual mechanisms they may be assessed to be rather weak. So, the result of the strength of functions/mechanisms analysis is to some extend pre-determined by the complexity of the functions or mechanisms identified on the Security Target level.

2.6 Case Study #6: Differential Power Analysis (DPA)

In July 1998 for instance, the new attack scenario Differential Power Analysis (DPA) [11] has been published.[3] Although it was known that an external observable like the power consumption or radiation contain information about the secrets being processed by a smart card or any other device, such kind of attacks have not thoroughly been considered before. In July 1998 our lab read out keys from smart cards using DPA the first time.

The principle (using the last round of the DES) is shown in Figure 5.

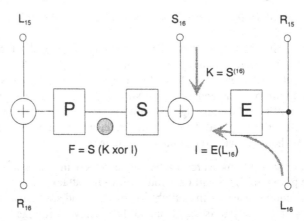

Fig. 5. Differential Power Analysis (DPA): Last Round of the DES

The pseudo code below shows the analysis process. The analysis is carried out for one or two bits (denoted by b) of each S-Box (denoted by i). K(i) and I(i) are the values of the lines associated with the input lines of that S-Box i.

```
locate "time interval" first
procedure start
choose: bit line b, key hypothesis K(i)
for (very much input values) do
    calculate F(b) = S { I(i) ⊕ K(i) }
    if F(b) = 0 then V(m) = -1 else V(m) = +1
    m = m+1
```

---

[3] The attack has been announced in spring 1998.

```
measure power consumption over time S(m,t)
for (each time in the interval) do
   calculate linear correlation
   between V(m) and S(m,t) giving CR(t)
   check if CR(t) shows significant "peaks"
end start.
for (each key hypothesis K(i) & other bit lines b) do
   execute "procedure start"
```

Results measured by our lab are shown in Figure 6. The upper curve is the co-variance CR(t), the power consumption over time S(m,t) is the curve below.

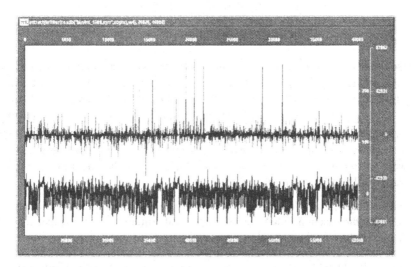

Fig. 6. Linear correlation (co-variance) CR(t) and power consumption over time S(m,t)

All manufacturers of smart card hardware and software were requested immediately by Zentraler Kreditausschuß (ZKA) to add countermeasures against the DPA (both in hardware and software) soon. ZKA has the knowledge and as the responsible organization is in the position to demand such improvement of the system. This again shows the immediate reaction of this evaluation schema.

2.7 Case Study #7: Hardware versus Software

The design hierarchy of a smart card is shown in Figure 7. Things like the process technology are changed not very often but the customer's software may change rather rapidly. The higher the level the more effort and time must be invested to make a change. So, one likes to assess the hardware and the application software separately.

Modern smart card hardware is equipped with special security mechanisms like detectors etc. For the mechanisms to be effective often the software has to properly take advantage of them. The mechanisms must be enabled or initialized. Register bits (flags) signaling a possible attack must be evaluated, interrupts must be served, and the software must properly respond to such events to bring the card into a secure state.

| Application |
| Additional Software (Patches etc.) |
| System Software (ROM Mask Part 2) |
| Firmware (ROM Mask Part 1) |
| Hardware Design (Basic Mask Set) |
| Process Technology, Libraries, and Paramaters |

Fig. 7. Design Hierarchy of a Smart Card

In some cases, just the evaluator of the *hardware* formulated requirements how to use the hardware's security characteristics best possible. Restrictions and conditions were discovered during the evaluation of the hardware. Or security relevant software had to be changed since the characteristics of the memories (especially E^2PROM) showed that a vulnerability might have been introduced. The other way around it was just the evaluator of the *software* who discovered that special characteristics of the hardware are required to maintain security. But of course he could not check if the hardware fulfills these requirements.

Such findings must be listed as guidelines for the evaluator of the other component. Zentraler Kreditausschuß (ZKA), as the Approval Authority, ensured that these lists are checked before final approval.

These issues could not have been discovered by locking at either hardware or software only. In addition, it is often not feasible to have such information on a security target (or requirements) level. In many cases, the guidelines required (restrictions or conditions) were outcomes of an evaluation. Communication between different Evaluation Facilities (if needed), mediated by an experienced Approval Authority or Certification Body, is needed when assessing different components which together built a secure system.

It is hard to force a company to disclose the details needed to the another company. Usually, they will not co-operate in order to have one evaluation for a composite product. In addition, this information comes often from an assessment the developer normally did not performed by himself. The information being published in evaluation reports are not sufficient. If such information would have been included secret information is disclosed. Therefore, the evaluation schema must support the technical communication between the evaluators and force them to look beyond his target of evaluation.

2.8 Case Study #8: Whom do you trust?

Security evaluation criteria like the ITSEC and the Common Criteria are designed to assess technical measures. But obviously, security can not be guaranteed by technical means alone. They have to be supplemented by organizational, personnel and other measures.

Obviously, if the developer and the Evaluation Facility collaborate backdoors and exploitable vulnerabilities may exist. More general it is up to the evaluators to guarantee assurance. Therefore, laboratories with long-term experience and good reputation shall be used.

Many organizations define requirements for the smart card design and production process. They consider traceability aspects as well as *technical* measures to proof the authenticity of the components being approved. Some of those concepts are reasonable but other ideas go too deep. It is always important to consider carefully before requiring technical measures. In smart card production processes this may introduce too much overhead and overtax the possibilities of the vendors. One should focus instead on the hostile actions for the card in the field and on the most risky actions like (i) the injection of keys and other critical data, (ii) the mechanisms needed to protect components not being ready to be issued and (iii) the security provided by the hardware and software of the card.

So, there is again a trade-off: Technical measures help to reduce the trust needed when services are delegated to other parties. But complex technical solutions can be too difficult and expensive to realize and may overtax the possibilities of the vendors.

3 Summary

Presently, too little attention is taken on the fact that the evaluation or approval process itself effects the evidence of the assurance and the value of evaluation verdicts. There are different evaluation schemas being designed for different purposes. After having introduced the elements of an evaluation schema trade-off situations typical of smart card evaluations were discussed. Studying these examples give value information to individuals, companies and organizations using evaluated products:

The criteria set out in the ITSEC or Common Criteria permit the sponsor to define the security functions. If the evaluations are directed by the developer, then the evaluation may often fail to give the evidence exactly needed by the user. Users must form consortiums to define functional requirements not only assurance requirements.

Software provides security against hostile accesses on a well-defined interface or external channel. Security function and threat agent are often well separated. Software provides logical security. Hardware modules with embedded software used in a hostile environment can be subject to tampering or other types of influencing its behavior. Physical security is required. Kerckhoff assumed that an attacker may have complete knowledge of the cryptographic algorithm and its implementation. For smart cards this assumption does not hold. Especially for hardware the secrecy of design information is important. But Kerckhoff's assumption does not always hold for software too. But restricting the availability of information reduces the number of experts having the possibility to approve, disapprove or improve the mechanisms considered. Differences between hardware and software measures have not been thoroughly taken into account. The ITSEC and the Common Criteria and their evaluation guidelines do not consider the study time needed to prepare the attack. The ZKA approach distinguishes between first and following attacks.

In fact all security mechanisms realized in the hardware can be disabled or bypassed by direct manipulation based on previous reverse-engineering. According to the ITSEC and the Common Criteria the countermeasures must be described in the Security Target. But this would discloses the details to the public. But especially card issuers taking the risk for frauds in payment systems have legitimate interests to be informed about the existence and the effectiveness of the security measures built in.

Different entities are involved when producing a smart card. For practical reasons and to prevent disclosure of details of the design to the other party, hardware and software are evaluated separately. Apart from such confidentiality aspects each company must take the responsibility for his own developments or services. And the organization of the personalization process turns out to affect the security functionality to be provided by the smart card. In principle, the evaluation criteria used in public markets (ITSEC and Common Criteria) cover all the aspects of a product's life-cycle. But sometimes the security objectives for a smart card component depend very much on the production and personalization process. A central authority is required being responsible to ensure that all the measures fit together. The published result of an ITSEC or Common Criteria evaluation (certification report) usually do not contain enough information to decide whether continual security is guaranteed.

Unfortunately, it is often difficult to define a threat or security objective for the hardware as required by the criteria. General statements like robustness against failures are hard to check since there are many ways to affect the chip and the effect of a malfunction caused by an attacker is difficult to rate without knowing the application context. For software it can rather easily be decided whether a solution is "secure". For hardware the rating is more difficult. The effort for a successful attack is often significantly smaller. In addition, the result of the strength of functions/-mechanisms analysis is to some extend pre-determined by the complexity of the functions or mechanisms identified on the Security Target level.

Zentraler Kreditausschuß (ZKA) has the knowledge and is in the position to demand improvement of the system. Examples given show the immediate reaction of this evaluation schema. A plan has been presented and worked out how to attack the Data Encryption Standard (DES). As a result, the German banks decided to move to the Triple-DES. Some years later all manufacturers of smart card hardware and software were requested by ZKA to add countermeasures against the DPA (both in hardware and software) soon. The criteria were extended.

In some cases, just the evaluator of the *hardware* formulated requirements how to use the hardware's security characteristics best possible. Restrictions and conditions were discovered during the evaluation of the hardware. The other way around it was just the evaluator of the *software* who discovered that special characteristics of the hardware are required to maintain security. Such findings must be listed as guidelines for the evaluator of the other component. Therefore, the evaluation schema must support the technical communication between the evaluators and force them to look beyond his target of evaluation.

Obviously, security can not be guaranteed by technical means alone. They have to be supplemented by organizational, personnel and other measures. Many organizations focus on technical measures for the smart card design and production process. In fact, technical measures help to reduce the trust needed when services are delegated to other parties. But complex technical solutions can be too difficult and expensive to realize and may overtax the possibilities of the vendors.

References

1. Common Criteria for Information Technology Security Evaluation; Part 1: Introduction and General Model; Version 2.0, May 22nd, 1998
2. Common Criteria for Information Technology Security Evaluation; Part 2: Security Functional Requirements, Part 2: Annexes; Version 2.0, May 22nd, 1998
3. Common Criteria for Information Technology Security Evaluation; Part 3: Security Assurance Requirements; Version 2.0, May 22nd, 1998
4. Information Technology Security Evaluation Criteria (ITSEC); Provisional Harmonised Criteria, Version 1.2, June 1991
5. Information Technology Security Evaluation Manual (ITSEM); Provisional Harmonised Methodology, Version 1.0, September 1993
6. ITSEC Joint Interpretation Library (JIL), Information Technology Security Evaluation Criteria; Version 2.0, November 1998
7. Department of Defense Trusted Computer System Evaluation Criteria (TCSEC), DoD 5200.28-STD, December 1985 ("Orange Book")
8. German Information Technology Security Criteria (ITSK), "Green Book"
9. Criteria for the Security of electronic cash Systems, Zentraler Kreditausschuß (ZKA)
10. Criteria for the Security of Smart Card based Payment Systems, Zentraler Kreditausschuß (ZKA)
11. Paul Kocher, Joshua Jaffe, and Benjamin Jun: Introduction to Differential Power Analysis and Related Attacks, Cryptography Research, July 31st, 1998
12. M. Wiener: Efficient DES Key Search, Manuscript, Bell-Northern Research, Ottawa, 1993 August 20
13. debis IT Security Services: Brute-Force-Attack on the Data Encryption Standard (DES), March 1996
14. Protection Profile Smart Card Integrated Circuit, Version 2.0, Issue September 1998, Registered at the French Certification Body under the number PP/9806
15. Act on Digital Signature (Digital Signature Act - Signaturgesetz - SigG), in: Article 3 Federal Act Establishing the General Conditions for Information and Communication Services - Information and Communication Services Act - (Informations- und Kommunikationsdienste-Gesetz - IuKDG); Federal Ministry of Education, Science, Research and Technology, 22 July 1997
16. Digital Signature Ordinance (Signaturverordnung - SigV), On the basis of § 16 of the Digital Signature Act of 22 July 1997 (Federal Law Gazette I S. 1870, 1872)
17. Guidelines for Implementing and Using the NBS Data Encryption Standard; FIPS PUB 74-1; April 1st, 1981

A Design of Reliable True Random Number Generator for Cryptographic Applications

Vittorio Bagini, Marco Bucci

Fondazione Ugo Bordoni, Via B. Castiglione 59, 00142 Roma-Italy
bagini, bucci@fub.it

Abstract. The scheme of a device that should have a simple and reliable implementation and that, under simply verifiable conditions, should generate a true random binary sequence is defined. Some tricks are used to suppress bias and correlation so that the desired statistical properties are obtained without using any pseudorandom transformation. The proposed scheme is well represented by an analytic model that describes the system behaviour both under normal conditions and when different failures occur. Within the model, it is shown that the system is robust to changes in the circuit parameters. Furthermore, a test procedure can be defined to verify the correct operation of the generator without performing any statistical analysis of its output.

Keywords: True random number generators, noise, cryptography, tests for randomness.

1 Introduction

Cryptographic systems should use only true random number generators for producing keys and other secret quantities. This paper aims at defining the scheme of a true random number generator that has a simple and reliable implementation and is not expensive in production. To ensure all these features, the generator must be able to stand large tolerances in its components without any calibration or compensation. Furthermore, possible malfunctions must be foreseen and tests to be made during prototype development, production and (possibly) operation must be defined. Since the generator is designed for cryptographic applications, the random source it uses must be suitable to be constructed in a protected and insulated environment. In this way the device can be certified to work under general and heavy operating conditions.

A popular way of generating truly random binary sequences is to sample analogical white noise after it has been quantized by means of a comparator. Because of offsets and bandwidth limitations, the generated sequence is typically affected by bias and symbol correlation, but some tricks are used to suppress both. The bias is eliminated by sending the quantized signal into a binary counter before sampling it, whereas the bit correlation is kept under a fixed value by choosing a suitably low sampling frequency [1-5]. Therefore, in this kind of generators, defects in the bit statistics are not masked (e.g. by means of a pseudorandom transformation) but simply suppressed. This can be considered the most correct solution since the device should generate a sequence whose entropy *is* the maximum possible, not a sequence whose entropy *looks like* the maximum possible. In a certification testing one is thus forced to conclude by an analysis of the scheme that, if the output sequence looks random, i.e., if it passes the statistical tests, it is actually random.

The generator proposed in this paper (see Fig. 1) follows this scheme, but its peculiarity is that the input noise is sampled and held. This solution ensures that the input noise does not change its value during the comparator response time so that the devices in the successive stages can operate under the conditions they are designed for [3]. The proposed scheme is then well represented by an analytic model that describes the device behaviour both under normal conditions and in presence of different failures. In this way the system insensibility to changes in the circuit parameters can be evaluated. Within the same model, a test procedure can be defined to verify the correct operation of the circuit without performing any statistical analysis of its output. It is shown that, if the random source is shielded (so that no external signal is injected) and does not sustain self-oscillations, the circuit operation can be tested by simply counting the transitions of an internal signal.

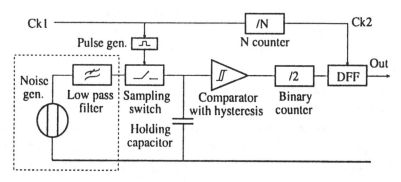

Fig. 1. Block design of the generator

The rest of the paper is organized as follows. In Section 2 each of the blocks that constitute the circuit is described and its role is explained. Furthermore, the generator self-testing procedure is proposed. In Section 3 an analytical model of the circuit is sketched and the autocorrelation function of the binary counter output, i.e., of the signal to be sampled for obtaining a binary random sequence, is given. Results of numerical simulations, which are in good agreement with the model, are also reported. A criterion for choosing the output sampling frequency, based on the form of the autocorrelation function, is then proposed. Some instructions for the practical design of the generator are given in Section 4 and conclusions of the work are presented in Section 5. The details of the calculation of the autocorrelation function are described in Appendix A and some numerical results supporting the self-testing procedure are reported in Appendix B.

2 Scheme of the Circuit

Our scheme uses a gaussian white noise source, e.g. shot noise in a directly polarized semiconductor junction. Shot noise is completely controlled by the polarization current, but its amplitude is typically very low and must therefore be strongly amplified. Since a high gain is required, some caution must be taken in the amplifier design so that external disturbances are shielded and coloured noises are not added [6]. In Fig. 1 the amplified real noise generator is represented by an ideal noise generator connected in series with a low-pass filter, whose cutoff frequency v_0 represents the bandwidth limitations of the real generator.

The sampling and holding operation ensures that the comparator works correctly and permits to sample the binary counter output in a synchronous way. All the statistical defects that could appear in the output binary sequence if it were generated by sampling an unstable signal are therefore avoided. It will be explained in the following how the holding time, i.e., the period of the clock Ck1, must be chosen for this purpose. Details of the sample-and-hold circuit will not be examined because it is well known that such devices, operating up to some GHz, can be implemented in a simple and economical way.

To obtain simple analytic results, in the following the sampled noise that enters the comparator is supposed to be white, i.e., uncorrelated. This hypothesis is reasonable, since the sampled noise correlation is fixed by the filter bandwidth and by the input sampling frequency, i.e., the frequency of Ck1. For instance, if $x(t)$ is the signal obtained by means of a first order Butterworth filtering [7] of white noise, its autocorrelation function is, see e.g. [8],

$$R_x(\tau) = \frac{\langle x(t)x(t+\tau)\rangle}{\langle x(t)^2\rangle} = \exp(-2\pi v_0|\tau|), \tag{1}$$

where brackets denote statistical average. If the input sampling frequency is v_1, the correlation between two consecutive samples of $x(t)$ is

$$\exp\left(-2\pi\, v_0/v_1\right) \tag{2}$$

and is controlled by the ratio of the two frequencies.

The comparator converts the analogical noise into a binary signal. Comparators with hysteresis are generally used to obtain a fast response time. Notice that the comparator is supposed to be the slowest circuit component, so that its response time τ_c determines the whole system operating frequency. Using current technologies, this is often the case.

The binary counter ensures that its output takes on both its possible values for the same average time[1], even if its input is biased because of the offsets introduced by the comparator and by the sample-and-hold [9]. An alternative way of eliminating bias is to control the comparator threshold by means of a feedback loop, see e.g. [10]. Anyway, it is well known that this solution may introduce some degree of correlation in the output bits [2]. Furthermore, the feedback circuit is critical and requires accurate calibration, which is not needed in our scheme.

The DFF (delay flip flop) samples the binary counter output at times corresponding to the edges of the clock Ck2[2] and generates the required binary sequence. The N counter produces Ck2 as a submultiple of the clock Ck1 at which the input noise is sampled. N is chosen to keep the output bit correlation lower than a fixed value.

Since Ck2 is synchronous with Ck1 by construction, if the period of Ck1 is larger than the comparator response time τ_c[3] it is ensured that the binary counter output is sampled when it is in a stable state. Any effect due to threshold offset, asymmetry in saturation output voltages and in rising/falling times, threshold dependence upon the state of the device and bandwidth limitation of the components is therefore avoided. These effects are very insidious, since they cause fluctuations of the time required by the binary counter output for crossing the DFF threshold and can reintroduce in this way a new bias to the produced bits [3]. In fact, as long as the comparator response time is small enough, both the binary counter and the DFF work on the usual binary signals they are designed for, so that the behaviour of these devices should be extremely reliable.

On the other hand one can be persuaded that an increase in τ_c, as well as any offset and any decrease in the amplifier gain and bandwidth, can be detected. In fact, while making the output statistics worse, all these effects result in a decrease of the

---

[1] Corresponding to the average time between two transitions of the comparator in the same direction.

[2] Notice that the output sampling may be triggered indifferently by negative or positive edges of Ck2.

[3] Response times of the following stages are supposed to be negligible with respect to τ_c.

number of circuit internal transitions.[4] In Appendix B it is shown by numerical results that such a decrease is noticeable before the output statistics is substantially damaged. Counting the internal transitions can therefore be a simple self-testing procedure for the generator. In Appendix A the expected number of transitions during a given time interval is calculated under ideal conditions. If the counted number shows a significant departure from this expected value, it is reasonable to suspect that some circuit component is faulty enough to spoil the statistics of the produced bits, that consequently have to be discarded.

3 Model of the Circuit and Output Correlation

The amplified noise $x(t)$ is assumed to be a stationary and ergodic stochastic process and the random variable x_n represents the value sampled at the instant t_n and held until t_{n+1}. The comparator output during this interval, if there is no hysteresis and the threshold value is 0, can be defined as

$$y_n = \text{sign}(x_n) = \begin{cases} +1 & \text{if } x_n \geq 0 \\ -1 & \text{if } x_n < 0 \end{cases}. \tag{3}$$

This transformation is known in literature as hard limiting or clipping [11]. Here the value -1 is chosen instead of 0 so that $\langle y_n \rangle = 0$ means that no bias occurs. This happens if there is no offset, i.e., the comparator threshold coincides with the sampled noise mean value, $\langle x_n \rangle = 0$. The following calculations are made under such hypothesis, that will be discussed at the end of this section. If the clipped noise produced by the comparator is unbiased, its autocorrelation function is

$$R_y(k) = \langle y_n y_{n+k} \rangle. \tag{4}$$

The sampled noise x_n is supposed to be δ-correlated, that is $R_x(k) = \delta_{k,0}$, where δ is the Kronecker symbol. As stated in the previous section, this hypothesis is not critical. In Appendix A it is shown that, as long as the comparator shows no hysteresis, $R_y(k)$ is δ-shaped too.

The binary counter output, denoted by z_n, takes on the values ± 1. For the very nature of this device, $\langle z_n \rangle = 0$ and this result holds even if there is any offset in the previous stages, causing $\langle y_n \rangle$ to differ from zero. The binary counter output autocorrelation function is

---

[4] This is not true for periodic disturbances, which are suppressed by a careful circuit shielding.

$$R_z(k) = \langle z_n z_{n+k} \rangle . \tag{5}$$

If no hysteresis is present, calculation of this function (see Appendix A) yields the result

$$R_z(k) = 2^{-|k|/2} \cos \frac{k\pi}{4} . \tag{6}$$

It must be remarked that, after passing through the binary counter, the noise is no longer δ-correlated.

When the comparator shows hysteresis, the relation (3) becomes

$$y_n = \begin{cases} \text{sign}(x_n - x_u) & \text{if } y_{n-1} = -1 \\ \text{sign}(x_n - x_d) & \text{if } y_{n-1} = +1 \end{cases}, \tag{7}$$

where x_u and x_d are two different threshold values and $x_u > x_d$. As it can be seen in Appendix A, the calculation of $R_y(k)$ and $R_z(k)$ is connected to the problem of counting the noise zero crossings, which in presence of hysteresis is usually considered difficult [1]. Nevertheless for discrete time evolution analytic results can be obtained if thresholds are symmetric with respect to the noise mean value, i.e., if $x_d = -x_u$. In this case, since the used input noise distribution $p(x)$ is symmetric too, the probability p of a comparator state change at any time step does not depend upon the change direction and it is given by

$$p = \int_{x_u}^{\infty} p(x)\,dx = \int_{-\infty}^{-x_u} p(x)\,dx < \frac{1}{2} . \tag{8}$$

In Appendix A the result

$$R_y(k) = (1 - 2p)^{|k|} , \tag{9}$$

which shows that hysteresis provides the comparator output with memory even if the input noise is white, is obtained. Furthermore in Appendix A it is shown that

$$R_z(k) = [r(p)]^{|k|} \cos[k\,\theta(p)] , \tag{10}$$

where

$$r(p) = \left[(1-p)^2 + p^2 \right]^{1/2} \tag{11}$$

(notice that $0 < r(p) < 1$) and

$$O(p) = \arctan\left(\frac{p}{1-p}\right). \tag{12}$$

Eq. (10) shows that the envelope of $R_z(k)$ decays exponentially for any value of the probability p. In particular, the fastest possible decay takes place for $p = 1/2$, i.e., when no hysteresis is present and Eq. (10) reduces to Eq. (6).

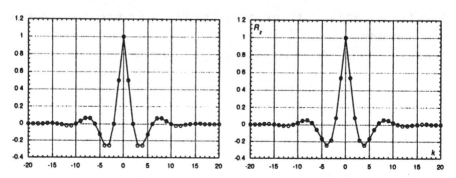

Fig. 2. Analytical form (*continuous line*) and numerical values (*circles*) of $R_z(k)$ without hysteresis (*left*) and with hysteresis (*right*). In the latter case the threshold values are ± 0.1

The circuit behaviour has been numerically simulated by means of the *Simulink* software. Gaussian white noise with standard normal distribution has been used and $R_z(k)$ has been estimated as a time average using 800000 samples of z_n. The plot on the left in Fig. 2 shows the result of a simulation where no hysteresis is present, together with the theoretical curve (6), whereas the plot on the right shows the result of a simulation with $x_u = 0.1^5$, together with the theoretical curve (10). In the latter case the value of p is

$$p = \frac{1}{\sqrt{2\pi}} \int_{0.1}^{\infty} \exp\left(-\frac{x^2}{2}\right) dx \cong 0.46 . \tag{13}$$

In both figures the agreement between theoretical values and numerical data (represented by circles) looks good. Indeed, the r.m.s. difference is about 10^{-3}.

The form of $R_z(k)$ provides us with a criterion for choosing the output sampling frequency. If a bit correlation lower than ε is required, the minimum value k_0 such that

$$[r(p)]^k < \varepsilon \quad \forall \quad k \geq k_0 \tag{14}$$

---

[5] Notice that thresholds are measured in units of the noise mean amplitude.

has to be determined. k_0 is the optimal ratio of the input sampling frequency to the output one and therefore the value $N = k_0$ must be chosen for the N counter.[6]

Throughout the calculations no offset has been supposed. If this were the case, the comparator output would be unbiased and the binary counter would not be needed at all. The analytical study of the correlation becomes difficult and cumbersome if offset is taken into account, but the results found here under simplifying hypotheses allow a conservative estimate of the output sampling frequency even in real circumstances.

Consider indeed a comparator affected by the offset s, with thresholds $s \pm x_u$. For a given input noise this device shows a larger transition rate with respect to a comparator with no offset and thresholds $\pm x_0$, where $x_0 = |s| + x_u$. An intuitive explanation can be gained by looking at Fig. 3, where the case $s > 0$ is represented and $x(t)$ is shown instead of its samples.

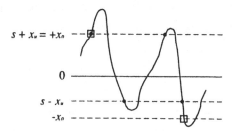

Fig. 3. Crossings of thresholds affected by offset (*dots*) and of broader thresholds with no offset (*squares*) by the same input noise

A smaller transition rate causes a slower decay of the correlation. Therefore a conservative estimate of the output sampling frequency can be obtained by considering the correlation calculated for the larger hysteresis band defined above to include offset.

4 Some Design Instructions

The designer of a random number generator of the type considered here should take into account the following set of instructions.

1) The input sampling frequency v_1, i.e., the clock frequency of the circuit, is determined by the comparator response time τ_c through the condition

---

[6] N could also be chosen in order to obtain $\cos[N0(p)]=0$, but such a condition is more critical than the one stated in Eq. (14).

$$v_1 < \frac{1}{\tau_c} . \tag{15}$$

2) The correlation of the sampled noise must be negligible with respect to the correlation introduced by the subsequent stages. If the maximum acceptable value for the latter is ε, the amplifier cutoff frequency v_0 must verify

$$\exp(-2\pi \, v_0/v_1) < \varepsilon \tag{16}$$

for the filter considered here, or a similar condition for a different filter. Eq. (16) gives

$$v_0 > v_1 \frac{|\ln \varepsilon|}{2\pi} . \tag{17}$$

In Appendix B it is shown that a practically white input noise can be obtained even if v_0 and v_1 are of the same order. A similar result is obtained in [9].

3) Once the input noise distribution $p(x)$ has been estimated, the probability

$$p = \int_{x_0}^{\infty} p(x)\,dx \tag{18}$$

is determined by x_0. This positive quantity has been defined in the previous section in terms of the actual hysteresis and offset, both measured in units of the noise mean amplitude. $r(p)$ is then calculated by means of Eq. (11).

4) Finally the condition

$$N \geq \frac{|\ln \varepsilon|}{\left|\ln[r(p)]\right|} , \tag{19}$$

which follows from Eq. (14) with $k_0 = N$, sets the value of N and therefore of the bit rate

$$v_2 = \frac{v_1}{N} . \tag{20}$$

Notice that, once the bit correlation ε has been fixed, v_2 increases with p, i.e., as it is intuitive, the bit rate grows as long as offset and comparator hysteresis, which cannot be totally suppressed, diminish with respect to the noise amplitude.

5 Conclusions

The complete unpredictability of the random numbers used by a cryptographic system is a necessary condition for the system security that can be satisfied only by means of a truly random source. On the other hand, sources of this kind often produce bit sequences whose statistics depend in a critical way on details of the implementation.

The circuit proposed in this paper belongs to a kind of true random number generators that are well known to produce unbiased bit sequences. It is designed to be insensitive as possible to fluctuations in the behaviour of the circuit components so that no calibration nor compensation is required. Furthermore, it is satisfactorily described by an analytical model that gives a relationship between the bit rate and the maximum expected bit correlation. The model gives also the expected value of the circuit internal transition rate. Since in our design phenomena that could spoil the bit statistics also slow down the circuit dynamics, counting the transitions and comparing their rate to its expected value can be a good self-testing procedure.

An actual circuit that verifies the hypotheses underlying our model generates binary sequences whose randomness is ensured by the circuit design. Such a system requires a small amount of time for its testing during production, since demanding statistical tests can be performed on prototypes only. Furthermore, true randomness of the generated bits can be controlled in a simple and effective way even while the system is operating.

Acknowledgements

This work has been carried out in the framework of the agreement between the Italian PT Administration and the Fondazione "Ugo Bordoni".

References

1. Murry, H.F.: A General Approach for Generating Natural Random Variables. IEEE Transactions on Computers C-19 (1970) 1210-1213
2. Vincent, C.H.: The Generation of Truly Random Binary Numbers. Journal of Physics E 3 No. 8 (1970) 594-598
3. Vincent, C.H.: Precautions for the Accuracy in the Generation of Truly Random Binary Numbers. Journal of Physics E 4 No. 11 (1971) 825-828
4. Maddocks, R.S., Matthews, S., Walker, E.W., Vincent, C.H.: A Compact and Accurate Generator for Truly Random Binary Digits. Journal of Physics E 5 No. 8 (1972) 542-544

5. Gude, M.: Concepts for a High Performance Random Number Generator Based on Physical Random Phenomena. Frequenz 39 No. 7-8 (1985) 187-190
6. Holman, W.T., Connelly, J.A., Dowlatabadi, A.B.: An Integrated Analog/Digital Random Noise Source. IEEE Transactions on Circuits and Systems - I 44 No. 6 (1997) 521-528
7. Terrell, T.J.: Introduction to Digital Filters. 2nd edn. Mac Millan, London (1988)
8. Bendat, J.S.: Principles and Applications of Random Noise Theory. Wiley, New York (1958)
9. Petrie, C.A.: An Integrated Random Bit Generator for Applications in Cryptography. Ph.D. Thesis, Georgia Institute of Technology (November 1997).
10. Yarza, A., Martinez, P.: A True Random Pulse Train Generator. Electronic Engineering 50 No. 614 (1978) 21-23
11. Kedem, B.: Binary Time Series. Lecture Notes in Pure and Applied Mathematics, Vol. 52. Marcel Dekker, New York (1980)
12. Papoulis, A.: Probability, Random Variables and Stochastic Processes. McGraw-Hill, New York (1965)

Appendix A: Calculation of the Autocorrelation Functions

In the following the probability $P_k(l)$ that the comparator change its state a number l of times in the interval $[t_n, t_{n+k}]$ will be needed. l is the number of noise zero crossings during the considered interval. Under the assumptions of discrete time evolution, white noise and no offset, if the distribution $p(x)$ of x_n is symmetric (not necessarily gaussian), the probability p of a comparator state change at any time step does not depend upon the change direction. Therefore $P_k(l)$ follows a binomial distribution,

$$P_k(l) = \binom{k}{l} p^l (1-p)^{k-l} . \tag{A.1}$$

When the comparator shows no hysteresis,

$$p = \int_0^\infty p(x)\,dx = \frac{1}{2} . \tag{A.2}$$

When hysteresis is present, the hypotheses leading to the binomial distribution $P_k(l)$ given by Eq. (A.1) still hold provided that thresholds are symmetric with respect to the sampled noise mean value, i.e., $x_d = -x_u$. In this case the value of p is given by Eq. (8).

Since the clipped noise y_n is represented by a sign function, its autocorrelation $R_y(k)$, defined as in Eq. (4), can be given the form

$$R_y(k) = \sum_{l\,even} P_k(l) - \sum_{l\,odd} P_k(l) \,. \tag{A.3}$$

It can be proven by simple algebra, using Eqs. (A.3) and (A.1), that

$$R_y(k) = (1 - 2p)^{|k|} \tag{A.4}$$

(here and in the following, the absolute value of k is used to generalize results to negative values of k). When no hysteresis is present, Eq. (A.2) holds and therefore

$$R_y(k) = \delta_{k,0} \,. \tag{A.5}$$

$R_z(k)$ can be evaluated by means of the probability $P_e(k)$ that the binary counter change its state an even number of times in $[t_n, t_{n+k}]$. Indeed $R_z(k)$ can be given a form analogous to Eq. (A.3), which is also equivalent to

$$R_z(k) = 2P_e(k) - 1 \,. \tag{A.6}$$

If at the instant t_n the comparator has changed its state an even number of times, in $[t_n, t_{n+k}]$ every transition of the counter corresponds to two transitions of the comparator. Therefore in this case the number l of comparator state changes must be equal to $4m$ or $4m + 1$, where m is an integer such that $l \in \{0...k\}$, to make the counter change its state $2m$ times. On the other hand, if at the instant t_n the comparator has changed its state an odd number of times, its first transition in $[t_n, t_{n+k}]$ coincides with the first counter transition. Therefore in this case one less comparator transition is needed for an even number of counter transitions to occur and l must be equal to $4m - 1$ or $4m$.

When there is no hysteresis, it follows from Eqs. (A.3) and (A.5) that the number of comparator transitions occurred before t_n has the same probability of being even or odd for every value of n. In presence of hysteresis this is no longer an exact result, but it is nevertheless a valid approximation, since $R_y(k)$ drops exponentially. In both cases thus

$$P_e(k) = \frac{1}{2}\left(\sum_{\substack{l=0\\mod 4}} P_k(l) + \sum_{\substack{l=1\\mod 4}} P_k(l) \right) + \frac{1}{2}\left(\sum_{\substack{l=-1\\mod 4}} P_k(l) + \sum_{\substack{l=0\\mod 4}} P_k(l) \right) \tag{A.7}$$

$$= \frac{1}{2}\left(1 + \sum_{\substack{l=0\\mod 4}} P_k(l) - \sum_{\substack{l=2\\mod 4}} P_k(l) \right) \,.$$

This result gives Eq. (A.6) the form

$$R_z(k) = \sum_{\substack{l=0 \\ \text{mod}\,4}} P_k(l) - \sum_{\substack{l=2 \\ \text{mod}\,4}} P_k(l) \,. \tag{A.8}$$

Substituting Eq. (A.1) into Eq. (A.8) gives

$$R_z(k) = \sum_n \binom{k}{2n} (-1)^n p^{2n} (1-p)^{k-2n} = \Re\left\{ \left[(1-p) \pm ip \right]^k \right\}, \tag{A.9}$$

where $\Re$ denotes the real part. This expression is generalized by taking the absolute value of k and it can be put in the form (10) using the polar representation of complex numbers. If there is no hysteresis, $p = 1/2$ and Eq. (6) is obtained.

Appendix B: Number of Internal Transitions vs. Output Correlation

Counting the internal transitions is a good self-testing procedure for the generator we designed, as long as the increase in output correlation is due to phenomena that slow down the circuit dynamics and not to periodic disturbances. The connection between the number of transitions and the output correlation has been confirmed by further numerical simulations of the circuit in which two different effects have been separately considered.

The first phenomenon taken into account has been the increase in the comparator hysteresis, which, in our model, can represent lowering input noise as well as increasing offset. In each simulation 100000 samples of z_n have been generated for a fixed value of the hysteresis band half width x_0. Some of the results are shown in Table 1. Eq. (18), where $p(x)$ is the standard normal distribution, holds for the probability p and the expected number of binary counter transitions,

$$\langle N_z \rangle = \frac{p}{2} N_{samples} = 50000 p \,, \tag{B.1}$$

is in good agreement with the counted number N_z.

In Table 1 theoretical and numerical values of $R_z(20)$ are also reported, since $N = 20$ can be a suitable value for the N counter. Theoretical values have been calculated by means of Eqs. (10-12). As the r.m.s. difference between theoretical and numerical values of $R_z(k)$ is about 5×10^{-3} in each experiment, simulations can be considered consistent with the model. Notice that data in parenthesis, whose absolute value is lower than the r.m.s. error, are shown only for the sake of completeness. It can be seen that a significant increase in correlation occurs when the number of transitions reduces to about one half of the initial value.

Table 1. Number of internal transitions and output correlation (both expected and numerical) for different comparator threshold values.

x_0	$\langle N_z \rangle$	N_z	$R_z(20)$ theor.	$R_z(20)$ num.
0	25000	25166	-0.0010	(0.0045)
0.1	23029	23149	3×10^{-5}	(10^{-5})
0.5	15515	15463	-0.0021	(0.0006)
0.7	12208	12133	0.0105	0.0204
1	7883	7894	-0.0355	-0.0353

In the second series of simulations the effect of a finite noise bandwidth, i.e., of a correlated input, has been studied. In each experiment 100000 samples of z_n have been generated for a fixed value of the frequency ratio v_0/v_1 always assuming no hysteresis, i.e., $x_0 = 0$. Some of the results are shown in Table 2.

Table 2. Number of internal transitions (expected and numerical) and numerical output correlation for different cutoff frequencies.

v_0/v_1	$\langle N_z \rangle$	N_z	$R_z(20)$ num.
∞	25000	25166	0.0045
0.5	24312	24384	-0.0010
0.1	16044	16071	-0.0058
0.05	11967	11953	-0.0156
0.01	5583	5706	0.1546

In this case the expected value $\langle N_z \rangle$, which looks in good agreement with the numerical value N_z, is still given by Eq. (B.1), but p has now the form

$$p = \frac{1}{\pi} \arccos\left[R_x(1) \right] = \frac{1}{\pi} \arccos\left[\exp\left(-2\pi \, v_0/v_1 \right) \right]. \tag{B.2}$$

according to the well known arcsine law [12] assuming first order Butterworth filtering. It can be seen in this case too that a significant increase in correlation occurs when the number of transitions reduces to about one half of the initial value.

Notice that numerical values only of $R_z(20)$ are reported in Table 2. Indeed, the model used throughout this paper for determining the function $R_z(k)$ considers input white noise. This hypothesis is crucial for the binomial distribution (A.1) to hold. As the frequency ratio decreases, the model loses its validity and, for $v_0/v_1 \leq 0.1$, it can be seen that it gives no longer account for the numerical results. On the other hand, Table 2 shows how larger values of v_0/v_1, e.g. 0.5, do not cause significant deviations from the ideal case of infinite v_0. This result confirms that the white noise hypothesis is not critical.

Random Number Generators Founded on Signal and Information Theory

David P. Maher[1], Robert J. Rance[2]

[1] Intertrust, Sunnyvale, CA, USA
dpm@intertrust.com

[2] Lucent Technologies, Bell Laboratories Innovations, No. Andover, MA, USA
rrance@lucent.com

Abstract. The strength of a cryptographic function depends on the amount of entropy in the cryptovariables that are used as keys. Using a large key length with a strong algorithm is false comfort if the amount of entropy in the key is small. Unfortunately the amount of entropy driving a cryptographic function is usually overestimated, as entropy is confused with much weaker correlation properties and the entropy source is difficult to analyze. Reliable, high speed, and low cost generation of non-deterministic, highly entropic bits is quite difficult with many pitfalls. Natural analog processes can provide non-deterministic sources, but practical implementations introduce various biases. Convenient wide-band natural signals are typically 5 to 6 orders of magnitude less in voltage than other co-resident digital signals such as clock signals that rob those noise sources of their entropy. To address these problems, we have developed new theory and we have invented and implemented some new techniques. Of particular interest are our applications of signal theory, digital filtering, and chaotic processes to the design of random number generators. Our goal has been to develop a theory that will allow us to evaluate the effectiveness of our entropy sources. To that end, we develop a Nyquist theory for entropy sources, and we prove a lower bound for the entropy produced by certain chaotic sources. We also demonstrate how chaotic sources can allow spurious narrow band sources to add entropy to a signal rather than subtract it. Armed with this theory, it is possible to build practical, low cost random number generators and use them with confidence.

Introduction

RNGs (Random Number Generators) are hardware and/or software sources that supply bits (or numbers) that ideally are statistically independent. In this paper we will talk solely about analog RNGs, that is, RNGs whose initial source of entropy is analog noise. As such, these RNGs are non-deterministic. In contrast, PRNGs (pseudo-random number generators) are deterministic in that their output is completely determined from their initial state or "seed".

RNGs are used to generate independent bits for cryptographic applications such as key generation or random starting states, where it is vital that the key or state cannot be predicted or inferred by the adversary. They are also used as hashing or blinding factors in various signature schemes, such as the Digital Signature Standard.

Our RNGs have employed either thermal or shot noise, and have been implemented in both discrete and integrated forms. Other RNGs have been driven by sources such as: vacuum tube shot noise, radioactive decay, neon lamp discharge, clock jitter, and PC hard drive fluctuations.

We further define these RNGs as hybrid RNGs since they comprise an analog noise source followed by digital post-processing. The post-processing greatly enhances the entropy (statistical independence) of the output, usually at the cost of an acceptable reduction in bit-rate. The post-processing could also be termed digital nonlinear filtering or lossy compression. Finally, we further class the RNGs as either chaotic or non-chaotic.

Reliable, high speed, and low cost analog random number generation of highly entropic bits is a hard problem. This is because practical noise sources are a few microvolts while other co-resident, typically fast-transitioning digital signals are several volts. Even greatly attenuated interference from these deterministic sources can rob the RNG output of most or all of the entropy that a cryptographic application may depend on. Amplification and sampling of noise signals can further degrade the entropy because the amplifiers are inevitably band-limited, and sampling thresholds are typically biased. By applying some results from chaotic processes and signal theory we have been able to overcome the problems mentioned above, producing reliable, high-speed, low-cost, highly entropic RNGs whose performance is supported by strong theory.

In particular, we study the effect of filtering on sampled analog noise sources. We demonstrate that under ideal conditions, a relatively narrow-band noise source can be used to produce a perfectly uncorrelated bit-stream. Seeking more practical solutions, we demonstrate how simple digital feedback processes can be used to improve RNG statistics and to nullify the effects of certain spurious noise sources. Finally, we demonstrate how digital feedback, directly interacting with a chaotic amplifier, can produce a noise source that coerces other spurious noise sources to contribute their entropy to the main source, rather than rob that source of its entropy. We prove a lower bound for the amount of entropy per bit that such a chaotic source will produce, we calculate the probability density function for the source, and we discuss how to use this source to compress n bits of entropy into a vector of length n.

We believe that the results provided here can help designers include high quality, stand-alone, non-deterministic RNGs in low-cost crypto-modules and ICs.

1 Signal Theory Applied to RNGs

We begin by showing that it is possible to produce very good, random bit sequences of completely independent values by carefully filtering and sampling a non-white natural noise source. We show how bandwidth limitations reduce to bit-rate limitations. A classic natural random number generator is modeled as follows:

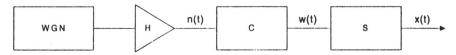

Fig. 1. Classic Random Number Generator

WGN is Stationary White Gaussian Noise that we assume is wide-band and low-power (such as thermal or shot noise). H is the transfer function of a band-limited amplifier. C is a comparator or quantizer function, and S is a sampling process that samples every interval of length τ.

Generally one can find wide-band, low-power white noise sources, but to work with them considerable amplification is needed. An inevitable limitation of bandwidth results. The signal x(t) is expressed in terms of the Dirac delta function δ(t):

$$x(t) = \sum_k w(t)\delta(t - k\tau) \tag{1}$$

summing over all integers k. The Fourier Transform of x(t) is then

$$X(f) = \int_R x(t)e^{-j2\pi ft}\,dt = \int_R \sum_k w(t)\delta(t - k\tau)e^{-j2\pi ft}\,dt = \sum_k w(k\tau)e^{-j2\pi k\tau f} \tag{2}$$

Using the Poisson summation formula on the last expression, we get:

$$X(f) = \frac{1}{\tau}\sum_k W(f - k/\tau) \tag{3}$$

For the moment, ignore the effect of the quantizer, and note that if the shape of W is completely determined by the filtering of the amplifier H, we can arrange that W(f) = 0 for $|f| > 1/\tau$ by selecting the sampling rate to match the amplifier's rolloff. In addition, if the amplifier characteristic is equalized so that W(τ/2-f) = W(τ/2+f), then the right hand side of Equation (3) is a constant, indicating white noise, and therefore x(t) will be completely uncorrelated. Again, if we ignore the effects of the quantizer, then by the Gaussian assumption we can conclude that the values x(nτ) are independent. We have shown that we can apply the Nyquist rule of thumb: "Make the sampling rate about twice the bandwidth," and we have shown that we can carefully filter a stationary Gaussian narrowband noise source to completely eliminate correlation. Nyquist theory refers to sampling theorems in hybrid analog and digital systems where the goal is to eliminate the effects of aliasing and to reduce intersymbol interference. We have shown that it applies just as well to sampled noise signals where the performance criterion is intersymbol correlation. It is also clear from the above analysis that if the original noise source is non-white, an equalizer W can be still be designed so that the sum in Equation (3) is constant.

Recall that the effect of the quantizer has thus far been ignored. The theory is much more involved for most common quantizers. We will treat one common and useful case in the next section.

2 Practical and Simple Examples

In order to economically manufacture a good natural random number generator, we have to use some simpler digital filtering techniques, shooting for less than Nyquist precision. We show how simple digital filtering and sampling techniques can reduce correlation. Some examples of RNGs with and without digital post-processing can be found in Murry [1], Bendat [2], Boyes [3], Castanie [4], and Morgan [5].

Referring to Figure 1, we assume w(t) has zero mean with a power spectral density function W (f), and we use a simple two-pole amplifier with non-Nyquist filtering. W (f) rolls off from a flat spectrum at f_a and f_b, the lower and upper cutoff frequencies of the amplifier. Let us also consider the effect of the quantizer function C. Here we assume that C is an infinite clipper. That is, C assigns the value +1 to a positive voltage and -1 to a negative voltage. We also assume the comparator has a bias with an offset voltage Δ. Let $x_n = x(n\tau)$. We are interested in values: $\mu_x = E[x_n]$, $\rho_x(i) = E[x_n \cdot x_{n+i}]$. The latter can be expressed in terms of $\rho_w(\tau) = E[w(t) \cdot w(t + \tau)]$. The mean of the process x is a straightforward error function approximated by $\mu_x \cong 0.4\ \Delta/\sigma$ (when the offset is small enough compared to the signal power), and with the aid of Price's theorem [6], the autocorrelation function can be expressed in closed form as:

$$\rho_w(\tau) = \frac{2}{\pi} \operatorname{Sin}^{-1}(\rho_n(\tau)) + \mu_x^2 \tag{4}$$

where

$$\rho_n(\tau) = \frac{f_b \exp(-2\pi f_b \tau) - f_a \exp(-2\pi f_a \tau)}{f_b - f_a} \tag{5}$$

For a typical selection of components for a low cost RNG as modeled above, we get unsatisfactory mean and correlation values even if we very carefully isolate the RNG components from spurious noise sources. Thus we are motivated to use some simple digital filtering techniques. First, suppose we follow the sampler in Figure 1 with a simple feedback loop, where the analog noise source below contains the sampler:

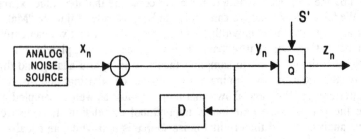

Fig. 2. Digital Processing

We note that if the delay D is one clock cycle then $y_i = \prod_{j=1}^{i} x_j$. After a short period of time, μ_y is going to be extremely small even if the x_i's are highly biased and strongly correlated. This is clearly true when the x_i are independent. More generally, we can show that $E[y_i] \to 0$ almost as quickly as $\left(\rho_x(1) + \mu_x^2\right)^{i/2} \to 0$, depending on the characteristics of w(t)'s autocorrelation function, $\rho_w(\tau)$, which we assume is asymptotically well-behaved and monotonically decreasing (as in the case of the two pole filter we have assumed). The difficulty here is that the autocorrelation $\rho_y(1) = \mu_x$ is unacceptable. Note that the effect of the feedback loop is just to shift values of lower order statistics to higher order statistics. Now consider the sequence z_n. We sample the output w_j at a rate $f_s = f_s/r$ where f_s is the sampling rate producing x_k. Let the function D delay the feedback by d clock cycles, then

$$E[z_j z_{j\,k}] = E[(x_{rj} x_{rj\,d} x_{rj\,2d} \cdots)(x_{r(j\,k)} x_{r(j\,k)\,d} \cdots)] \qquad (6)$$

We choose d to be relatively prime to r. Then when d does not divide k, we see that there are no duplications in the subscripts in the above expression, and so there are no symbolic cancellations of the x values, and so $\rho_z(k)$ is the expectation of the product of a large number of samples of x which grows larger as n grows large. As for the case when d divides k: If we set m=k/d, then

$$E[z_j z_{j\cdot k}] = E[y_{rj} y_{r(j\cdot k)}] = E[x_{rj} x_{rj\cdot d} \cdots x_{rj\cdot(rm-1)d}] \qquad (7)$$

Therefore, this correlation is the expectation of the product of rm bits each spaced d apart. This works very well with a decreasing autocorrelation function for w(t). In cases where the acf $\rho_w(\tau)$ decreases slowly, then the value of d. should be increased to compensate. Heuristically, we are taking the expectation of the product of many ± 1 values spaced far apart in time. For a typical acf, increasing the spacing will effect an exponential reduction in expectation, and increasing the number of bits will also cause an exponential drop. Thus increasing d and r serve to reduce the expectation synergistically and powerfully. Both of these techniques are novel. With reasonable assumptions on n(t) and w(t), we can show that $|\rho_z(1)| \le |B_n(\rho_x(1), \mu_x^2)|$ given B_n:

$$B_n(\rho, \mu^2) \equiv \sum_{k=0}^{n/2} a_k \mu^{n-2k} \rho^k \qquad (8)$$

$$a_k \equiv \binom{n/2}{k} k! / \left((k/2)! \pi^{k/2}\right) \qquad (9)$$

The actual closed form expressions for $\rho_z(\tau)$ are difficult to analyze asymptotically. We use Price's theorem [6] to calculate the autocorrelation function of the output of the infinite clipper, and our expressions include a determinant of an autocorrelation matrix whose entries are values of the autocorrelation function for the process n(t) at times $n\tau$ (see Figure 1 again). For the example where n(t) is flat noise filtered by a two-pole filter the autocorrelation function magnitude, $|\rho_n(t)|$, is eventually monotonically decreasing, and thus we can estimate bounds for $\rho_z(k)$. Overall, the key to improving the statistics is selection of the sampling rates S and S' as we show next.

Thus, we can produce a sequence z_k with zero mean (after a brief transient), and unnoticeable correlation. In fact, the bound given above also provides a measure of independence in that $|\log(1 - E[x_1 x_2 \ldots x_{rk}])|$ bounds the average mutual information between bits z_i and z_{i+k}. Thus, this system approximates a sequence of equiprobable mutually independent (Bernoulli) samples of events from a sample space $\{1, -1\}$ which can be produced using simple components with very modest performance characteristics.

Suppose we use a Western Electric WE-459G noise diode as the WGN source. It's output is 0.8 μV/sqrt(Hz) with a power spectral density that is flat $\pm$ 2 dB from 100 Hz to 500 kHz. We use a comparator with offset of $\Delta = 10$ mV, and an amplifier with voltage gain of 100 and a flat transfer function from 100 Hz to 10 kHz. The mean will then be μ_x = 0.05. For a sampler S with frequency $f_s = 10$ kHz, then the two-pole amplifier model above predicts all covariances between x_i and x_{i+j} for j between 1 and 50 to be on the order of 10^{-2} down to 10^{-3}. Applying the loop model, we "oversample" with $f_s = 30$ kHz, and then set the second sampler frequency to $f_{s'} = 10$ kHz. Thus r=3, and we choose a delay of 256 bits. The output sequence z_i then has zero mean, and all correlations are 0, except $\rho_k(k) \approx (0.05)^{3m}$ for $k = 256m$.

In general, our experience shows that the above structure works extremely well under the assumption that we are reasonably faithful to the model, and we are careful to isolate the analog source from coupling effects from the digital filter components and other on-board components. This latter requirement is either difficult or expensive to satisfy, but it turns out that the same techniques mentioned above employing a double-loop topology will mitigate the effects of such coupling.

3 Reducing Coupling Effects by a Double Loop

Measured statistics on the output of an implemented single-loop RNG showed a small mean bias. This result violated the above theory and we attributed this effect to coupling from the high-level digital output into the analog noise source. Coupling between the digital output and analog input is denoted by ε. Since the digital output is 5 to 6 orders of magnitude larger than the analog noise levels, the coupling effect will be significant in practical designs and will place a bias on the y_n, independent of the sub-sampling ratio. This places a fundamental upper limit on the output entropy. However, this limit can be surmounted by placing two loops in tandem:

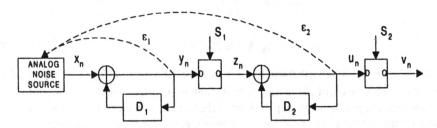

Fig. 3. Double-Loop Coupling

The second loop exponentially mitigates the ε_1 coupling effect and that the first loop will similarly reduce the ε_2 coupling effect. In the first case, we can model the noise source and first loop as a single noise source with some mean bias induced by ε_1. The second loop will greatly enhance the bit independence as shown in the single-loop RNG analyses in the preceding sections. In the second case, the mean bias on the noise source output induced by ε_2 will just be treated as another noise source bias by the first loop. In fact, we have found that two loops are enough; three or more loops produce similar results. Again, this result is novel.

No mean bias was observed at the output of a dual-loop RNG. We have implemented this dual-loop RNG in a single IC that includes fault-tolerance and testability features. The performance of this device bears out the theory.

4 Chaotic RNG

This section departs from the above theory in that it treats a radically different type of RNG termed a chaotic RNG. Due to the great disparity between analog noise and digital signal levels, it is difficult (expensive) to ensure that interference of undesirable (low entropic) character will not dominate the analog noise source output. This dominance would nullify the beneficial effects of the various techniques described above. To free ourselves of this constraint, and other constraints imposed by other low-entropic interferers such as 1/f noise, we developed a chaotic RNG. The chaotic RNG has the advantageous property of accepting all noises, good and bad, and extracting their entropies. We discovered this idea by observing that the LSBs of high-resolution A/Ds tend to yield independent bits, regardless of the statistical nature and amplitude of the "desired" signal being converted. High resolution A/Ds require much hardware; this hardware can be sharply minimized by implementing the A/D in a loop with a 1-bit quantizer:

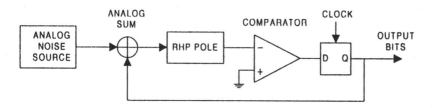

Fig. 4. Chaotic RNG

The selection of the RHP (right-half-plane) pole and the clock frequency determine the loop gain, A, of the "A/D". A standard, radix-2 A/D can be implemented by setting the loop gain at 2 and by setting the "analog noise source" to a fixed voltage.

Setting A to unity and replacing the "fixed voltage" by a time-varying signal implements a sigma-delta modulator. Finally, increasing A to somewhere between 1 and 2, and replacing the "time varying signal" with an analog noise source creates a chaotic RNG.

Electrical engineers are taught, almost from birth, to avoid poles in the right half plane. However, the loop's negative feedback creates a stable overall response that circumvents the RHP instability. We chose an RHP pole design since it provided the simplest implementation using discrete parts. In particular, the RHP pole comprises an OP-AMP (operational amplifier), capacitor, and a few resistors:

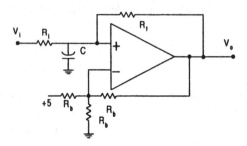

Fig. 5. RHP Pole Circuit

Referring back to Figure 4, we can immediately draw up an equation describing the evolution of voltage at the OP-AMP's output. We will call the normalized RHP pole output voltage, at sample time nT, b_n. Note that this voltage at time $(n+1)T$ is a linear combination of the voltage at time n, b_n, the sign of the voltage at time n, $sgn[b_n]$, thermal noise, g_n and interference, s_n. The first term is the initial state for the RHP filter, and the latter three terms are inputs accumulated by the RHP filter during the $[nT, (n+1)T]$ period. The RHP pole will increase its initial state over this time period by a factor of A. After normalization, the voltage is described as

$$b_{n+1} = Ab_n - sgn[b_n] + g_n + s_n \qquad (10)$$

For convenience, we will often use c_n as shorthand for $sgn[b_n]$.

4.1 What is Chaos?

Chaos can be described as a response that grows exponentially larger with time due to an arbitrarily small perturbation. A good introductory description to chaos is Schuster [7]. Note that its title is "Deterministic Chaos". In fact, it is the marriage of (analog) deterministic chaos with an analog noise source that engenders a potent random number generator.

Mathematically, a positive Lyapunov exponent defines chaos. In discrete time, the Lyapunov exponent is defined as the averaged logarithm of the absolute value of gain each cycle. For our chaotic RNG, the gain A is constant, so the Lyapunov exponent is $ln(A)$. Since $A > 1$, the exponent is positive, verifying chaos.

4.2 Other Chaotic RNGs

Bernstein [8] and Espejo-Meana [9] describe two (of many) possible implementations of chaotic RNGs. Of the two, Espejo-Meana is the most similar to the implementation described here.

4.3 Why is Chaos Good for Random Number Generation?

Chaos guarantees that any noise contribution, no matter how small and how buried in deterministic interference, will ultimately significantly effect the output bits since the noise's effect increases exponentially. This means that we can greatly relax the isolation requirements on the analog noise source. As long as the deterministic or low-entropic interference does not lower the Lyapunov exponent by causing OP-AMP saturation, chaotic operation will occur. In fact, we built the above circuit with both analog and digital circuitry powered by the same +5V supply (which also powered much other digital circuitry). We observed no interference with chaotic operation.

This RNG employs the simplest possible topology for a chaotic RNG implemented in discretes, has a constant Lyapunov exponent, and is therefore (relatively) easily analyzed. In Appendix A, we calculate a lower bound on the output bit entropy, expressed in bits:

$$H_{lb} = (N-1)\log_2(A) - \frac{1}{2}\log_2(1 - A^{-2}) + \log_2(\sigma_s) + 1.77 \tag{11}$$

Here N denotes the number of successive output bits collected and σ_s denotes the noise standard deviation. Note that for large N,

$$H_{lb} \approx N\log_2(A) \tag{12}$$

We have implemented this chaotic RNG and have verified that its output entropy approximates this lower bound to the precision we could measure. The output entropy is *guaranteed* in the sense that it will always be greater than the bound expressed in Equation 11, independent of any non-saturating interfering signal. This is a very important property for an RNG. In contrast, typical non-chaotic designs are plagued with very difficult isolation issues, tight tolerance on parameter matches, or clock phase-locking. Chaotic designs are often plagued with regions with negative Lyapunov exponents. Finally, both chaotic and non-chaotic designs can often be very difficult to analyze.

4.4 Output Whitening

The derivation in Appendix A suggests a particular form of post-processing to provide independence. Specifically, we proved that the two sides of Equation 13 are asymptotically equal:

$$\sum_{k=0}^{N-1} A^{-k} c_k \approx \sum_{k=0}^{N-1} A^{-k} g_k + \sum_{k=0}^{N-1} A^{-k} s_k \tag{13}$$

where c_k = sgn[b_k] and the s_k denote the interfering signal(s). The LHS comprises a quantizer that represents accurately a Gaussian random variable with some mean arising from the interference. The post-processor can then re-express the LHS as a binary number, which will comprise a standard binary-weighted A/D converter. Selecting a (large) subset of the bits of this binary number will yield a nearly independent bit-stream. Heuristically, the MSBs are not independent since they are heavily influenced by the signal's distribution. Also, the LSBs after some point cannot convey any additional entropy, since only $N \log_2(A)$ bits of entropy are available. Thus at this point, these LSBs become deterministically related to the prior bits. This leaves us with a mid-range of bits that *are* independent. Of course other whitening methods such as post-processing via a hash function or DES are always valid.

4.5 Concluding Remarks on Chaotic RNG

We believe that the following are novel with regard to this type of chaotic RNG: use and implementation of RHP pole, calculation of entropy lower bound, realization that this lower bound is independent of external interference, form of whitening filter, and the derivation of a probability distribution.

References

[1] H. F. Murry, "A General Approach for Generating Natural Random Variables," IEEE Trans. Computers, Vol. C-19, pp. 1210-1213, December 1970.

[2] J. S. Bendat, Principles and Applications of Random Noise Theory, John Wiley and Sons, Inc., 1958.

[3] J. D. Boyes, "Binary Noise Sources Incorporating Modulo-N Dividers," IEEE Trans. Computers, Vol. C-23, pp. 550-552, May 1974.

[4] F. Castanie, "Generation of Random Bits with Accurate and Reproducible Statistical Properties," Proc. IEEE, Vol. 66, pp. 807-809, July 1978.

[5] D. R. Morgan, "Analysis of Digital Random Numbers Generated from Serial Samples of Correlated Gaussian Noise", IEEE Trans. on Info. Theory, Vol. IT-27, No. 2, March 1981, pp. 235-239.

[6] R. Price, "A Useful Theorem for Non-linear Devices Having Gaussian Inputs," IRE PGIT, Vol. IT-4, 1958.

[7] H. G. Schuster, Deterministic Chaos, VCH, 1989.

[8] G. M. Bernstein, M. A. Lieberman, "Secure Random Number Generation Using Chaotic Circuits", IEEE Trans. on Circuits and Systems, Vol. 37, No. 9, September 1990, pp. 1157-1164.

[9] S. Espejo-Meana, J. D. Martin-Gomez, A. Rodriguez-Vazquez, J. Huertas, "Application of Piecewise-Linear Switched Capacitor Circuits for Random Number Generation", Proc. Midwest Symp. Circuits and Systems, August 1989, pp. 960-963.

Appendix: Entropy Calculation

Equation (10) is cast into an equivalent form by applying the filter $\sum_{k=0}^{N-1} A^{-k}(\cdot)$:

$$A^{-(N-1)}b_N = Ab_0 + \sum_{k=0}^{N-1} A^{-k}\left(g_k + s_k - c_k\right) \tag{14}$$

Due to the negative feedback via the $\{c_n\}$, $|b_n| < 1$ for all n. Thus the RHS above is bounded in amplitude by $A^{-(N-1)}$. In other words, with maximum error $A^{-(N-1)}$,

$$\sum_{k=0}^{N-1} A^{-k}c_k \approx Ab_0 + \sum_{k=0}^{N-1} A^{-k}s_k + \sum_{k=0}^{N-1} A^{-k}g_k \tag{15}$$

For fixed N, the RHS of Equation (A2) is the sum of:
1. The initial condition, Ab_0
2. A possibly large term due to the extraneous interference called S: $S = \sum_{k=0}^{N-1} A^{-k}s_k$

3. A zero-mean Gaussian random variable: $G = \sum_{k=0}^{N-1} A^{-k}g_k$

We cannot rely on Ab_0 and S to supply entropy, at least entropy that is unknown to an adversary who is attempting to break a cryptosystem. The initial condition b_0 may be largely deterministic if it is defined as the initial value of b_n just after the circuitry has been powered-up or supplied with a clock. Moreover, b_0 may be correlated to the previous exercise of the RNG, thereby reducing its entropy. Since we cannot specify what entropy that S will have that is unknown to the adversary, we will assume conservatively that S is deterministic. Therefore, for the remainder of this argument, we conservatively model the RHS of Equation (15) as a Gaussian random variable with mean $(Ab_0 + S)$ and standard deviation

$$\sigma_G \equiv \frac{\sigma_g}{\sqrt{1 - A^{-2}}} \tag{16}$$

The LHS of Equation (15) is the (scaled) quantized value of the RHS in the sense that the $\{c_k\}$ assume values of $\{-1,1\}$ which can be mapped into binary ones and zeros. The quantizer defined by the set $\{c_k\}$ has the property that it represents the RHS of this equation with a maximum error of $A^{-(N-1)}$. There is an infinite set of quantizers that have this same property. Generally, these quantizers would have different entropies. The minimum-entropy quantizer with this property is one with as few quantization steps as possible, namely one that uniformly spans $[-1, +1]$ with step size $2A^{-(N-1)}$. The entropy of this minimum-entropy quantizer is thus a lower bound on entropy of the $\{c_k\}$ quantizer. We calculate this lower bound here:
Divide the $[-1, +1]$ range into 2M levels, M positive and M negative. Then, to a high accuracy since M is very large,

$$M \approx \frac{A^{N-1}}{2} \tag{17}$$

The entropy (in nats) of this quantizer is

$$
H_{nats} = -2\sum_{i=1}^{M} \frac{\exp\left(-\dfrac{(i-\bar{i})^2}{2\sigma_{G}^2 M^2}\right)}{\sqrt{2\pi}\,\sigma_{G} M} \ln\left(\frac{\exp\left(-\dfrac{(i-\bar{i})^2}{2\sigma_{G}^2 M^2}\right)}{\sqrt{2\pi}\,\sigma_{G} M}\right) \tag{18}
$$

$\bar{i}$ denotes the mean value of the RHS of Equation (A2): $\bar{i} = (Ab_0 + S)M$. The tails of the Gaussian pdf have been ignored since σ_G is small. Expanding the natural logarithm in Equation (18) gives

$$
H_{nats} = 2\sum_{i=1}^{M} \frac{\exp\left(-\dfrac{(i-\bar{i})^2}{2\sigma_{G}^2 M^2}\right)}{\sqrt{2\pi}\,\sigma_{G} M} \left[\ln(M) + \ln\left(\sqrt{2\pi}\,\sigma_{G}\right) + \frac{(i-\bar{i})^2}{2\sigma_{G}^2 M^2} \right] \tag{19}
$$

The first two terms in the brackets are independent of i and equivalently pre-multiply the summation operator. The weighting function, $\exp(\cdot)$ is just a pdf which sums to one[1]. The last term in the brackets, when summed, very closely approximates the following integral where $\bar{x} = \dfrac{\bar{i}}{M} = Ab_0 + S$:

$$
2\int_{0}^{1} \frac{\exp\left(-\dfrac{(x-\bar{x})^2}{2\sigma_{G}^2}\right)}{\sqrt{2\pi}\,\sigma_{G}} \left(\frac{(x-\bar{x})^2}{2\sigma_{G}^2}\right) dx \tag{20}
$$

The integral approximates unity due to the small σ_G. Therefore, the entropy is

$$
H_{nats} = \ln(M) + \ln\left(\sqrt{2\pi}\,\sigma_G\right) + 1 \tag{21}
$$

Substituting for σ_G and M from Equations (16) and (17), and converting to bits yields

$$
H_{lb} = (N-1)\log_2(A) - \frac{1}{2}\log_2\left(1 - A^2\right) + \log_2\left(\sigma_z\right) + 1.77 \tag{22}
$$

---

[1] Again ignoring the tails of the Gaussian since σ_G is small

A High-Performance Flexible Architecture for Cryptography

R. Reed Taylor[1] and Seth Copen Goldstein[2]

[1] Department of Electrical and Computer Engineering, Carnegie Mellon University,
Pittsburgh, PA, 15213, USA
rt2i@ece.cmu.edu,
WWW home page: http://ece.cmu.edu/~rt2i
[2] Computer Science Division, School of Computer Science, Carnegie Mellon
University, Pittsburgh, PA, 15213, USA
seth@cs.cmu.edu,
WWW home page: http://cs.cmu.edu/~seth

Abstract. Cryptographic algorithms are more efficiently implemented in custom hardware than in software running on general-purpose processors. However, systems which use hardware implementations have significant drawbacks: they are unable to respond to flaws discovered in the implemented algorithm or to changes in standards. In this paper we show how reconfigurable computing offers high performance yet flexible solutions for cryptographic algorithms. We focus on PipeRench, a reconfigurable fabric that supports implementations which can yield better than custom-hardware performance and yet maintains all the flexibility of software based systems. PipeRench is a pipelined reconfigurable fabric which virtualizes hardware, enabling large circuits to be run on limited physical hardware. We present implementations for Crypton, IDEA, RC6, and Twofish on PipeRench and an extension of PipeRench, PipeRench+. We also describe how various proposed AES algorithms could be implemented on PipeRench. PipeRench achieves speedups of between 2x and 12x over conventional processors.

1 Introduction

Most cryptographic algorithms function more efficiently when implemented in hardware than in software. This is largely because customized hardware can take advantage of bit-level and instruction-level parallelism that is not accessible to general-purpose processors. Hardware implementations, lacking flexibility, can only offer a fixed number of algorithms to system designers. In this paper we describe a reconfigurable fabric which delivers high performance hardware implementations with the flexibility of general-purpose processors.

The efficiency of an implementation is directly related to the degree to which it is customized to perform a given task. Hardware implementations are even more efficient when they are customized for a specific instance of an algorithm. For example, a hardware multiplier with one constant operand will generally take much less area than a general-purpose two operand multiplier.

Of course implementing circuits with such a high degree of specificity in VLSI is generally infeasible because the cost of development and manufacturing must be offset by the chip's applicability. Furthermore, to be responsive, a system must have some control over its embedded algorithms. For example, if a particular algorithm is discovered to be insecure, the system is rendered useless unless a different algorithm can be implemented. Reconfigurable hardware strikes a balance between customization and performance on the one hand and flexibility and cost on the other hand by permitting any algorithm to be highly customized.

Reconfigurable hardware is a general term that applies to any device which can be configured, at run-time, to implement a function as a hardware circuit. Reconfigurable devices occupy a middle ground between traditional computing devices, e.g., microprocessors, and custom hardware. Microprocessors compute a function over time by multiplexing a limited amount of hardware using instructions and registers. They are thus general-purpose and can compute many different functions. At the other end of the spectrum, custom hardware is used to implement a single function, fixed at chip fabrication time. A reconfigurable device, of which the most common is a Field Programmable Gate Array (FPGA), has sufficient logic and routing resources that it can be configured, or programmed, to compute a large set of functions in space. Later, it can be re-programmed to perform a different set of functions. It shares attributes of microprocessors, in that it can be programmed post-fabrication, and of custom hardware, in that it can implement a circuit directly; avoiding the need to multiplex hardware.

The primary ways in which reconfigurable devices are tailored to an application are by matching application parallelism with as many function units as needed, by sizing function units to the word size of the application, by creating customized instructions, by introducing pipelining, and, by eliminating control overhead associated with the multiplexing of function units as in a microprocessor.

In the next section, we describe how reconfigurable computing devices can achieve the efficiency of highly customized designs while maintaining both cost-effectiveness and security. Section 3 focuses on how the components of typical cryptographic algorithms map to reconfigurable devices. Section 4 describes a pipelined reconfigurable device called PipeRench which overcomes many of the problems of using commercial FPGAs to implement datapaths. In particular PipeRench supports hardware virtualization which, like virtual memory, allows designs that do not fit on the physical device to run. Section 5 describes our implementations of several algorithms on PipeRench and our support of on-the-fly customization even in embedded systems. Related work is covered in Section 6. We conclude in Section 7.

2   Reconfigurable Computing

Functions for which a reconfigurable fabric can provide a significant benefit exhibit one or more of the following features:

1. The function operates on bit-widths that are different from the processor's basic word size.
2. The data dependencies in the function allow multiple function units to operate in parallel.
3. The function is composed of a series of basic operations that can be combined into a single specialized operation.
4. The function can be pipelined.
5. Constant propagation can be performed, reducing the complexity of the operations.
6. The input values are reused many times within the computation.

These functions take two forms. *Stream-based functions* process a large data input stream and produce a large data output stream, while *custom instructions* take a few inputs and produce a few outputs. Notice that cryptographic algorithms possess many of the features described above. They can be implemented as stream-based functions which run completely on a reconfigurable device, or, when impractical to implement completely on the a reconfigurable device, pieces of them can be implemented on the reconfigurable device as custom instructions. After presenting a simple example of a custom instruction to illustrate how a reconfigurable fabric can improve performance, we discuss the ways in which a fabric can be integrated into a complete system.

2.1 Custom Instructions: The q_x permutation from TwoFish

In Twofish [27], in order to generate the key dependent S-boxes, multiple invocations of the q function are required. This function combines XOR, rotation, bit truncation, and table lookups. One way to accelerate the creation of the key dependent S-boxes is to implement a custom instruction, the q-instruction, on a reconfigurable fabric. This instruction takes an 8-bit operand and produces an 8-bit result. The custom instruction exploits the ability of the reconfigurable fabric to operate on small bit-width operands (4-bits), to execute many operations in parallel, and to combine a sequence of operations into a single operator (through the use of lookup tables).

2.2 A System Architecture

Reconfigurable fabrics enhance performance mainly by providing the computational datapath with more flexibility. Their utility and applicability is thus influenced by the manner in which they are integrated into the datapath. We recognize three basic ways in which a fabric may be integrated into a system: as an attached processor on the I/O or memory bus, as a co-processor, or as a functional unit on the main CPU. They are most widely useful when integrated into the processor as a reconfigurable function unit (RFU). The RFU has access to both the register file and the primary cache. The main reason for this is that the they may be used to implement custom instructions which can operate on data in the processor registers. Furthermore, the bandwidth between the fabric and the processor (and the data in the processor's cache) is highest when the fabric can directly access the cache. As we will show in the rest of the paper, this organization leads to a system which can significantly enhance the performance of all cryptographic algorithms.

A fourth possible system organization is the system-on-a-chip approach used in embedded computing systems. In such an organization the fabric is closely coupled with a processor, but not so tightly coupled as to be on the processors datapath.

3 Cipher Components

Most ciphers can be specified as dataflow graphs consisting of a few different components. In this section we will enumerate the most common of these components and discuss how they map onto reconfigurable hardware.

- Simple Arithmetic Operations
  Simple operations such as addition and subtraction appear frequently in cryptographic algorithms. These operations map easily to hardware, but due to their simplicity they offer no real gain for reconfigurable systems.
- Narrow and Unusual Bit-widths
  Operations involving narrow bit-widths appear often in stream ciphers, and they are important in highly customized ciphers of any type. Standard microprocessors are notoriously bad at performing narrow bit-width operations, particularly if the values are not multiples of the natural word length of the architecture. Customized hardware supports operations on values of any width, avoiding the computation of unneeded values and the costly masking of undesired bits. Implementing a highly customized design with a constant key allows all datapaths to be reduced to their minimum widths, eliminating the need for paths wide enough to support all possible key combinations.
- Multiplication
  Multiplication is a difficult task to perform in hardware, in that simple hardware multipliers consume a large amount of hardware and compute their results very slowly. Because there are many different ways to improve their performance, multipliers are a prime candidate for optimization and acceleration. Here we consider three different types of multiplier.
  - General-Purpose
    General-purpose multipliers (where both operands may take any value) are costly to implement in hardware. However, in many cryptographic algorithms, the result of $n x n$ multiplies is often only n bits wide. On a reconfigurable device, the size and number of the adders can be reduced accordingly, eliminating the need to compute bits which are later ignored.*
  - Multiplication by a Constant
    Implementing highly customized cryptographic hardware, for example when the key has been set to a constant value, can serve to change many (or all) of the general-purpose multipliers in a design into constant multipliers

(multipliers where one operand is a constant). Constant multipliers can be made considerably smaller and faster in hardware than general-purpose multipliers.

Suppose that one operand of a multiplier set to a constant. The multiplier requires only as many partial products as there are 1's in the constant operand. On average, single-operand multipliers of this type are half the size and twice as fast as their general-purpose counterparts.

- Multiplication Using a Redundant Coding Scheme
  A great deal of space can be saved when performing constant multiplication through the use of a redundant coding scheme. For example, it is straightforward to transform a constant into canonical signed digit, or CSD, form. CSD vectors reduce the number of partial products needed for multiplication by permitting bits in the constant operand to take on negative values. For example, the number 7 in binary is 0111, or $2^2 + 2^1 + 2^0$. Multiplication by this constant requires three partial products; one for each 1 in the binary representation. The CSD representation of 7, however, is $100(-1)$, or $2^3 - 2^0$. Multiplication by this constant vector requires only two partial products.
  As long as addition and subtraction take the same amount of time, no hardware overhead is incurred in implementing this type of multiplier. On average, a constant CSD multiplier will be about 75% smaller than a general-purpose multiplier because the number of partial products in constant CSD multipliers scales with the number of *sequences* of ones in the original constant.
- Parallel Logical Operations
  Hardware allows many logical operations to be performed in parallel. This instruction level parallelism is one of the fundamental advantages of hardware over software in computation. Reconfigurable devices can be programmed to perform such complex logical operations in parallel, harnessing all the parallelism available to a hardware implementation. Furthermore, since the number and kind of function units needed at any point in the computation is configured for the application, the parallelism is never artificially constrained by a lack of function units (as might happen in a VLIW architecture, for example).
- Sequences of Logical Operations
  Most reconfigurable architectures, including standard commercial Xilinx FPGAs the PipeRench architecture discussed later in this paper, implement function units using lookup tables. Thus a sequence of operators can often be combined into a single operator by setting the lookup-table appropriately.
- Table Lookup
  Most block ciphers include a substitution box, or S-box. S-boxes are generally not easily expressible as linear transformations and are therefore implemented as table look-ups. Many reconfigurable architectures can implement tables of this kind, while others may need external scratch memory to store the S-box values.
- Rotation and Shifting
  Lastly, two very common operations in cryptography are bitwise shifts and rotations. Microprocessors, particularly if programmed in C, are very inefficient at performing operations of this type.

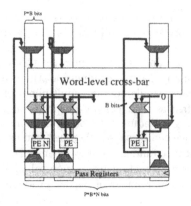

Fig. 1. *Hardware virtualization in PipeRench overlaps computation with reconfiguration and provides the illusion of unlimited hardware resources.*

Hardware, on the other hand, can shift and rotate numbers easily. Variable shifts and rotates can be accomplished with barrel shifters, and constant shifts and rotates do not require any resources at all, as they can be achieved by simply reordering the actual wires.

Reconfigurable hardware can accomplish all of the benefits associated with hardware while providing even more opportunities for optimization. In highly customized designs such as fixed-key implementations, variable shifts and rotations may become fixed, reducing running time and freeing resources.

4   PipeRench

PipeRench is a reconfigurable fabric being developed at CMU. It is an instance of the class of pipelined reconfigurable fabrics [26]. From the point of view of implementing cryptographic algorithms the three most important characteristics of PipeRench are: it supports hardware virtualization, it is optimized to create pipelined datapaths for word-based computations, and it has zero apparent configuration time. Hardware virtualization allows PipeRench to efficiently execute configurations larger than the size of the physical fabric, which relieves the compiler or designer from the onerous task of fitting the configuration into a fixed-size fabric. PipeRench achieves hardware virtualization by structuring the fabric (and configurations) into pipeline stages, or *stripes*. The stripes of an application are time multiplexed onto the physical stripes (see Figure 1). This requires that every physical stripe be identical. It also restricts the computations it can support to those in which the state in any pipeline stage is a function of the current state of that stage and the current state of the previous stage in the pipeline. In other words, the dataflow graph of the computation cannot have long cycles.

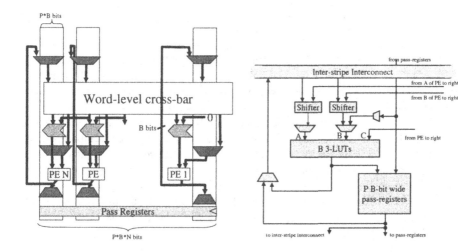

Fig. 2. *The interconnection network between two adjacent stripes. All switching is done at the word level. All thick arrows denote B-bit wide connections.*

Fig. 3. *The structure of a processing element. There are N PEs in each stripe. Details about the zero-detect logic, the fast carry chain and other circuitry are left out.*

Each stripe in PipeRench is composed of N processing elements (PEs). In turn, each PE is composed of B identically configured 3-LUTS, P B-bit pass registers, and some control logic. The three inputs to the LUTS are divided into two data inputs (A and B) and a control input similar to [8]. Each stripe has an associated *inter-stripe interconnect* used to route values to the next stripe and also to route values to other PEs in the same stripe. An additional interconnect, the *pass-register interconnect*, allows the values of all the pass registers to be transferred to the pass registers of the PE in the same column of the next stripe.

The structure of the interconnect is depicted in Figures 2 and 3. Both the inter-stripe interconnect and the pass-register interconnect switch B-bit wide buses, not individual bits. A limited set of bit permutations are supported in the interconnect by barrel shifters, which can left shift any input coming from the inter-stripe interconnect. Currently, the inter-stripe interconnect is implemented as a full crossbar.

From the perspective of cryptographic algorithms, the current version of PipeRench has one significant drawback: It cannot perform large table lookups. Thus S-boxes with more than a few entries cannot be efficiently supported directly in the current version of PipeRench. One proposed extension to PipeRench, PipeRench+, allows the individual stripes to make memory accesses. This would allow PipeRench to efficiently support S-boxes and thus all the operations listed in Section 3. Without the memory extension, algorithms with S-box operations would be best supported by decomposing the algorithm into pieces where the non-S-box portions are implemented as custom instructions and the S-box lookups are performed in the processor core.

The performance numbers we use in this paper are for an implementation in a 0.25 micron process. After an analysis described in [12] we determined that each stripe will have 16 8-bit PEs, yielding a 128-bit wide stripe. Each PE contains 8 pass registers. The final chip will use $100mm^2$ for 28 stripes and an on-chip cache capable of holding more than 512 virtual stripes.[1]

Along with the development of the PipeRench fabric, a fast compilation framework was built [6]. Except where noted, all performance numbers are on simulations of a 28-stripe $N = 16$, $B = 8$, $P = 8$ instance of PipeRench running configurations created automatically by the compiler. For PipeRench+, we consider PipeRench to be augmented by a small scratchpad memory of 1K bytes. We consider two versions, PipeRench+16, which allows up to 16 simultaneous reads, and PipeRench+4, which supports up to 4 simultaneous reads. They increase the total area by 5% to 20%.

5 Applications

In this section we describe how IDEA, Crypton, RC6, and Twofish can be implemented on PipeRench, yielding high performance. We also describe how these algorithms would be aided by PipeRench+. As an example of the flexibility of reconfigurable systems we describe how key-specific instances of IDEA can be easily created on the fly, without a compiler. We then evaluate reconfigurable implementations for the proposed algorithms of AES.

5.1 IDEA

The IDEA block cipher [28] is comprised entirely of three fundamental operations described in Section 3: addition modulo 2^{16}, 16-bit XOR, and 16x16 multiplication modulo $2^{16} + 1$. The 128-bit key is used to generate 52 16-bit subkeys. Throughout the algorithm there are no backwards paths for data. In addition, one operand of every multiplication operation in the algorithm is a subkey, and for a highly-customized implementation it may be treated as a constant.

This means that the algorithm maps exceptionally well onto PipeRench. The forward-only datapath permits the entire application to be constructed as a single, long virtual pipeline. PipeRench is sufficiently wide to receive one complete 64-bit cleartext block and to return one 64-bit ciphertext block per cycle.

Multipliers are the best candidates for optimization in this algorithm. If implemented as general-purpose two-operand shift-and-add multipliers, they require 16 partial products each. The modulo $2^{16} + 1$ operation can be packed into one stripe and computed using only three operations [18].

A simple key-specific implementation can be created if the compiler is given the subkeys as constants. The compiler performs constant propagation reducing

---

[1] First silicon for a prototype of PipeRench implemented in 0.35 micron technology is expected in October 1999. It will have 16 stripes.

Processor	Clock Speed	Clocks per Block	Throughput (MBytes/sec)
PipeRench (template)	100 MHz	6.3	126.6
PipeRench (compiler)	100 MHz	12	66.3
Pentium-II using MMX [21]	450 MHz	358	10.0
Pentium [23]	(scaled) 450 MHz	590	6.1
IDEACrypt Kernel [22]	100 MHz	3	90.0

Table 1. *Comparison of IDEA implementations.*

the number of partial products to an average of 8 per multiplier. Further optimization can be performed by transforming the shift-and-add multiplier into a constant CSD multiplier.

In Table 1 we compare both the template- and compiler-generated IDEA to optimized software implementations running on state-of-the-art processors, and to custom VLSI designs. PipeRench outperforms the processors listed by over 10x.

Somewhat surprisingly, PipeRench outperforms the .25 micron IDEACrypt Kernel from Ascom [22]. This is due to several factors: first, the PipeRench implementation of IDEA does not include the time taken to generate keys. This is because PipeRench targets streaming media applications, in which key generation comprises only a small preprocessing step. Secondly, because of the pipelined nature of PipeRench, IDEA has effectively been pipelined into 177 stages. If a custom silicon implementation were built with such a high degree of pipelining, the circuit would allow a fast clock (at the cost of silicon area.) Lastly, there is a 177-cycle latency through the pipeline. Nonetheless, it is noteworthy that the raw throughput of PipeRench is 40% faster than full-custom silicon.

5.2 IDEA in Embedded Systems

One of the challenges of placing a PipeRench fabric running IDEA in an embedded system is to reduce the time to generate a single-key configuration. While the compiler for PipeRench can compile a complete, single-key optimized IDEA application in less than one minute, this is too long for an embedded system. One method for accomplishing this is the use of precompiled CSD multiplier templates.

An 8-round IDEA pipeline (with the output transformation) contains 34 16x16 bit constant multipliers. The vast majority of the compilation time in generating a single-key IDEA pipeline is spent propagating constants through these multipliers, reducing them to the minimum required number of partial products. This operation is very important since nearly all the efficiency gained by fixing the key is a result of this reduction.

The task of compilation can be separated into two components: optimization and generation of the multipliers, and generation of the rest of the pipeline. If an interface is agreed upon by the multipliers and the rest of the pipeline, the two tasks can be performed independently. We have developed a system for creation of IDEA with template-based multiplication.

The non-multiplier portions of the pipeline were hand-compiled beforehand to give maximum performance. They interact with the template-generated multipliers according to a pre-defined interface, and they do not place any important data in the registers which are used by the multiplier. The multipliers are thus treated as black boxes, and the non-multiplier operations are wrapped tightly around the multipliers to perform all the necessary computations in the minimum possible number of stripes.

To generate the multipliers themselves, a system is required that rapidly returns configuration bits for the stripes that perform the multiplication. Rather than expending a tremendous amount of effort and silicon area on hardware which actually computes these bitstreams, it is preferable to simply construct a lookup table which converts constant multiplicands into the necessary configuration bits.

Such an implementation would consist of nothing more than a ROM preloaded with the appropriate values. To reduce the size of the ROM, each 16-bit constant is broken into two 8-bit constants. The 8-bit constant is used as an index into a table of stripe configurations which implement that portion of the multiplier. The ROM would need approximately 256 120-bit entries. Although there is some overhead when recombining the two portions of the constant, using CSD representations we can still build the entire multiplier in two or three stripes. (Three stripes are required only for certain "bad" CSD vectors, which occur in only 1/16 of the entries. The multiplier interface is maintained whether the multiplier needs two or three stripes.)

Because the design of the template-generated multipliers and the logic placed between them only needs to be done once, great care can be taken to make the design very efficient. A single round of IDEA generated by this system is generally only 20 or 21 stripes long, resulting in a complete IDEA pipeline of only 177 stripes. This is a tremendous improvement over the 338-stripe compiler-generated pipeline. This improvement is primarily due to the compiler's not using registered feedback within a stripe.

5.3 Crypton

The Crypton [20] cipher can be implemented as a complete stream-function on PipeRench, with reasonable speedup. There are, however, operations within the Crypton cipher which are difficult to accomplish on the PipeRench architecture.

Most parts of the cipher map easily onto PipeRench. The byte transposition τ can be implemented entirely in the interconnect of PipeRench and does not require any computational resources at all. The bit permutations, π_o or π_e, can each be completed in four stripes. The key addition, σ_k, takes only one stripe.

The nonlinear S-box substitution, γ, however, is not easy to implement on PipeRench. Each of the three small 4x4 P_x s-boxes can be implemented either as logic or as a look-up table. In either case, because the PEs on PipeRench operate only on 8-bit quantities, 7/8 of each PE is wasted. Each PE generates

Cipher	System	method	Clocks/block	Throughput (Mbytes/sec)	Speedup
RC6 [25]	PipeRench	A	28	58.8	4.7x
Crypton [20]	PipeRench	A	65	24.8	1.3x
	PipeRench+4	A	50	32.5	1.8x
	PipeRench+6	A	19	86.8	4.7x
Twofish [27]	PipeRench+4	C	51	15.6	2.8x
	PipeRench+16	C	15	54.3	9.7x
	PipeRench+4	A	36	36.0	3.9x
	PipeRench+16	A	9.7	164.7	14.6x

Table 2. *Comparison of different PipeRench systems and the speedups they achieve over the best versions in the cited papers. The "method" column indicates how the code was created: 'C' indicates it was automatically generated by the compiler. 'A' indicates a hand-coded version in CVHASM [15].*

only a single bit, which is later combined by other PEs into a 4-bit quantity. When implemented by hand, this process requires three stripes per P_x.

A single round of Crypton uses 16 S-boxes, with three P_x units in each S-box. As a result, a single round of Crypton requires about 150 stripes, causing the entire 12-round cipher to occupy 1800 virtual stripes. PipeRench may have difficulty handling an application of that size due to limitations on the storage space for virtual stripes. The application can be re-pipelined on the inside in order to re-use the S-box in each round on all four 32-bit words. This cuts the length of the virtual pipeline by a factor of four, but it incurs a considerable amount of overhead in re-pipelining, and so reduces the overall throughput of the application.[2] Even with this large number of stripes, the application still gives 24.8 MByte/sec of throughput (with tremendous latency), compared with 18.46 MByte/sec on a 450 MHz Pentium Pro (coded with in-line assembly).

When implemented on PipeRench+, Crypton is significantly smaller and faster. Each of the S-boxes requires only a single stripe—the stripe that contains the load. Thus, the entire round takes only 24 stripes yielding a total of 288 stripes. Due to the limit on memory accesses per cycle this implementation yields 87 MByte/sec on PipeRench+16.

5.4 RC6

RC6 [25] is easy to implement as a stream function, but is only 5x faster than a 200Mhz Pentium-Pro due to the general purpose 32-bit multiplies (2 in each round). Unlike IDEA, neither operand of the multiplier is a constant. However, because the multiplier result is only 32-bits, the size of the multiplier is reduced by half. The variable rotates also require a significant amount of hardware: six stripes to do both rotates in parallel.

---

[2] Another solution is to chain multiple PipeRench chips together making a bigger pipeline. This solution also doubles performance.

5.5 Twofish

The Twofish [27] algorithm, like many others makes substantial use of S-boxes which is unsuitable for PipeRench. However, as described in Section 2.1, pieces of the algorithm can be mapped to custom instructions. For example, the q-function can be mapped to ten stripes on PipeRench. The space-consuming part of the function is again the four table lookups. In spite of this, using PipeRench to compute the S-boxes reduces the time for key setup making "full keying" a viable option even when very few blocks are encrypted with a single key.

However, on PipeRench+, the S-box lookups are easily handled. Each round requires 16 loads (from the g_0 and g_1 functions) and some rotating, XORing, and addition all of which are easy to accomplish on PipeRench+. Both the compiler and hand-coded versions achieve a speedup of about 3x on PipeRench+4 over the fastest assembly version running on a 200Mhz Pentium Pro/II. Interestingly, this is in spite of the fact that the compiler version is twice as large. This is because the compiler version spaces out the loads so there are very few stalls. The extra memory bandwidth of PipeRench+16 allows the hand-coded version to get more than 14x speedup.

5.6 Other AES Algorithms

We now discuss the AES algorithms which cannot be fully implemented on PipeRench. However, certain operations in the algorithms can be implemented as custom instructions, resulting in faster overall performance. In addition, most could be implemented on PipeRench+.

Cast-256 [1], like many of the proposed AES algorithms is heavily based on S-boxes which are not amenable to the current version of PipeRench. However, for each of the three keyed-round operations there are key-specific left rotations which can be optimized to constant rotations. Thus custom instructions can be created based on the subkeys which reduce the time to compute the S-box input to one cycle.

Performance of DFC [33] would improve by implementing the necessary multiprecision arithmetic as custom instructions.

The Hasting Pudding Cipher [29] gains significant performance as a series of custom instructions. The basic operations are all easily performed on PipeRench, but the entire cipher cannot be implemented as a stream function due to the large tables needed to hold the key expansion.

Both the key schedule and encryption/decryption in LOKI97 [5] can be sped up with a custom instruction which implements all but the S-box of the g-function and the data routing.

Deal [17], E2 [31], FROG [11], MAGENTA [3] and MARS [7] appear to be unsuitable for implementation on PipeRench due to the use of large table lookups in both key formation and encryption/decryption.

When encrypting or decrypting using Rijndael [10] each round consists of 4 basic functions of which three can be implemented as custom instructions.

SAFER+ [9], like HPC, DFC, Cast-256, Rijndael, and LOKI97, can benefit from using custom instructions, although the entire cipher cannot be implemented on PipeRench due to the large M matrix.

Serpent [2] uses S-boxes for the entire process except for the initial and final permutations which could be custom instructions.

6 Related Work

There is a growing body of work on using reconfigurable devices to implement cryptographic algorithms. Reconfigurable implementations of DES [32, 14] and RSA [30] have all achieved significant speedups over general-purpose processors. However, in none of these cases were key-specific hardware implementations generated. The impact on the hardware size and throughput of key-specific implementations of DES using Xilinx FPGAs is discussed in [19].

In [16] FPGA-based implementations of DES are described that make use of many of the helpful attributes of reconfigurable devices which are used on PipeRench, including loop unrolling (which PipeRench requires) and pipelining (which PipeRench does implicitly). The acceleratation of modular multiplication and exponentiation (as used in RSA) using arithmetic architectures which have been optimized for use on FPGAs is described in [4].

More generally, PipeRench is one of several approaches towards making reconfigurable hardware more applicable to computation of the sort needed by cryptographic applications. PRISC [24] is among the earliest work on integrating a reconfigurable function unit with a processor. GARP [14], Chimaera [13], and One-Chip [34] are more recent examples of such work. The main difference between these systems and PipeRench is that PipeRench supports virtual hardware, freeing the application designer from fixed hardware constraints.

7 Conclusions

The use of reconfigurable hardware in cryptographic systems has many advantages. Reconfigurable implementations benefit from the hardware-based performance of custom VLSI while maintaining the flexibility and adaptability of software. Unlike fixed hardware, reconfigurable devices deliver highly customized, efficient solutions that are adaptable and robust to changing system needs.

The PipeRench reconfigurable architecture is well suited to many cryptographic tasks. Because it supports hardware virtualization, it can implement designs which are larger than the amount of physical hardware available. This is important, as many of the algorithms which can be mapped entirely to reconfigurable devices require tremendous amounts of physical hardware. Some algorithms which cannot be mapped completely onto PipeRench can still be accelerated by building custom instructions. PipeRench has two drawbacks; both relating primarily to table lookups. For small tables, such as the P tables in Crypton, the 8-bit PEs are very inefficient. For larger tables, such as S-boxes found in many of the algorithms, PipeRench does not have any facility for performing memory accesses. We explore an extension to PipeRench, PipeRench+, which overcomes this second drawback.

Finally, key-specific circuits for PipeRench can be generated in embedded systems: with the use of a simple table lookup operation, the full performance of PipeRench can be obtained without waiting for software-based compilation. In the case of IDEA, we are able to exceed the performance of custom hardware.

Acknowledgements

This work was supported by DARPA contract DABT63-96-C-0083. We would like to thank the PipeRench group for tools and ideas, especially Mihai Budiu and Hari Cadambi for their help with the compiler.

References

1. C. Adams. The CAST-256 encryption algorithm.
   www.entrust.com/resources/pdf/cast-256.pdf.
2. R. Anderson, E. Biham, and L. Knudsen.
   SERPENT. www.cl.cam.ac.uk/~rja14/serpent.html.
3. E. Biham, A. Biryukov, N. Ferguson, L. Knudsen, B. Schneier, and A. Shamir. Cryptanalysis of MAGENTA. www.ii.uib.no/~larsr/papers/magenta.pdf.
4. T. Blum and C. Paar. Montgomery modular exponentiation on reconfigurable hardware. In *Proceedings of the 14th IEEE Symposium on Computer Arithmetic (ARITH-14)*, Adelaide, Australia, April 1999.
5. L. Brown and J. Pieprzyk. Introducing the new LOKI97 Block Cipher. www.adfa.oz.au/~lpb/research/loki97/.
6. M. Budiu and S.C. Goldstein. Fast compilation for pipelined reconfigurable fabrics. In *Proceedings of the 1999 ACM/SIGDA Seventh International Symposium on Field Programmable Gate Arrays*, Feb. 1999.
7. Burwick, Coppersmith, D'Avignon, Gennaro, Halevi, Jutla, Matyas Jr., O'Connor, Peyravian, Safford, and Zunic. MARS - a candidate cipher for AES. www.research.ibm.com/security/mars.html.
8. D. Cherepacha and D. Lewis. A datapath oriented architecture for FPGAs. In *Second Int'l ACM/SIGDA Workshop on Field Programmable Gate Arrays*, 1994.
9. Cylink Corporation. SAFER+. www.cylink.com/SAFER.
10. J. Daemen and V. Rijmen. AES Proposal: Rijndael. www.esat.kuleuven.ac.be/~rijmen/rijndael/.
11. D. Georgoudis, D. Leroux, and B.S. Chaves. The "FROG" encryption algorithm. www.tecapro.com/aesfrog.htm.
12. S.C. Goldstein, H. Schmit, M. Moe, M. Budiu, S. Cadambi, R.R. Taylor, and R. Laufer. Piperench: A coprocessor for streaming multimedia acceleration. In *Proceedings of the 26th Annual International Symposium on Computer Architecture*, pages 28–39, May 1999.
13. S. Hauck, T. W. Fry, M. M. Hosler, and J. P. Kao. The Chimaera reconfigurable functional unit. In *IEEE Symposium on FPGAs for Custom Computing Machines (FCCM '97)*, pages 87–96, April 1997.

14. J.R. Hauser and J. Wawrzynek. Garp: A MIPS processor with a reconfigurable coprocessor. In *IEEE Symposium on FPGAs for Custom Computing Machines*, pages 24–33, April 1997.
15. Kevin Jaget. Cvhasm: An assembler for pipelined reconfigurable architectures. Master's thesis, Carnegie Mellon University, 1998.
16. J.-P. Kaps and C. Paar. Fast DES implementation for FPGAs and its application to a universal key-search machine. In *Selected Areas in Cryptography '98*, volume 1556 of *Lecture Notes in Computer Science*, Kingston, Ontario, Canada, August 1998. Springer-Verlag.
17. L. Knudsen. DEAL: A 128-bit block cipher. Technical Report 151, Department of Informatics,University of Bergen, Norway, Feb 1998.
18. X. Lai and J.L. Massey. A proposal for a new block encryption standard. In *Advances in Cryptology Eurocrypt '90*, pages 389–404, 1991.
19. J. Leonard and W. H. Mangione-Smith. A case study of partially evaluated hardware circuits: Key-specific DES. In *Field-programmable Logic and Applications: 7th International Workshop (FPL'97)*, London, UK, September 1997.
20. C. H. Lim. "CRYPTON". crypt.future.co.kr/~chlim/crypton.html.
21. H. Lipmaa. IDEA: A cipher for multimedia architectures? In *Selected Areas in Cryptography '98*, volume 1556 of *Lecture Notes in Computer Science*, pages 248–263, Kingston, Ontario, Canada, August 1998. Springer-Verlag.
22. Ascom Systec Ltd. IDEACrypt Kernel. www.ascom.ch/infosec/idea/kernel.html.
23. B. Preneel, V. Rijmen, and A. Bosselaers. Recent developments in the design of conventional cryptographic algorithms. In *Computer Security and Industrial Cryptography*, volume 1528 of *Lecture Notes in Computer Science*, pages 106–131. Springer-Verlag, 1998.
24. R. Razdan and M.D. Smith. A high-performance microarchitecture with hardware-programmable functional units. In *MICRO-27*, pages 172–180, November 1994.
25. R. Rivest, M.J.B. Robshaw, R. Sidney, and Y.L. Yin. The RC6 Block Cipher. theory.lcs.mit.edu/~rivest/rc6.ps.
26. Herman Schmit. Incremental reconfiguration for pipelined applications. In J. Arnold and K. L. Pocek, editors, *Proceedings of IEEE Workshop on FPGAs for Custom Computing Machines*, pages 47–55, Napa, CA, April 1997.
27. Schneier, Kelsey, Whiting, Wagner, Hall, and Ferguson. Twofish: A 128-bit block cipher. www.counterpane.com/twofish.html.
28. Bruce Schneier. *Applied cryptography: Protocols, algorithms, and source code in C*, chapter 13, pages 319–325. John Wiley and Sons, Inc., 1996.
29. R. Schroppel. The Hasty Pudding Cipher. www.cs.arizona.edu/~rcs/hpc.
30. M. Shand and J. Vuillemin. Fast implementations of RSA cryptography. In *11th IEEE Symposium on COMPUTER ARITHMETIC*, 1993.
31. Nippon Telegraph and Telephone Corporation. The 128-bit block cipher E2. info.isl.ntt.co.jp/e2/.
32. K. W. Tse, T. I. Yuk, and S. S. Chan. Implementation of the data encryption standard algorithm with FPGAs. In W. Moore and W. Luk, editors, *More FPGAs: Proceedings of the 1993 International workshop on field-programmable logic and applications*, pages 412–419, Oxford, England, September 1993.
33. S. Vaudenay. DFC. www.dmi.ens.fr/~vaudenay/dfc.html.
34. R. Wittig and P. Chow. OneChip: An FPGA processor with reconfigurable logic. In *IEEE Symposium on FPGAs for Custom Computing Machines*, 1996.

CryptoBooster:
A Reconfigurable and Modular Cryptographic Coprocessor

Emeka Mosanya[1], Christof Teuscher[1], Héctor Fabio Restrepo[1], Patrick
Galley[2], and Eduardo Sanchez[1]

[1] Logic Systems Laboratory, Swiss Federal Institute of Technology
CH-1015 Lausanne, Switzerland.
http://lslwww.epfl.ch
E-mail: {name.surname}@epfl.ch,

[2] LIGHTNING Instrumentation SA
Av. des Boveresses 50, CH-1010 Lausanne, Switzerland.
http://www.lightning.ch,
E-mail: Patrick.Galley@lightning.ch

Abstract The CryptoBooster is a modular and reconfigurable cryptographic coprocessor that takes full advantage of current high-performance
reconfigurable circuits (FPGAs) and their partial reconfigurability. The
CryptoBooster works as a coprocessor with a host system in order to
accelerate cryptographic operations. A series of cryptographic modules
for different encryption algorithms are planned. The first module we implemented is IDEACore, an encryption core for the International Data
Encryption Algorithm ($IDEA^{TM}$).

Keywords: **Cryptography, Coprocessor, Reconfiguration, FPGA, IDEA.**

1 Introduction

In this paper we describe a novel cryptographic coprocessor, the CryptoBooster, optimized for reconfigurable computing devices (e.g., Field Programmable Gate Arrays, or
FPGAs). Our implementation is modular, scalable, and helps to resolve the trade-off
between device size and data throughput. The implementation is designed to support
a large number of different encryption algorithms and includes appropriate session
management.

The first CryptoBooster implementation we propose implements $IDEA^{TM1}$, a
symmetric-key block cipher algorithm [7]. IDEACore is the first of a series of cryptographic modules for the CryptoBooster coprocessor. A simple reconfiguration of the
reconfigurable computing device will suffice to replace IDEACore by another block cipher module (e.g., DES). The other modules of the CryptoBooster generally remain
unchanged.

---

[1] IDEA is patented in Europe and the United States [11, 12]. The patent is held by
Ascom Systec Ltd., http://www.ascom.ch/systec.

In section 2 we describe the CryptoBooster architecture. Section 3 gives a short introduction to the $IDEA^{TM}$ encryption algorithm and reviews existing implementations of this algorithm in hardware. Section 4 describes the implementation of IDEACore, the first cryptographic module for the CryptoBooster. We conclude in section 5 the paper with the synthesis and performance results of our implementation.

2 CryptoBooster

CryptoBooster is a modular coprocessor dedicated to cryptography. It is designed to be implemented in Field Programmable Gate Arrays (FPGAs) and to take advantage of their partial reconfiguration features. The CryptoBooster works as a coprocessor with a host system in order to accelerate cryptographic operations. It is connected to a session memory responsible for storing session information (Figure 1). A session is characterized by a set of parameters describing the cryptographic packets, the algorithm used, the key(s), the initial vector(s) for block chaining, and other information.

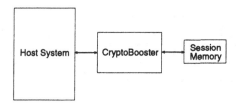

Figure 1. The CryptoBooster works as a coprocessor together with a host system. Typically, the host system is a PC. The CryptoBooster needs additional memory to store session information.

Our design is motivated by the following objectives: (1) The main goal is to have maximum data throughput to provide a design able to cope with ever-increasing network speed. This justifies hardware implementation in place of a software solution. Physical security may, however, also be an argument for hardware implementation. (2) Our requirements include the ability to easily configure different algorithm sub-blocks and the associated subkey generation. We clearly need a highly modular architecture, allowing us to easily change building blocks. The modularity has been pushed far enough to allow partial reconfiguration of the coprocessor. Partial reconfiguration allows to offer several algorithms for one and the same physical chip with limited resources.

2.1 Modular Architecture

A block diagram of the CryptoBooster architecture is shown in Figure 2. The InterfaceAdapter module is a technology-dependent interface to the host system. Typically, this is a PCI or a VME interface, but one may also imagine a networking interface like Ethernet. The HostInterface is the software interface to the host system. It offers read/write registers and interruptions to configure and control the coprocessor. The SessionMem module allows to interface different types and configurations of physical memories. A separate session memory has been chosen mainly to limit the communication between host and coprocessor and in order to change rapidly between different sessions.

The CryptoCore module itself is subdivided into three parts:

- *CypherCore*: encryption algorithm,
- *SessionAdapter*: session parameter management (specific to each CypherCore),
- *SessionControl*: central controller for session management.

The CypherCore and SessionAdapter modules are intelligent modules and can be queried by the SessionControl module. They respond with the implemented features available. It is therefore possible to exchange these modules without changing the control mechanism. All these modules communicate together and with the modules outside the CryptoCore using unidirectional point-to-point links called *CoreLink*. These links are designed to transmit control or data packets. This homogeneity at the interconnection level strongly enforces the modularity of the system.

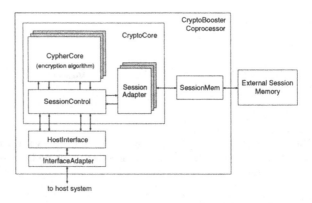

Figure 2. Block diagram of the CryptoBooster architecture. The architecture is highly modular and uses standardized interconnections.

2.2 Advantages of an FPGA-based Implementation and Reconfiguration Features

An FPGA circuit is an array of logic cells placed in an infrastructure of interconnections [14]. Each cell is a universal function or a functionally complete logic device, which can be programmed to realize a certain function. Interconnections between the cells are also programmable. The versatility allowed by logic blocks and the flexibility of the interconnections provide high freedom of design during the utilization of FPGAs.

The CryptoBooster is implemented using the VHDL hardware description language. The design can thus be synthesized without major problems for FPGAs as well as for VLSI technology. A VLSI solution results in general in higher performance than an FPGA implementation but the latter has several important advantages: (1) Reconfigurability of the FPGA allows the developer to easily provide specific solutions to the customer as it is often needed in cryptography; (2) A VLSI multi-algorithm co-processor requires all corresponding CypherCores to be implemented in the chip which demands a huge amount of transistors; (3) On the other hand, an FPGA only contains one encryption algorithm at a given time. Other algorithms are available in the form of a configuration bitstream. Thus the maximum area required corresponds to the area used by the largest algorithm.

One can distinguish between full and partial FPGA reconfiguration. Full reconfiguration is the common method currently used. The configuration is replaced by a new one each time the algorithm is changed (Figure 3). Partial reconfiguration allows to reconfigure parts of the FPGA, i.e., only the part where the algorithm is implemented on the FPGA has to be reconfigured. This normally results in a much shorter interrupt of service compared to full reconfiguration. The CryptoBooster is designed to take full advantage of partial reconfiguration.

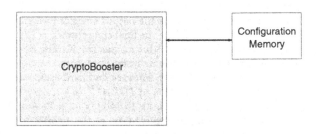

Figure 3. Full reconfiguration requires to reprogram all the chip including common parts.

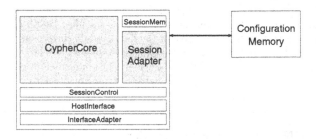

Figure 4. Partial reconfiguration requires only to reprogram the parts specific to a particular algorithm.

3 The $IDEA^{TM}$ Block Encryption Algorithm

The International Data Encryption Algorithm ($IDEA^{TM}$) was developed by Xuejia Lai and James Massey of the Swiss Federal Institute of Technology in Zurich. The original version—called PES (Proposed Encryption Standard)—was first published in 1990 [8]. The following year, after Biham and Shamir's demonstrated differential cryptanalysis, the authors strengthened their cipher against the attack and called the new algorithm IPES (Improved Proposed Encryption Standard) [9]. IPES changed its name to IDEA in 1992 [7].

$IDEA^{TM}$ is one of a number of conventional encryption algorithms that have been proposed in recent years to replace DES. However, there has been no rush to adopt it as a replacement to DES, partly because it is patented and must be licensed for commercial applications, and partly because people are still waiting to see how well the algorithm fares during the upcoming years of cryptanalysis.

$IDEA^{TM}$ is a 64-bit block cipher that uses a 128-bit key to encrypt data (DES also uses 64-bit blocks but only a 56-bit key). The same algorithm is used for encryption and decryption. It consists of 8 computationally identical rounds followed by an output transformation. Round r uses six 16-bit subkeys $K_i^{(r)}$, $1 \leq i \leq 6$, to transform a 64-bit input X $= (X_1, X_2, X_3, X_4)$ into an output of four 16-bit blocks, which are input to the next round. The round 8 output enters the output transformation, employing four additional subkeys $K_i^{(9)}$, $1 \leq i \leq 4$ to produce the final ciphertext Y $= (Y_1, Y_2, Y_3, Y_4)$. All 52 16-bit subkeys are derived from the 128-bit master key K. The key is long enough to withstand from exhaustive key searches well into the future.

$IDEA^{TM}$ is easy to implement in both software and hardware, even in embedded systems. A typical software implementation of $IDEA^{TM}$ in C (using the Ascom Systec source code) performs data encryption at 16 Mbit/s on a PentiumPro 180 MHz machine. It is clear that a hardware implementation of the algorithm may essentially speed up the throughput. To the best of our knowledge, the first VLSI implementation was developed at the Swiss Federal Institute of Technology in Zurich by Bonnenberg et al. [1] and reached a throughput of 44 Mbit/s at 25 MHz.

Current hardware implementations have stressed the importance of the combinatorial delay and area consumption of the multiplication modulo $(2^{16} + 1)$ units which are crucial to the entire system. These units are the limiting factor to obtain high data throughput. Various methods of implementing such a multiplication are investigated in [3, 5, 10, 15, 16]. The VINCI implementation uses a modified Booth recording multiplication and fast carry select additions for the final modulo correction [17]. In general, for larger words, ROM-based solutions using lookup-tables require large ROMs. In a recent paper, Zimmermann [16] presents an efficient VLSI implementation of the modulo $(2^{16} + 1)$ multiplication. Several different hardware implementations of the $IDEA^{TM}$ algorithm are given below.

3.1 The VINCI VLSI–Implementation

A new VLSI implementation—called VINCI [4, 17]—was developed in 1993 at ETH Zurich. The chip consists of 250,000 transistors on a total area of $107.8 mm^2$ and attained 177.8 Mbit/s @ 25 MHz. The data path was optimized using an eight-stage pipeline and full-custom modulo $(2^{16} + 1)$ multipliers. The VINCI chip was the first chip that could be used for on-line encryption in high-speed networking like ATM or FDDI. Figure 5 shows the pipeline of the VINCI data path for one round. The multiplication modulo $(2^{16} + 1)$ operation is distributed in two pipeline stages to reduce the critical path.

Note that the output round is identical to the first section (3 stages) of a regular round as shown in Figure 5.

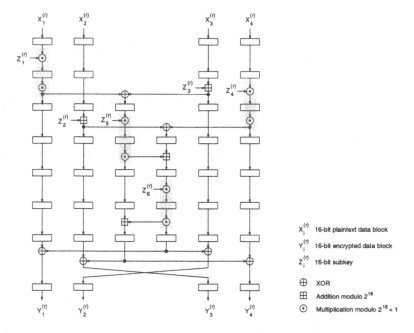

Figure 5. Pipeline of the VINCI [17] data path for one round.

3.2 Ascom IDEACrypt Coprocessor

Ascom Systec Ltd., holder of the $IDEA^{TM}$ patents, offers a high-speed implementation of the $IDEA^{TM}$ algorithm as an embedded ASIC core. The IDEACrypt kernel implements data encryption and decryption in all common operating modes for block ciphers (ECB, CBC, CFB, OFB) [6]. IDEACrypt provides flexible key management for both master-session key and asymmetric key and stores the keys in a RAM which may be either on-chip or off-chip. Therefore, with a RAM, a bus interface, and a global controller, a complete $IDEA^{TM}$ cipher may be implemented. The entire IDEACrypt coprocessor is implemented in synthesizable VHDL code. Ascom Systec lists a complexity of approximatively 35k gates.

With 0.25 micron technology using a 3-stage pipeline, the chip provides a throughput of 300 Mbit/sec (@ 40 MHz) in ECB mode and a throughput of 100 Mbit/sec in the other modes. At 100 MHz, the throughput goes up to 720 Mbit/sec in ECB mode.

3.3 Existing FPGA-Implementations

Caspi and Weaver [2] proposed IDEA as a benchmark for reconfigurable computing. Their implementation on Xilinx XC4005 achieves a throughput of 0.477 Mbit/s. A high space-conserving design was used because of the limited resources of this FPGA.

Mencer et al. [13] compared different implementations of the $IDEA^{TM}$ algorithm. They compared a Digital Signal Processor (DSP) from Texas Instruments to Xilinx XC4000 series FPGAs and pointed out the benefits and limitations of FPGA and DSP technologies for $IDEA^{TM}$. The FPGA implementation has a throughput rate of 528

Mbit/s @ 33 MHz using a fully pipelined version of $IDEA^{TM}$ distributed on four XC4000XL FPGAs. Taking into account the powerful programming capabilities of the FPGA technology and the performance of the new families (e.g., Xilinx Virtex), this kind of circuit is becoming an excellent option to implement $IDEA^{TM}$.

4 The First CypherCore: IDEACore Encryption Module

The first CypherCore module—called IDEACore—implements the $IDEA^{TM}$ algorithm. It is composed of a scalable pipeline and an associated block chaining module (Figure 6).

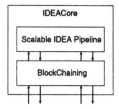

Figure 6. The IDEACore module is composed of a scalable pipeline and a block-chaining module.

4.1 A Scalable $IDEA^{TM}$ Pipeline

We adopted a highly scalable solution for our $IDEA^{TM}$ pipeline: the length of the pipeline can be chosen at compilation time. The regular round is inspired by the VINCI datapath (see Figure 5). Figure 7 shows the regular round consisting of seven pipeline stages. The first three stages of a regular round simply form the output round (Fig. 8).

As Figure 9 shows, the minimal pipeline length is one regular round followed by one output round. Data has to be fed eight times through the regular round before passing through the output round. The longest pipeline (full-length pipeline) consists of eight rounds and one output round. In each configuration, the data needs 59 clock cycles to pass through the pipeline. A longer pipeline has a smaller latency and thus a higher throughput.

We use a fully self-controlled pipeline with a control pipeline associated in parallel to the data pipeline. The control pipeline addresses the key memories attached to each stage that needs an encryption key. Every 64-bit data block has an associated counter that indicates the current round. Data is automatically feed to the output stage if it was sent the correct number of times through the regular rounds or it is fed back through the block of regular rounds. Pipeline bubbles (data marked by a non-valid bit) are automatically inserted into the pipeline if the module preceding the pipeline (block chaining) is not able to deliver new data packets. This mechanism allows us to avoid pipeline stalls.

Multipliers and Modulo ($2^{16} + 1$) We currently use simple bit-parallel multipliers optimized for FPGAs and the low-high algorithm [7] for the modulo ($2^{16} + 1$) calculation. As stated in section 3, the combinatorial delay and area consumption of the multiplication modulo ($2^{16} + 1$) units are crucial to the entire design and are limit the data path.

Bit-parallel multipliers are perhaps not the best choice, but we were surprised by the performance they achieved and the area they used in our FPGA-based implementation. In the near future, we intend to optimize the multiplication modulo ($2^{16} + 1$) units to achieve yet higher throughput.

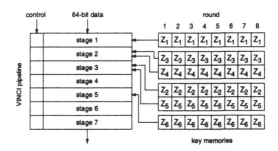

Figure 7. One round with associated encryption key memories of the CryptoBooster pipeline.

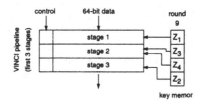

Figure 8. Output round with associated encryption key memories of the CryptoBooster pipeline.

4.2 Block-Chaining

The block-chaining module implements the commonly used block-chaining algorithms like ECB (Electronic Codebook Mode), CBC (Cipher Block Chaining Mode), CFB (Cipher Feedback Mode), and OFB (Output Feedback Mode). To prevent the pipeline from stalling, the block-chaining module always disposes of enough initial vectors (the number depends on the number of regular rounds used in the pipeline).

As with all other modules in the CryptoBooster architecture, the block-chaining module is connected to the other modules by CoreLink unidirectional point-to-point interconnections.

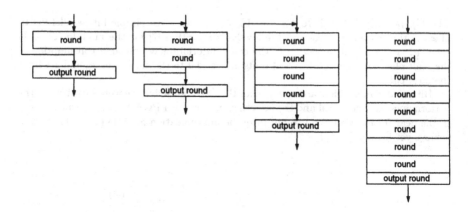

Figure 9. The four different pipeline configurations: 1+1, 2+1, 4+1, and 8+1 rounds. Data has to be feed 8, 4, 2, and 1 time through the regular rounds.

4.3 Performance of the IDEACore CryptoCore

The CryptoBooster is designed to achieve maximum throughput for a given area in the FPGA. Our current implementation allows pipeline lengths of 1, 2, 4, or 8 regular rounds, followed by one output round. A full-length pipeline consists of 59 (8 regular rounds + 1 output round) stages with a latency of 1 clock cycle when using bit-parallel multipliers (Figure 9).

The peak performance of our current implementation is estimated at 200 Mbit/s for a 1-round pipeline (1 regular round + 1 output round) and it easily fits into a state-of-the-art FPGA. Performance for a full-length pipeline (8 regular rounds + 1 output round) is estimated at more than 1500 Mbit/s. However, the area needed in terms of reconfigurable logic blocks in the FPGA is quite important.

Session initialization and key calculation slightly decrease the overall performance over a complete session. Depending on the block-chaining mode used, performance may, however, significantly decrease.

5 Conclusions

The CryptoBooster is a modular and reconfigurable cryptographic coprocessor taking full advantage of current high-performance reconfigurable circuits (FPGAs). Reconfigurable circuits can be reconfigured within a few milliseconds and they provide speed rates close to ASIC designs. Our main goal is to have maximum data throughput so as to provide a design able to cope with ever-increasing network speed. This justifies hardware implementation in place of a software solution. Physical security may, however, also be an argument for hardware implementation.

As our results show, the throughput of the CryptoBooster allows the coprocessor to be used in today's high-speed networks like ATM, Sonet, and GigaEthernet; moreover, it is competitive with full-custom circuits or DSP implementations. $IDEA^{TM}$ was chosen as a first CryptoCore module. More modules with different algorithms (e.g., DES) are planned.

Acknowledgments

The CryptoBooster project is a joint project between the Swiss Federal Institute of Technology in Lausanne and Lightning Instrumentation SA, with funding from the Swiss Federal Office for Education and Technology. The authors are grateful to Moshe Sipper for his careful reading of this work and his helpful comments.

References

1. H Bonnenberg, A. Curiger, N. Felber, H. Kaeslin, and X. Lai. VLSI implementation of a new block cipher. In *Proceeding of the International Conference on Computer Design: VLSI in Computer and Processors*, pages 510–513, Washington, 1991. IEEE, IEEE Computer Society Press.
2. E. Caspi and N. Weaver. IDEA as a benchmark for reconfigurable computing. Technical report, BRASS Research Group, University of Berkeley, December 1996. Report available form http://www.cs.berkeley.edu/projects/brass/projects.html.
3. A. Curiger, H. Bonnenberg, and H. Kaeslin. Regular VLSI-architectures for multiplication modulo $(2^n + 1)$. *IEEE Journal of Solid-State Circuits*, 26(7):990–994, July 1991.
4. A. Curiger, H. Bonnenberg, R. Zimmermann, N. Felber, H. Kaeslin, and W. Fichtner. VINCI: VLSI implementation of the new block cipher IDEA. In *Proceedings of the IEEE CICC'93*, pages 15.5.1–15.5.4, San Diego, CA, May 1993. IEEE.
5. A. Hiasat. New memoryless modulo $(2^n + 1)$ residu multiplier. *Electronic Letters*, 41(3):314–315, January 1992.
6. ISO. Information technology-security techniques-modes of operation for an n-bit block cipher. International standard ISO/IEC 10116:1997(E), ISO/IEC, 15 April 1997.
7. X. Lai. *On the Design and Security of Block Ciphers*. Number 1 in ETH Series in Information Processing. Hartung-Gorre Verlag Konstanz, 1992.
8. X. Lai and J. Massey. A proposal for a new block encryption standard. In *Advances in Cryptology–EUROCRYPT'90*, pages 389–404, Berlin, 1990. Springer-Verlag.
9. X. Lai, J. Massey, and S. Murphy. Markov ciphers and differential cryptanalysis. In *Advances in Cryptology–EUROCRYPT '91*, pages 8–13, Berlin, 1991. Springer-Verlag.
10. Y. Ma. A simplified architecture for modulo $(2^n + 1)$ multiplication. *IEEE Transactions on Computers*, 47(3):333–337, March 1998.
11. J. L. Massey and X. Lai. Device for convertig a digital block and the use thereof, 28 Nov 1991. International Patent PCT/CH91/00117.
12. J. L. Massey and X. Lai. Device for the conversion of a digital block and use of same, 25 May 1993. U.S. Patent #5,214,703.
13. O. Mencer, M. Morf, and M. J. Flynn. Hardware software tri-design of encryption for mobile communication units. In *Proceedings of International Conference on Acoustics, Speech, and Signal Processing*, Seattle, Washington, USA, May 1998.
14. S. Trimberger. *Field-Programmable Gate Array Technology*. Kluwer Academic Publishers, 1994.
15. Z. Wang, G. A. Jullien, and W. C. Miller. An efficient tree architecture for modulo $(2^n + 1)$ multiplication. *Journal of VLSI Signal Processing Systems*, 14(3):241–248, December 1996.

16. R. Zimmermann. Efficient VLSI implementation of modulo $(2^n + 1)$ addition and multiplication. In *Proceedings IEEE Symposium on Computer Arithmetic*, Adeleide, Australia, April 1999. IEEE.

17. R. Zimmermann, A. Curiger, H. Bonnenberg, H. Kaeslin, N. Felber, and W. Fichtner. A 177 Mbit/s VLSI implementation of the international data encryption algorithm. *IEEE Journal of Solid-State Circuits*, 29(3):303–307, March 1994.

Elliptic Curve Scalar Multiplier Design Using FPGAs

Lijun Gao, Sarvesh Shrivastava and Gerald E. Sobelman

Dept. of Electrical & Computer Engineering, University of Minnesota, Twin-Cities
200 Union Street S.E., Minneappolis, MN 55455, U.S.A
Email: {lgao,sobelman}@ece.umn.edu
URL: http://www.ece.umn.edu/users/lgao

Abstract. A compact fast elliptic curve scalar multiplier with variable key size is implemented as a coprocessor with a Xilinx FPGA. This implementation utilizes the internal SRAM/registers of the FPGA and has the whole scalar multiplier implemented within a single FPGA chip. The compact design helps reduce the overhead and limitations associated with data transfer between FPGA and host, and thus leads to high performance. The experimental data from the mappings over small fields shows that the carefully constructed hardware architecture is regular and has high CLB utilization.

Keywords. elliptic curves, public-key cryptography, scalar multiplication, Galois field, reconfigurable hardware, FPGA, coprocessor

1 Introduction

The motivation of this work is to develop high-speed elliptic curve scalar multipliers with the least development time, the lowest hardware cost and maximal flexibility. Recently, elliptic curve (EC) cryptosystems have become attractive due to their small key sizes and varieties of choices of the curves available. Their low cost and compact size are critical to some applications, including smart cards and hand-held devices[1]. In all those applications, an elliptic curve scalar multiplier serves as a basic building block for secret key exchange, authentication and certification. Most microprocessors have hardware-supported integer multiplication and logic functions like AND, OR or XOR, so an elliptic curve cryptosystem can be implemented on them. However, this is not efficient because of word size mismatch, less parallel computation, no hardware supported wire permutation and algorithm/architecture mismatch. As a result, such systems have low performance/cost ratios.

The solution to this problem is to build a coprocessor as a dedicated computing unit. Moreover, using an FPGA, the coprocessor can be reconfigured for different application instances or for different computation stages of one particular application. Thus, the total hardware utilization can be kept at a very high rate and the computation is speeded up.

Previous work in this area is based on a coprocessor for arithmetic operations over $GF(2^{155})$ using a gate array[11]. To accomplish an elliptic curve operation, the host controller and the coprocessor have to transfer data between each other frequently. The control of elliptic curve operations and the storage of intermediate variables are provided by the host controller. Therefore, the communication cost is large and may be a bottleneck for an elliptic curve cryptosystem.

In this paper, a compact fast elliptic curve scalar multiplier coprocessor is introduced which utilizes the internal SRAM/registers in an FPGA and is implemented within a single FPGA chip. The normal bases for the underlying finite field are chosen because the field squarings can be done with the bit shifts in hardware and are virtually free[2]. A pipelined digit-serial modified Massey-Omura multiplier is constructed and is used in the design. The scalar multiplier is implemented with a parameterized (in term of key size) VHDL description and is synthesized/mapped to a Xilinx FPGA. By changing the parameter for key size and re-doing synthesis, a different instance can be acquired. The architecture and algorithms adopted here are suitable for massively parallel computation. Therefore, with larger capacity FPGA chips, higher performance can be easily obtained with few changes in the underlying design.

2 Algorithm of EC Scalar Multiplier

The basic operation in an EC cryptosystem is the scalar multiplication over the elliptic curve and the most efficient method for computing EC scalar multiplications is to use an addition/subtraction method[2][4][5]. With this method, the scalar (or the integer) is decomposed in a non-adjacent format(NAF) and one scalar multiplication is done with a series of additions/subtractions of elliptic curve points. In turn, each addition/subtraction of EC points consists of a series of underlying field additions, squarings, multiplications and inversions. When the elliptic curve is defined over $GF(2^m)$ with an optimal normal basis, these underlying field operations have the least complexity. The elliptic curve used in this implementation is defined by Weierstrass equations as:

$$y^2 + xy = x^3 + ax^2 + b \tag{1}$$

where $a, b \in GF(2^m)$ and $b \neq 0$. The algorithms of EC scalar multiplication and EC addition/subtraction are shown below respectively[2].

Algorithm 1: EC scalar multiplication
Input:

      P−EC point
      $n−(e_{l-1}, e_{l-2}, ..., e_1, e_0)$ non-adjacent format integer and $e_{l-1}=1$

Output:

      $Q = nP$

Computation:

      1. $Q = P$;
      2. For $i = l - 2$ downto 0 do

Set $Q = 2Q$;
If $e_i = 1$ then set $Q = Q + P$;
If $e_i = -1$ then set $Q = Q - P$;
3. Output Q;

Algorithm 2: EC addition
Input:

$$P_0 = (x_0, y_0)$$
$$P_1 = (x_1, y_1)$$

Output:

$$(x_2, y_2) = P_2 = P_1 + P_0$$

Computation:

1. If $P_0 = 0$, then output $P_2 = P_1$ and stop;
2. If $P_1 = 0$, then output $P_2 = P_0$ and stop;
3. If $x_0 = x_1$ then

   If $y_0 = y_1$ then

   $$\lambda = x_1 + y_1/x_1;$$
   $$x_2 = \lambda^2 + \lambda + a;$$
   $$y_2 = x_1^2 + (\lambda + 1)x_2;$$

   else output O;

   else

   $$\lambda = (y_0 + y_1)/(x_0 + x_1);$$
   $$x_2 = \lambda^2 + \lambda + x_0 + x_1 + a;$$
   $$y_2 = (x_1 + x_2)\lambda + x_2 + y_1;$$

4. Output(x_2, y_2);

Since $-P_0 = (x_0, x_0 + y_0)$ for $P_0 = (x_0, y_0)$ and $P_1 - P_0 = P_1 + (-P_0)$, EC subtraction is as simple as EC addition and can be computed with one EC addition. The average and maximal number of non-zero bits among NAFs are about $m/3$ and $m/2$, respectively[2]. Therefore the average cost of Alg.1 is about m point doubles and $m/3$ point additions/subtractions[2], and the worst case cost is about m point doubles and $m/2$ point additions/subtractions. This is much better than binary decomposition, which has $m/2$ non-zero bits on average and m non-zero bits in the worst case.

3 Hardware Architecture

3.1 System structure

Since the word size (or key size) for a typical elliptic curve cryptosystem is large, the above algorithm can not be unfolded. Therefore, a folded hardware architecture is constructed with a controller to sequence the computation. The underlying field multiplier, a $GF(2^m)$ multiplier with optimal normal basis, can be implemented as either a serial multiplier or a digital-serial multiplier or a parallel multiplier, depending on the amount of available hardware resources. In

Fig. 1, the two FIFOs serve as input/output buffers and the dual-port register file is used to save input parameters and intermediate data. This is realizable because Xilinx FPGAs have a large amount of internal SRAM and registers. Alternatively, if we were to use external SRAM, it would take either many external user pins with multiple SRAM chips, or multiple cycles to read in one single word. This would result in low bandwidth data transfers due to the large word size. The hardware provides $GF(2^m)$ arithmetic units GF_adder, GF_squarer, GF_multiplier and GF_inverter. With the finite field of characteristic 2 as the underlying field, addition is just a bit-wise XOR, and with normal bases representation, squaring is a simple cyclic right shift. The internal structures of the GF_multiplier and GF_inverter are given in the following sections.

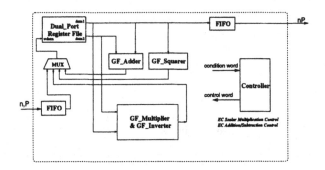

Fig. 1. Hardware architecture of EC coprocessor

3.2   GF_multiplier structure

The structure of GF_multiplier is a modified form of the Massey-Omura multiplier. Compared to the implementation in [9], the modified structure reduces the number of AND gates and the wire permutation by 50% in the AND Plane

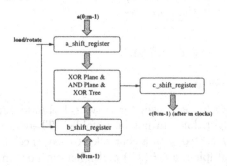

Fig. 2. Structure of $GF(2^m)$ serial multiplier

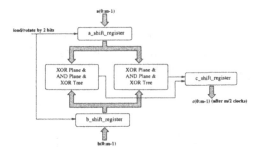

Fig. 3. Structure of $GF(2^m)$ digit-serial multiplier

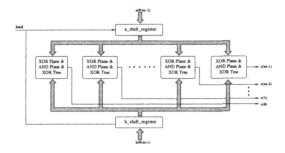

Fig. 4. Structure of $GF(2^m)$ parallel multiplier

without changing the total number of XOR gates. A serial multiplier of such a structure is shown in Fig. 2, which can be simply unfolded to a digital-serial multiplier in Fig. 3 or a parallel multiplier in Fig. 4. In this serial multiplier, at each cycle, a_shift_register and b_shift_register make a cyclic right shift, and one bit of the product is computed and shifted into the product register, c_shift_register. Therefore, each multiplication takes m clock cycles. If the serial multiplier is unfolded to a parallel multiplier, then each multiplication only takes one clock cycle. However, the required hardware will be m times that of the serial multiplier. Pipeline techniques are also applied to the XOR Plane—AND Plane—XOR Tree of the multiplier to reduce the clock cycle time. The modified Massey-Omura serial multiplier takes m AND gates, $2m$ XOR gates and $3m$ flip-flops, and has a latency of $m \times (T_{AND} + T_{XOR}\lceil log_2(m-1)\rceil)$ when it is not pipelined. However when it is pipelined, the serial multiplier has a latency of $(m + \lceil log_2(m-1)\rceil) \times Max(T_{AND}, T_{XOR})$ and a total cost of m AND gates, $2m$ XOR gates and $5m$ flip-flops. It is obvious that the pipelined multiplier is much faster than non-pipelined ones when m is large. The same techniques can also apply to a digit-serial multiplier. The digit-serial multiplier with a digit size of k will generate k bits of the product simultaneously and thus one multiplication can be done in kth fold time taken by a serial multiplier. The digit-serial multiplier makes a trade-off of the speed and the hardware between a serial multiplier and a parallel multiplier.

3.3 GF_inverter structure

The structure of GF_inverter is derived from the method introduced by T. Itoh et al[8] and is shown in Fig. 5. The inverse takes $\lfloor log_2(m-1) \rfloor$ recursive iterations and a total of $\lfloor log_2(m-1) \rfloor + HW(m-1) - 2$ underlying field multiplications, where $HW(m-1)$ is the Hamming weight of $(m-1)$.

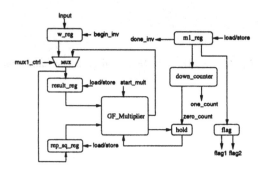

Fig. 5. Structure of $GF(2^m)$ inverter

3.4 Controller structure

The controller takes advantage of the abundance of internal SRAM and registers in Xilinx FPGAs. The controller is built up as a finite state machine with table look-up to implement the logic functions. Since the whole look-up table consists of small look-up tables from each CLB (configurable logic block), the controller can be pipelined to have a clock cycle time equal to one CLB delay. A pipelined structure of the controller is illustrated in Fig. 4. At each cycle, a selector associated with the value of PC is generated. It then selects the appropriate bit of the condition word to make PC either increment or load a new value from the Branch PC look-up. In case of a branch hazard, the pipeline register is cleared.

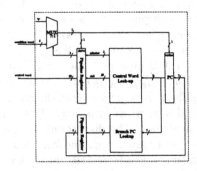

Fig. 6. Pipelined structure of the controller

4 Dataflow of EC Scalar Multiplication

The algorithm and hardware architecture leads to the computation dataflow chart shown in Fig. 7 and Fig. 8. Each bit of the decomposed scalar is encoded by 2 bits to represent $\{1,0,-1\}$. A total of 9 intermediate storage elements are needed and it leads to a 4-bit address space for the dual-port register file. Then each data bit of the dual-port register file can be implemented with one CLB and the dual-port register file has one CLB delay. It is obvious that the dataflow has many conditional branches. Therefore, the branch hazard problem has to be taken care. From the dataflow chart, the schedule and control of computation in the scalar multiplier is worked out and the corresponding VHDL description is implemented.

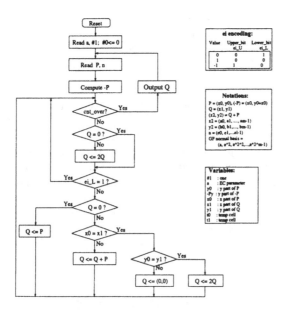

Fig. 7. Dataflow chart I of EC scalar multiplier

5 Development of EC Scalar Multiplier

All components in the scalar multiplier are implemented according to above algorithms and hardware architectures. A pipelined digit-serial multiplier is implemented because a serial or a parallel multiplier is only a special case for digit size of 1 or m respectively. The controller is a finite state machine according to Fig. 6. The implementation of the controller follows the dataflow charts I & II in Fig. 7 and Fig. 8, respectively. All operations in the dataflow can be categorized as one of the following atomic operations: unconditional jump, conditional

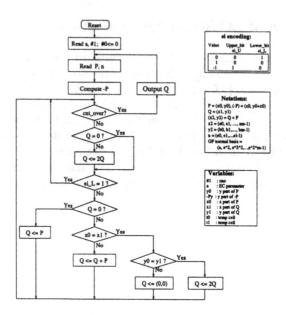

Fig. 8. Dataflow chart II of EC scalar multiplier

jump, operand load, operand store, finite field addition, finite field squaring, finite field multiplication and finite field inversion. Then, each state of the controller consists of one or more such atomic operations because addition, squaring, multiplication/inversion and load/store can be executed concurrently. The execution schedule is optimized to provide the shortest computing time. These atomic operations are represented as macros and are re-used in the VHDL code. One example of the VHDL simulation results is shown in Fig. 9 for $m = 39$ and type II optimal normal basis[14]. The example shows two scalar multiplications which compute

$$7 \times (1A0C3EB323, 2EE60CF558) = (7527E64FAD, 34A9265CF1)$$

$$7 \times (1A0C3EB323, 34EA32467B) = (7527E64FAD, 418EC0135C)$$

for $a = 1A28CE01DD$ and $b = 1200569A44$ in equation (1). The hex encoding follows the method in [14].

6 Mapping onto Xilinx XC4000XL-Series FPGA

6.1 Synthesis/mapping results

After VHDL code simulations, the design is setup with a pipelined $GF(2^m)$ serial multiplier (or digit size of 1) and is mapped onto a Xilinx XC4000XL-series FPGA for some small values of m. The synthesis is done with Exemplar synthesis tools and the mapping/layout is done with Xilinx Design Manager.

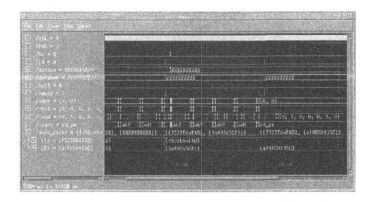

Fig. 9. VHDL simulation waveform for $m = 39$

The results are shown in Table 6.1 and one example of the layout for $m = 29$ is shown in Fig. 10. The mapping efficiency is represented with the percentage of total CLBs in a FPGA that is used by the design and the throughput is represented with the scalar multiplications per second. Both the throughput and clock cycles in Table 6.1 represent the worst case performance. It is shown that the architecture of the prototype is regular and the designed coprocessor has very high CLB utilization. The dominant operation in the EC scalar multiplier is $GF(2^m)$ multiplication. Therefore, if the $GF(2^m)$ serial multiplier is unfolded with a factor of 2, then the throughput, in terms of scalar multiplications per second, will be doubled.

Value of m	XC4000XL Device	CLB Usage	Clock Cycles	Throughput (scalar mul/sec)
5	4010XL	$272/400 = 68\%$	126	179856
11	4013XL	$478/576 = 83\%$	825	19230
29	4028XL	$962/1024 = 93\%$	7158	1653
53	4044XL	$1626/1936 = 84\%$	26753	417

Table 1. FPGA chip area utilization and throughput

6.2 Analysis of the mapping results

Since the proposed hardware architecture is regular and simple, the expected mapping results can also be obtained with an estimation formula. Then, a comparison can be made to analyze the mapping results. In order to derive the formula, two summaries are listed below:

Fig. 10. FPGA Layout for m=29 with XC4028XL

1. CLB (Configurable Logic Block) structure of Xilinx XC4000XL-series:
   - 2 flip-flops(FFs) per CLB
   - 2 function generators (FGs) per CLB (4 input/single output logic unit)
   - 2 single-port 16x1 RAMs per CLB (using two logic units)
   - 1 dual-port 16x1 RAM per CLB (using two logic units)
2. Cost of implementing basic components:
   - two m-bit registers take $2m$ FFs.
   - three m-bit shift registers take $3m$ FFs and FGs.
   - four m-bit 2:1 MUXs take $4m$ FGs.
   - one m-bit GF_Adder takes m FGs.
   - one m-bit dual-port 9 word RAM takes $2m$ FGs.
   - one m-bit 6 word FIFO takes $6(m+1)$ FFs and FGs.
   - one m-bit 2 word FIFO takes $2(m+1)$ FFs and FGs.
   - one m-bit GF_Multiplier takes $5m$FFs and $3m$ FGs (pipelined bit-serial multiplier).
   - one m-bit GF_Inverter takes $3(m+log_2 m)$FFs and FGs (excluding GF_Multiplier).

From above basic facts, the cost of one EC scalar multiplier with key size m is derived as:

- Total FFs = $21m + 3log_2 m + 48$
- Total FGs = $24m + 3log_2 m + 308$
- Minimal value of Total CLBs
  = max(Total FFs, Total FGs)/2
  = $12m + (3log_2 m)/2 + 154$
- Maximal value of Total CLBs
  = (Total FFs + Total FGs)
  = $45m + 6log_2 m + 356$

Table 2 is constructed with above estimation formula and the data in Table 6.1. In Table 2, the actual CLBs means the CLB count of the mapping obtained with Xilinx Design Manager. The Min/Max CLBs comes from the estimation formulas and puts a lower/upper limit for the total number of CLBs needed for the design. Using regression, a polynomial curve of degree 2 is fitted with the data of the actual CLBs and the CLB usage for larger fields are predicted through extrapolation with the fitted curve. However routing resources are not taken into account in the extrapolation and they could become a bottleneck for larger fields. For m=160, the estimated number of CLBs ($\approx$ 4000) is larger than the capacity of the XC4085XL. However, the design should easily fit onto the Virtex XCV1000 chip.

Value of m	Expected Device	Actual CLBs	Min CLBs	Max CLBs
5	XC4010XL	272	219	600
11	XC4013XL	478	292	876
29	XC4028XL	962	510	1692
53	XC4044XL	1626	799	2778

Table 2. Mapping analysis

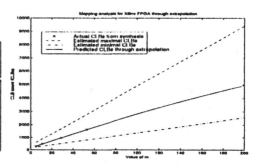

Fig. 11. Extrapolated mapping analysis

7 Conclusions and Future Work

The experimental results from the mappings over small fields show that the hardware architecture is regular and achieves high CLB utilization and high speed. The use of an FPGA in the development of an elliptic curve scalar multiplier demonstrates many advantages:

- Reduced development time and cost.
- Tailorable design for a particular application.
- Reduced hardware overhead and high performance.
- Increased chip area utilization.
- Hardware performance with advantages of software development.
- Simplified hardware architecture and ability to easily add new functions.

The effectiveness and eventual performance/cost ratio of applying reconfigurable hardware to cryptography depends on many factors and research in this area is highly experimental. Therefore, future work remains in many areas and is summarized as follows:

- To map the design onto more types of FPGA chips to show the usefulness of the design and to reveal the relationship between the algorithm and the architecture/resources of FPGAs.

– To build some EC application systems, such as an EC digital signature or an EC Diffie-Hellman key exchange[1][14][15], by using reconfigurable hardware, such that a direct comparison can be made with other implementations using microprocessors[12][13].

Acknowledgments

The authors would like to thank Professor Keshab Parhi for valuable conversations. This research was supported by Defense Advanced Research Project Agency under contract number DA/DABT63-96-C-0050. The authors are grateful to the anonymous reviewers for their constructive comments and suggestions.

References

1. A. J. Menezes, *"Elliptic curves and cryptography"*, Dr. Dobb's Journal, April, 1997.
2. J. A. Solinas, *"An Improved algorithm for arithmetic on a family of elliptic curves"*, Journal of Cryptography, 1997.
3. N. Kobliz, *"CM curves with good cryptographic properties"*, Proc, Crypto '91, Springer-Verlag, 1992, pp.279-287.
4. F. Morain and J. Olivos, *"Speeding up the computations on an elliptic curve using addition-subtraction chains"*, Information Theory Application 24(1990), pp.531-543.
5. D. M. Gordon, *"A survey of fast exponentiation methods"*, Journal of Algorithms 27, 129-146, 1998.
6. G. B. Agnew, t. Beth, R. C. Mullin and S. A. Vanstone, *"Arithmetic operations in $GF(2^m)$"*, Journal of Cryptology, Vol 6, pp.3-13, 1993.
7. D. W. Ash, I. F. Blake and S. A. Vanstone, *"Low complexity normal bases"*, Discrete Applied Mathematics 25, 1989, pp.191-210.
8. T. Itoh and S. Tsujii, *"A fast algorithm for computing multiplicative inverses in $GF(2^m)$ using normal bases"*, J. Society for Electronic Communications(Japan)44(1986), 31-36.
9. C. C. Wang, T. K. Truong, H.M.Shao, L.J.Decutsch, J.K.Omura and I.S.Reed, *"VLSI architectures for computing multiplications and inverses in $GF(2^m)$"*, IEEE Tran. Comp., Vol c-34, No 8, August 1985.
10. G. B. Agnew, R. C. Mullin, I. M. Onyszchuk and S. A. Vanstone, *"An implementation for a fast public-key cryptosystem"*, Journal of Cryptology, Vol 3, pp.63-79, 1991.
11. G. B. Agnew, R. C. Mullin and S. A. Vanstone, *"An implementation of elliptic curve cryptosystems over $F_{2^{155}}$"*, Journal of Cryptology, Vol 3, pp.63-79, 1991.
12. M. J. Wiener, *"Performance comparison of public-key cryptosystems"*, RSA Laboratories's CryptoBytes, Vol. 4, No. 1 - Summer 1998.
13. H. Handschuh and P. Paillier, *"Smart card crypto-coprocessor for public-key cryptography"*, RSA Laboratories's CryptoBytes, Vol. 4, No. 1 - Summer 1998.
14. American Bankers Association, *X9.62-199x, Public Key Cryptography For The Financial Services Industry: The Elliptic curve digital signature algorithm (ECDSA)*, Working Draft, Nov. 1997
15. IEEE, *"IEEE P1363 standard specification for public key cryptography"*, draft, 1998.
16. Xilinx, *"The Programmable logic data book"*, Xilinx, Inc., 1998.

Highly Regular Architectures for Finite Field Computation Using Redundant Basis

Huapeng Wu[1], M. Anwarul Hasan[2], and Ian F. Blake[3]

[1] Dept of ECE, IIT, Chicago IL 60616
hpwu@ece.iit.edu
This work was done when he worked for his Ph.D degree
with the Dept of ECE, University of Waterloo.
[2] Dept of ECE, Univ. of Waterloo, Waterloo, ON N2L 3G1 Canada
ahasan@ece.uwaterloo.ca
Currently, he is with the Motorola Lab on a
sabbatical leave from the University of Waterloo.
[3] HP Lab., Mail Stop 3U-4, 1501 Page Mill Road, Palo Alto, CA 94304
ifblake@hpl.hp.com

Abstract. In this article, an extremely simple and highly regular architecture for finite field multiplier using redundant basis is presented, where redundant basis is a new basis taking advantage of the elegant multiplicative structure of the set of primitive n^{th} roots of unity over $\mathbb{F}_2$ that forms a basis of $\mathbb{F}_{2^m}$ over $\mathbb{F}_2$. The architecture has an important feature of implementation complexity trade-off which enables the multiplier to be implemented in a partial parallel fashion. The squaring operation using the redundant basis is simply a permutation of the coefficients. We also show that with redundant basis the inversion problem is equivalent to solving a set of linear equations with a circulant matrix. The basis appear to be suitable for hardware implementation of elliptic curve cryptosystems.

1 Introduction

Efficient computations in finite field and their architectures are very important to many cryptosystems, *e.g.*, elliptic curve systems. There are mainly three types of bases over finite fields, namely, polynomial basis (PB), normal basis (NB), and dual basis (DB)[12], which are commonly used to represent the field elements. The main advantage of using the normal basis is that the squaring operation in NB is simply a cyclic shift of the coordinates of the element, and thus this basis has found application in computing exponentiations and multiplicative inverses [10, 8, 1]. However, the computations of exponentiations and inverses require not only squaring but also multiplications. Massey and Omura devised an NB multiplier known as Massey-Omura multiplier [13]. Alternative bit-serial multiplications using the normal basis can be found in [5, 2]. The bit-parallel NB multipliers were proposed in [17, 9]. PB and DB have been also used for implementing bit-parallel multiplier [14, 8, 16, 6, 11, 19].

In this article a new basis – redundant basis (RB), is proposed. The redundant basis takes advantage of the elegant multiplicative structure of the set of primitive n^{th} roots of unity over $\mathbb{F}_2$ that forms a basis of $\mathbb{F}_{2^m}$ over $\mathbb{F}_2$. It is shown that finite field arithmetic operations using the redundant basis have extremely simple and highly regular structures.

Some similar work to ours using polynomial ring basis was proposed recetly [4]. We believe that the polynomial ring basis is a subset of the redundant basis proposed here.

The organization of this paper is as follows: Redundant basis is introduced in Section 2. In Section 3, multiplication operation using RB is discussed and then architectures of bit-serial and bit-parallel multipliers are proposed. Relation between RB and other types of bases is analyzed in Section 4. Squaring and inverse operation using RB are discussed in Section 5 and Section 6, respectively. Finally, a few examples are given in Section 7.

2   Redundant Basis

Definition 1. [12] *Let K be a field of characteristic p and n be a positive integer. The splitting field of $x^n - 1$ over a field K is called the nth cyclotomic field over K and denoted by $K^{(n)}$. The roots of $x^n - 1$ in $K^{(n)}$ are called the nth roots of unity over K and the set of all these roots is denoted by $E^{(n)}$. Then a generator of the cyclic group $E^{(n)}$ is called a primitive n^{th} root of unity over K if n is not divisible by p.*

Redundant basis uses the set of primitive n^{th} roots of unity over $\mathbb{F}_2$ that forms a basis of $\mathbb{F}_{2^m}$ over $\mathbb{F}_2$. Let β be a primitive n^{th} root of unity in $\mathbb{F}_{2^m}$ or some extension field of $\mathbb{F}_{2^m}$, then we have

$$\beta \cdot \beta^i = \begin{cases} \beta^{i+1} & i \neq n-1, \\ 1 = \sum_{j=1}^{n-1} \beta^j & i = n-1. \end{cases}$$

By adding the element '1' to the set of primitive n^{th} roots of unity, we have[1] $\langle 1, \beta, \beta^2, \ldots, \beta^{n-1} \rangle \triangleq I_1$, which can be used as a basis in $\mathbb{F}_{2^m}$ over $\mathbb{F}_2$ and it is referred to as a redundant basis. Note that the base elements are in the cyclotomic field $\mathbb{F}_2^{(n)}$ and they may not belong to the field $\mathbb{F}_{2^m}$. Clearly, any field $\mathbb{F}_{2^m}$ has a redundant basis if there is a cyclotomic field over $\mathbb{F}_2$ that contains $\mathbb{F}_{2^m}$ as a subfield. Thus one redundant basis can be the set of $(2^m - 1)$st roots of unity. To efficiently represent the field elements, the redundant basis should be chosen such that its size is as small as possible. Now the question is: Given $\mathbb{F}_{2^m}$, what is the smallest cyclotomic field $\mathbb{F}_2^{(n)}$ that contains $\mathbb{F}_{2^m}$ as a subfield? An algorithm for computing such an n is given below.

Algorithm 1 Computing the smallest cyclotomic field $\mathbb{F}_2^{(n)}$ that includes $\mathbb{F}_{2^m}$ as a subfield

---

[1] We denote a set by $\{\cdots\}$ and an orderly set by $\langle\cdots\rangle$.

1. Find all the factors d_i of $2^m - 1$ that are greater than m and list them in an increasing order: $d_1, d_2, \ldots, d_k = 2^m - 1$;
2. DO WHILE($i \leqslant k$)
   IF $m \mid \phi(d_i)$ AND $j = m$ is the smallest integer such that $2^j = 1 \bmod d_i$,
   THEN $t \leftarrow d_i$, and BREAK; ELSE $i \leftarrow i + 1$.
3. Let $n \leftarrow t$ and let h be the largest positive integer such that $t > hm$.
   IF $h > 1$ THEN
       FOR $i = 2$ TO h DO
         (a) Find all the factors d_i of $2^{im} - 1$ that are greater than im and list them in an increasing order: $d_1, d_2, \ldots, d_{k_i} = 2^{im} - 1$;
         (b) DO WHILE($i \leqslant k_i$)
           IF $im \mid \phi(d_i)$ AND $j = im$ is the smallest integer such that $2^j = 1 \bmod d_i$, THEN $n \leftarrow \min\{n, d_i\}$, and BREAK; ELSE $i \leftarrow i + 1$.

$\square$

Since the $(2^m - 1)^{\text{st}}$ cyclotomic field has a degree of $\phi(2^m - 1)$ and contains the field $\mathbb{F}_{2^m}$ as a subfield, we have that m divides $\phi(2^m - 1)$, where ϕ is the Euler Phi function.

3 Redundant Basis Multiplier

3.1 Multiplication operation

Consider the redundant basis in $\mathbb{F}_{2^m}$ over $\mathbb{F}_2$: $I_1 = \langle 1, \beta, \beta^2, \ldots, \beta^{n-1} \rangle$. Let field element $A \in \mathbb{F}_{2^m}$ and be represented with I_1:

$$A = a_0 + a_1\beta + a_2\beta^2 \cdots + a_{n-1}\beta^{n-1}, \tag{1}$$

where $a_i \in \mathbb{F}_2, i = 0, 1, \ldots, n-1$. Note that $n \geq m+1$ and the set of coefficients $\{a_i\}$ is not unique.

Now let us look at multiplication operation under the redundant basis I_1. Let $B \in \mathbb{F}_{2^m}$ be given as $B = b_0 + b_1\beta + b_2\beta^2 + \cdots + b_{n-1}\beta^{n-1}$. Then we have

$$\beta \cdot B = b_0\beta + b_1\beta^2 + b_2\beta^3 + \cdots + b_{n-2}\beta^{n-1} + b_{n-1}\beta^n$$
$$= b_{n-1} + b_0\beta + b_1\beta^2 + b_2\beta^3 + \cdots + b_{n-2}\beta^{n-1}.$$

Obviously, the coordinates of βB is a cyclic shift of those of B, with respect to I_1. From

$$\beta^i \cdot B = b_0\beta^i + b_1\beta^{i+1} + b_2\beta^{i+2} + \cdots + b_{n-1-i}\beta^{n-1} + b_{n-i} + b_{n-i+1}\beta + \cdots + b_{n-1}\beta^{i-1}$$
$$= b_{n-i} + b_{n-i+2}\beta + \cdots + b_{n-1}\beta^{i-1} + b_0\beta^i + b_1\beta^{i+1} + \cdots + b_{n-i-1}\beta^{n-1}$$
$$= \sum_{j=0}^{n-1} b_{(j-i)}\beta^j, \tag{2}$$

where $(j - i) = (j - i) \bmod n$ denotes that $j - i$ is to be reduced modulo n, we have

$$A \cdot B = \sum_{i=0}^{n-1} a_i(\beta^i \cdot B) = \sum_{i=0}^{n-1} a_i \sum_{j=0}^{n-1} b_{(j-i)}\beta^j = \sum_{j=0}^{n-1}\left(\sum_{i=0}^{n-1} a_i b_{(j-i)}\right)\beta^j.$$

If we define $AB = C = \sum_{j=0}^{n-1} c_j\beta^j$, then it follows

$$c_j = \sum_{i=0}^{n-1} a_i b_{(j-i)}, \quad j = 0, 1, \ldots, n-1. \tag{3}$$

3.2 Bit-serial multiplier architecture

Figure 1 shows the multiplier structure to realize multiplication using redundant basis. The coordinates of B with respect to the redundant basis I_1 are loaded into a register of length n bits whose contents can be shifted cyclically. The binary tree of $n - 1$ adders in $\mathbb{F}_2$ takes n terms of $a_i b_k$ as its inputs and generates a c_j term as output every clock cycle. All c_j's, $j = 0, 1, \ldots, n-1$, which are represented using I_1, are computed and obtained in n clock cycles. It can be seen that n AND gates, $n - 1$ XOR gates and n 1-bit registers are required for constructing the multiplier. The clock period should not be less than $T_A + \lceil \log_2 n \rceil T_X$, where T_A and T_X denote the time delays of an AND gate and an XOR gate, respectively.

Table 1 shows the complexity of the bit-serial multipliers using redundant basis and normal basis when there is a type I optimal normal basis. While Table 2 shows comparison of the complexities between RB multiplier and NB multiplier when there is a type II optimal normal basis or no optimal basis.

Multiplier	#AND	#XOR	#1-bit reg.	# clk cycles	Basis
Massey-Omura [13]	$2m - 1$	$2m - 2$	$2m$	m	normal
Feng [5]	$2m - 1$	$3m - 2$	$3m - 2$	m	normal
Agnew *et al* [2]	m	$2m - 1$	$3m$	m	normal
presented here	$m + 1$	m	$m + 1$	$m + 1$	redundant

Table 1. Comparison of bit-serial multipliers using type I ONB and RB (here $n = m + 1$).

It can be seen that the bit-serial redundant basis multiplier has lower complexity only when there is a type I optimal basis. When there is a type II optimal NB or no optimal basis, then the redundant basis multiplier will have a long time delay. In this case, partially parallel architecture can be employed and it is discussed in the next section.

Multipliers	#AND	#XOR	#1-bit reg.	# clk cycles	basis
Massey-Omura [13]	C_N	$C_N - 1$	$2m$	m	normal
Feng [5]	$2m - 2$	$C_N + m - 1$	$3m - 2$	m	normal
Agnew *et al* [2]	m	C_N	$3m$	m	normal
presented here	$km + 1$	km	$km + 1$	$km + 1$	redundant

In the example presented in [5], a technique of reusing partial sum was used to reduce the complexity. Thus the number of XOR gates should be not greater than $C_N + m - 1$ if a non-optimal normal basis is used.

Table 2. Comparison of bit-serial multipliers using NB and RB (where $n = km + 1$).

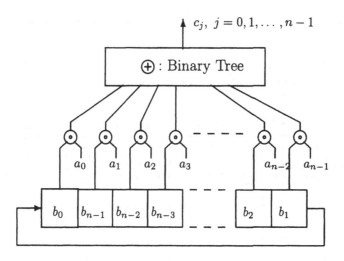

Fig. 1. Bit serial multiplier using the redundant basis.

3.3 Bit-parallel multiplier architecture

A parallel version of the multiplier using a redundant basis is shown in Figure 2. On the left side of the figure inputs $\{a_i\}$ and $\{b_i\}$ are fed into n blocks (Block B). The detailed structure of Block B is shown on the right side of the figure. It can be seen that n^2 AND gates and $n(n-1)$ XOR gates are required. The time delay is $T_A + \lceil \log_2 n \rceil T_X$.

Trade-off with complexities or partial-parallel architecture The proposed bit-parallel architecture can be easily made for trade-offs between size and time complexities: If t Block B's are used to construct a multiplier and thus in one clock cycle t c_j's are computed and output, then one multiplication operation can be completed in $\lceil \frac{n}{t} \rceil$ clock cycles. This feature has great significance for hardware implementation since it might be difficult to implement a full-scale bit-parallel multiplier in hardware if the field is very large.

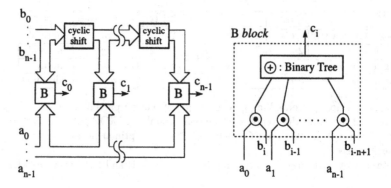

Fig. 2. Parallelization of the bit-serial multiplier using the redundant basis.

Table 3 adn Table 4 show the comparisons between bit-parallel redundant basis multipliers and bit-parallel normal basis multipliers.

Multipliers	#AND	#XOR	Time delay	Partial-parallel Arch.
Hasan et al [9]	m^2	$m^2 - 1$	$T_A + (1 + \lceil \log_2 m \rceil)T_X$	not avail.
Koc and Sunar [11]	m^2	$m^2 - 1$	$T_A + (2 + \lceil \log_2 m \rceil)T_X$	not avail.
New proposal	$(m+1)^2$	$m(m+1)$	$T_A + \lceil \log_2(m+1) \rceil T_X$	available

Table 3. Comparison of bit-parallel multipliers using type I ONB and RB, here $n = m + 1$.

Multipliers	#AND	#XOR	Time delay	Partial-parallel Arch.
Massey-Omura	$2m^2 - m$	$2m^2 - m$	$T_A + \lceil \log_2(2m - 1) \rceil T_X$	available
New proposal	m^2	$2m^2 - m$	$T_A + (1 + \lceil \log_2 m \rceil)T_X$	available

Table 4. Comparison of bit-parallel multipliers using type II ONB and RB (where $n = 2m + 1$).

3.4 Complexity

Clearly the complexity of the RB multipliers in $\mathbb{F}_{2^m}$ over $\mathbb{F}_2$ depends on the size n of the cyclotomic field $\mathbb{F}^{(n)}$. There seems no easy way to give a general relation between n and m. In Table 1, we have computed values of n for certain small values of m using Algorithm 1.

m	2	3	4	5	6	7	8	9	10	11	12	13	14	15	18	19
n	3	7	5	11	9	29	17	19	11	23	13	53	29	31	19	37

Table 5. Smallest cyclotomic field $\mathbb{F}^{(n)}$ that includes $\mathbb{F}_{2^m}$ as a subfield.

For a subset of redundant basis, which can be derived from certain normal basis (optimal normal basis), the complexity can be easily solved which is discussed in the next section. Also, for the field in which there exists an equally spaced polynomial (ESP), a small value of n can be found.

4 Relation/Conversion between Redundant Basis and Other Bases

4.1 Redundant basis and normal basis

Some redundant bases can be easily introduced by the normal basis generated with Gauss period, which also reveals the relation/conversion between the redundant basis and the normal basis.

Gauss period, normal basis and redundant basis The Gauss period (GP) was discovered by Gauss and is defined as follows: Let $m, k \geq 1$ be integers such that $r = mk + 1$ is a prime, and let q be a prime power with $\gcd(q, r) = 1$. Let $\mathcal{K}$ be the unique subgroup of order k of the multiplicative group of $\mathbb{Z}_r = \mathbb{Z}/r\mathbb{Z}$, then for any primitive rth root β of unity in $\mathbb{F}_{q^{mk}}$, the elements

$$\alpha = \sum_{\gamma \in \mathcal{K}} \beta^\gamma \tag{4}$$

is called a *Gauss period of type* (m, k) over $\mathbb{F}_q$. It can be checked that $\alpha \in \mathbb{F}_{q^m}$.

Gauss periods have been used to construct normal bases with low complexity [15, 3]. A Gauss period of type (m, k) over $\mathbb{F}_2$ naturally introduces a normal basis $I_2 = \langle \alpha, \alpha^2, \ldots, \alpha^{2^{m-1}} \rangle$ in $\mathbb{F}_{2^m}$ over $\mathbb{F}_2$ if and only if $\gcd(e, m) = 1$, where e is the index of 2 modulo r. Furthermore, such a normal basis has complexity at most $mk' - 1$ with $k' = k$ if k even and $k + 1$ otherwise [3, 18, 7]. Clearly, Gauss periods of type $(m, 1)$ and $(m, 2)$ generate optimal normal bases with complexity $2m - 1$, which are usually called type-I and type-II optimal normal bases (ONB), respectively [15].

On the other hand, a redundant basis in this case can be given as $I_1 = \langle 1, \beta, \beta^2, \beta^3, \ldots, \beta^{mk} \rangle$. Consider two sets of km elements in $\mathbb{F}_{2^{km}}$: $S_1 = \{\beta^{2^i \gamma^j}, i = 0, 1, \ldots, m-1; j = 0, 1, \ldots, k-1\}$ and $S_2 = \{\beta, \beta^2, \ldots, \beta^{km}\}$. For any element $\beta^{2^i \gamma^j} \in S_1$, we have $\beta^{2^i \gamma^j} = \beta^{2^i \gamma^j \bmod (mk+1)} \in S_2$, and thus, $S_1 \subseteq S_2$. Let $G = \mathbb{F}_{km+1}^*$ then $G = \langle 2, \gamma \rangle$. For any integer $l \in \{1, 2, \ldots, km\}$, there exist integers $i \in \{0, 1, \ldots, m-1\}$ and $j \in \{0, 1, \ldots, k-1\}$, such that $l = 2^i \gamma^j \bmod (km + 1)$. Therefore, $S_2 \subseteq S_1 \Rightarrow S_2 = S_1$.

Since $I_2 = \langle \alpha, \alpha^2, \ldots, \alpha^{2^{m-1}} \rangle = \langle \sum_{i=0}^{k-1} \beta^{\gamma^i}, \sum_{i=0}^{k-1} \beta^{2\gamma^i}, \ldots, \sum_{i=0}^{k-1} \beta^{2^{m-1}\gamma^i} \rangle$ and each element in I_2 is a sum of k elements in S_1, it can be seen that elements in $S_1 (= S_2)$ can serve as a basis in $\mathbb{F}_{2^m}$ and which is a permutation of $\langle \beta, \beta^2, \ldots, \beta^{mk} \rangle \triangleq I_3$. Obviously, the redundant basis can be obtained by adding element '1' to the basis I_3.

Conversion between normal basis and redundant basis Now let us look at the conversion from the normal basis $I_2 = \langle \alpha, \alpha^2, \ldots, \alpha^{2^{m-1}} \rangle$ to the redundant basis I_1. As we have seen before, the conversion between redundant basis I_1 and the basis consisting of elements from S_1 is simple. If $A = (a'_0, a'_1, \ldots, a'_{m-1})$ with the normal basis, then with the basis from S_1,

$$A = (a''_{0,0}, a''_{0,1}, \ldots, a''_{0,k-1}, \ldots, a''_{m-1,k-1}),$$

where $a''_{i,j} = a'_i$ for $j = 0, 1, \ldots, k-1$ and $i = 0, 1, \ldots, m-1$.

4.2 Redundant basis and polynomial basis

Given a basis I in $\mathbb{F}_{2^m}$, the general case of basis conversion between I and the redundant basis I_1 may not be trivial. If I is a normal basis generated with the Gauss period of type (m, k), then how to obtain I_1 has been discussed in the last section. If $I = \langle 1, \alpha, \ldots, \alpha^{m-1} \rangle$ is the polynomial basis, and if we know that the order of element α is $\text{ord}(\alpha)$, then the redundant basis I_1 can be obtained using the following algorithm:

Algorithm 2 Computing the redundant basis from a polynomial basis $\langle 1, \alpha, \ldots, \alpha^{m-1} \rangle$

1. Compute n using Algorithm 1;
2. Compute the order of the irreducible polynomial $\text{ord}(\alpha)$;
3. Let $t = \text{ord}(\alpha)/n$, then the redundant basis is given by $\langle 1, \alpha^t, \alpha^{2t}, \ldots, \alpha^{(n-1)t} \rangle$.

$\square$

It can be shown that for the field that there exists an ESP, the value of n is always between $m+1$ and $2m$.

5 Squaring Operation Using Redundant Basis

Let $\langle 1, \beta, \beta^2, \ldots, \beta^{n-1} \rangle$ be a redundant basis for $\mathbb{F}_{2^m}$ over $\mathbb{F}_2$. For a field element represented in the redundant basis:

$$A = a_0 + a_1\beta + \cdots + a_{n-1}\beta^{n-1}$$

its square is given by

$$A^2 = a_0 + a_1\beta^2 + \cdots + a_{n-1}\beta^{2(n-1)}.$$

Since $\beta^n = 1$, we have that $a_j\beta^{2j} = a_j\beta^{2j-n}$ if $2j > n-1$. It can be seen that a squaring operation using the redundant basis is equivalent to a permutation of the element coefficients.

6 Inversion with Redundant Basis

The problem of inversion in redundant basis is as follows: Given a field element

$$A = a_0 + a_1\beta + \cdots + a_{n-1}\beta^{n-1} \in \mathbb{F}_{2^m},$$

find

$$B = b_0 + b_1\beta + \cdots + b_{n-1}\beta^{n-1} \in \mathbb{F}_{2^m}$$

which is the inverse of A. Clearly, the methods proposed by Itoh and Tsujii [10] and by Agnew *et al* [1] can be used. With their methods, about $\frac{3}{2}\log(m-1)$ multiplications on average and $(m-1)$ squaring operations are required. Since squaring operation in the redundant basis is a permutation of lines and free, while the multiplication can be efficiently implemented in hardware, it is expected that with this method inversion using the redundant basis can be as good as using normal basis.

Another method for inversion is to solve a set of linear equations. From $AB = 1$, we have

$$1 = \sum_{i=0}^{n-1} a_i\beta^i \sum_{j=0}^{n-1} b_j\beta^j$$

$$= \sum_{j=0}^{n-1} b_j \sum_{i=0}^{n-1} a_i\beta^{i+j}$$

$$= \sum_{j=0}^{n-1} b_j \sum_{i=0}^{n-1} a_i\beta^{(i+j)}$$

$$= \sum_{j=0}^{n-1} \left(\sum_{i=0}^{n-1} a_{(j-i)}b_i \right)\beta^j,$$

where $(x) \overset{\triangle}{=} x \bmod (n)$, or,

$$\begin{bmatrix} a_0 & a_{n-1} & a_{n-2} & \cdots & a_1 \\ a_1 & a_0 & a_{n-1} & \cdots & a_2 \\ a_2 & a_1 & a_0 & \cdots & a_3 \\ \vdots & \vdots & \vdots & \vdots & \vdots \\ a_{n-1} & a_{n-2} & a_{n-3} & \cdots & a_0 \end{bmatrix} \begin{bmatrix} b_0 \\ b_1 \\ b_2 \\ \vdots \\ b_{n-1} \end{bmatrix} = \begin{bmatrix} 1 \\ 0 \\ 0 \\ \vdots \\ 0 \end{bmatrix} \tag{5}$$

The circulant matrix is always singular and the equations allow many solutions, all of which is a representation of the inverse B in the redundant basis. Note that the circulant matrix is a special case of Toplitz matrix and any algorithm for solving Toplitz system can also be used to solve (5).

7 Examples

Example 1. For the field $\mathbb{F}_{2^{10}}$, we can compute the smallest cyclotomic field that includes it as a subfield is $\mathbb{F}^{(11)}$. Highly regular architectures for bit-serial and bit-parallel multipliers using redundant basis can be built as discussed in Section 3. Clearly, a bit-serial multiplier requires 11 AND gates, 10 XOR gates, and 11 1-bit registers. It takes 11 clock cycles to accomplish a multiplication operation. The complexities for a fully bit-parallel multiplier are: 121 AND gates, 110 XOR gates and a propagation delay of $T_A + \lceil \log_2(m+1) \rceil T_X$.

Example 2. From Algorithm 1, we find the smallest cyclotomic field that includes $\mathbb{F}_{2^6}$ as a subfield is $\mathbb{F}^{(9)}$. Let the redundant basis in $\mathbb{F}_{2^6}$ over $\mathbb{F}_2$ be given by $\langle 1, \beta, \beta^2, \dots, \beta^8 \rangle$, where β is a primitve 9th root of unity in $\mathbb{F}_{2^6}$. In fact, β is a root of irreducible polynomial $x^6 + x^3 + 1$. The complexities of the bit-serial redundant basis multiplier are 9 AND gate, 8 XOR gate, 9 1-bit registers and 9 clock cycles for performing a multiplication operation.

Example 3. It can be computed that the redundant basis in $\mathbb{F}_{2^8}$ has 17 elements ($m = 8$ and $n = 17$). Then the redundant basis multipliers can be built and their complexities can be decided.

8 Summary

In this paper, we have presented redundant bases and their applications to the construction of multipliers. It has shown the new constructions are advantageous over other normal basis constructions when bit-parallel or partial-parallel structures are required. The comparisons have been made between redundant basis and normal basis, since the squaring operation using redundant basis is also a simple cyclic shift of lines. The inversion using the new basis has also been discussed. It can be shown that the polynomial ring basis proposed in [4] is a subset of the redundant basis.

References

1. Agnew, G.B., Beth, R., Mullin, R.C., Vanstone, S.A.: Arithmetic operations in GF(2^m). J. Cryptology **6** (1993) 3-13
2. Agnew, G.B., Mullin, R.C., Onyszchuk, I., Vanstone, S.A.: An implementation for a fast public key cryptosystem. J. Cryptology **3** (1991) 63-79
3. Ash, D.W., Blake, I.F., Vanstone, S.A.: Low complexity normal bases. Disc. Appl. Math. **25** (1989) 191-210
4. Drolet, G.: A New Representation of Elements of Finite Fields GF(2^m) Yielding Small Complexity Arithmetic Circuits. IEEE Trans. Comput. **47** (1998)
5. Feng, M.: A VLSI architecture for fast inversion in GF(2^m). IEEE Trans. Comput. **38** (1989) 1383-1386
6. Fenn, S.T.J., Benaissa, M., Taylor, D.: GF(2^m) multiplication and division over the dual basis. IEEE Trans. Comput. **45** (1996) 319-327

7. Gao, S., Vanstone, S.A.: On orders of optimal normal basis generators. Math. Comp. **64** (1995) 1227-1233

8. Hasan, M.A., Wang, M., Bhargava, V.K.: Modular construction of low complexity parallel multipliers for a class of finite fields $GF(2^m)$. IEEE Trans. Comput. **41** (1992) 962-971

9. Hasan, M.A., Wang, M., Bhargava, V.K.: A modified Massey-Omura parallel multiplier for a class of finite fields. IEEE Trans. Comput. **42** (1993) 1278-1280

10. Itoh, T., Tsujii, S.: A fast algorithm for computing multiplicative inverse in $GF(2^m)$ using normal bases. Inform. and Comput. **78** (1988) 171-177

11. Koc, C.K., Sunar, B.: Low-complexity bit-parallel canonical and normal multipliers for a class of finite fields. IEEE Trans. Comput. **47** (1998) 353-356

12. Lidl, R., Niederreiter, H.: Finite Fields. Addison-Wesley Publishing Company, 1983, Reading, MA

13. Massey, J.L., Omura, J.K.: Computational method and apparatus for finite field arithmetic. U.S. Patent No.4587627, 1984.

14. Mastrovito, E.D.: VLSI Architectures for Computations in Galois Fields. Ph.D Thesis, Linköping University, 1991, Linköping, Sweden

15. Mullin, R., Onyszchuk, I., Vanstone, S.A., Wilson, R.: Optimal normal bases in $GF(p^n)$. Disc. Appl. Math. **22** (1988) 149-161

16. Paar, C.: Efficient VLSI Architectures for Bit-Parallel Computation in Galois Fields. Ph.D Thesis, VDI-Verlag, Düsseldorf, 1994

17. Wang, C.C., et al: VLSI architectures for computing multiplications and inverses in $GF(2^m)$. IEEE Trans. Comput. **34** (1985) 709-717

18. Wassermann, A.: Konstruktion von Normalbasen. Bayreuther Mathematische Schriften **31** (1990) 155-164

19. Wu, H., Hasan, M.A.: Low complexity bit-parallel multipliers for a class of finite fields. IEEE Trans. Comput. **47** (1998) 883-887

Low Complexity Bit-Parallel Finite Field Arithmetic Using Polynomial Basis

Huapeng Wu

Dept of ECE, IIT, Chicago IL 60616
hpwu@ece.iit.edu

Abstract. Bit-parallel finite field multiplication in $\mathbb{F}_{2^m}$ using polynomial basis can be realized in two steps: polynomial multiplication and reduction modulo the irreducible polynomial. In this article, we prove that the modular polynomial reduction can be done with $(r-1)(m-1)$ bit additions, where r is the Hamming weight of the irreducible polynomial. We also show that a bit-parallel squaring operation using polynomial basis costs not more than $\left\lfloor \dfrac{m+k-1}{2} \right\rfloor$ bit operations if an irreducible trinomial of form $x^m + x^k + 1$ over $\mathbb{F}_2$ is used. Consequently, it is argued that to solve multiplicative inverse in $\mathbb{F}_{2^m}$ using polynomial basis can be as good as using normal basis.

1 Introduction

The increasing use of cryptographic techniques in computer and communication network systems has inspired many researchers to find ways to perform fast or bit-parallel algorithms and architectures over finite fields of characteristic two. Besides the discrete logarithm cryptosystems over $\mathbb{F}_{2^n}$, the elliptic curve cryptosystems, which utilize the group of points on an elliptic curve over a field, can also be realized using finite fields of characteristic two. These groups are generally used to take advantage of their efficiency over multiprecision arithmetic for large prime fields. The elliptic curve cryptosystems also have the advantage of their high cryptographic strength relative to the key size, and thus they are especially attractive in applications such as the financial industry, smart cards and wireless areas where power and bandwidth are limited.

There are generally three types of basis in finite field, namely, normal basis (NB), polynomial basis (PB) and dual basis (DB). Normal basis is often chosen in cryptographic application, since squaring operation is only a cyclic shift of the lines and thus inversion and exponentiation can be efficiently performed. Massey and Omura first found a regular architecture for normal basis multiplication [13], while the use of the optimal normal basis further reduces the complexity of multiplication [16]. Polynomial basis has long been used for finite field arithmetic. Polynomial basis multiplication based on the irreducible trinomial $x^m + x^k + 1$ with $1 \le k \le \left\lfloor \frac{m}{2} \right\rfloor$ are attractive because they require fewer bit operations for modular reduction. Mastrovito has proposed a bit-parallel multiplication algorithm and architecture when $f(x)$ is a trinomial [14]. He has shown that the number of both bit multiplications and bit additions needed is proportional to $2m^2$ when the degree of $f(x)$ is no greater than 15 and not equal to 8. The Karatsuba-Ofman (KOA) algorithm has also been considered for building bit-parallel finite field multipliers [1, 17]. An implementation of KOA [17] has shown that bit-parallel multiplication architectures in certain composite fields can have significantly lower complexity, compared to those proposed in [14]. However, the time delay of the architectures using the KOA can be longer. Polynomial dual basis and normal dual basis have also been considered for efficient multiplication [21, 19].

In this article, we prove that bit-parallel reduction modulo the irreducible polynomial costs only $(r-1)(m-1)$ when the irreducible polynomial $f(x)$ has the Hamming weight of r. Consequent work can be shown that a bit-parallel multiplier in $\mathbb{F}_{2^m}$ over $\mathbb{F}_2$ can be built with at most m^2 AND gates and m^2-1 XOR gates for any integer m when an irreducible trinomial of degree m exists. Polynomial basis bit-parallel squaring is also discussed. When the irreducible polynomial is chosen as a trinomial of form $x^m + x^k + 1$, then bit-parallel squaring operation can realized with no more than $\left\lfloor \frac{m+k-1}{2} \right\rfloor$ bit additions. Consequently, it is argued that to solve multiplicative inverse in $\mathbb{F}_{2^m}$ using polynomial basis can be as good as using normal basis.

The organization of this paper is as follows: Polynomial basis bit-parallel multiplication and squaring are discussed in Section 2 and Section 3, respectively. In Section 4, we argue that to solve multiplicative inverse using polynomial basis can be as good as using normal basis. Finally, a few concluding remarks are given in Section 5.

2 Bit-Parallel Polynomial Basis Multiplication in $\mathbb{F}_{2^m}$

Let the finite field $\mathbb{F}_{2^m}$ be generated with an irreducible r-term polynomial $f(x) = x^m + \sum_{i=0}^{r-2} x^{e_i}$, where $0 = e_0 < e_1 < \cdots < e_{r-2} < m$. Let $A(x) = \sum_{i=0}^{m-1} a_i x^i$ and $B(x) = \sum_{i=0}^{m-1} b_i x^i$ be any two elements in $\mathbb{F}_{2^m}$. Then, $C(x) = \sum_{i=0}^{m-1} c_i x^i \in \mathbb{F}_{2^m}$, the product of $A(x)$ and $B(x)$ can be obtained in two steps:

1. Polynomial multiplication:

$$S(x) = A(x)B(x), \tag{1}$$

where $S(x) = \sum_{k=0}^{2m-2} s_k x^k$, and s_k is given by

$$s_k \stackrel{\triangle}{=} \sum_{\substack{i+j=k \\ 0 \leqslant i,j \leqslant m-1}} a_i b_j, \quad k = 0, 1, 2, \ldots, 2m-2.$$

2. Reduction modulo the irreducible polynomial:

$$C(x) = S(x) \bmod f(x), \tag{2}$$

where $C(x) = \sum_{i=0}^{m-1} c_i x^i$, $c_i \in \mathbb{F}_2$.

Obviously, the complexities of polynomial basis bit-parallel multiplication in $\mathbb{F}_{2^m}$ are determined by these two parts. The complexity of the first step (polynomial multiplication) is independent of choice of the irreducible polynomial $f(x)$, and it has been shown to be $O(m \log m \log \log m)$ in bit operations [18]. We will show that the second step (modular reduction) requires at most $(r-1)(m-1)$ bit operations, where r is the Hamming weight of the irreducible polynomial $f(x)$.

2.1 Polynomial multiplication

In the first step of PB multiplication (1), if $S(x)$ is computed from $A(x)$ and $B(x)$ by the conventional polynomial multiplication method, it requires m^2 multiplications and $(m-1)^2$ additions in the ground field and the time delay is $T_A + \lceil \log_2 m \rceil T_X$. However, there are some asymptotically faster methods for polynomial multiplication over finite fields [3], such as, the Fast Fourier Transform method [3,11] and the Karatsuba-Ofman algorithm [10,1,17]. They can result in asymptotically fewer bit operations at the expense of longer time delay and/or certain costly pre- and post-computations. Another technique for polynomial basis multiplication that can combine polynomial multiplication with modulo reduction into one single step is called the Montgomery method [15,12].

2.2 Reduction modulo a polynomial

For modular reduction $C(x) = S(x) \bmod f(x)$, where $\deg f = m$, $\deg S \leqslant 2m-2$ and $\deg C \leqslant m-1$, if the conventional polynomial division method is used, the complexity is $O(m^2)$ in ground field operations. Mastrovito [14] has found that if the irreducible polynomial is chosen properly for $m \leqslant 15$, $m \neq 8$, the complexity of modulo reduction can be greatly reduced by using some partial sums. Paar [17] has also discussed this issue for certain small values of m. However, their methods are based on computer based exhaustive search and available for only moderately small size fields. In the following, we will present a new algorithm that can perform modulo reduction in $(r-1)(m-1)$ ground field operations for any irreducible polynomial $f(x)$ with the Hamming weight r.

Theorem 1. If the Hamming weight of the irreducible polynomial $f(x)$ is r, then the modular polynomial reduction (2) can be done with $(r-1)(m-1)$ bit operations.

Proof: Define

$$\sum_{i=0}^{m+l} s_i x^i \bmod f(x) \stackrel{\triangle}{=} \sum_{i=0}^{m-1} t_i^{(l)} x^i, \quad l = -1, 0, 1, \ldots, m-2. \tag{3}$$

$t_i^{(l)}$'s have the initial values $t_i^{(-1)} = s_i$, and, we try to solve for the 'final' values $t_i^{(m-2)} = c_i$, $i = 0, 1, 2, \ldots, m - 1$.

In the following, we shall prove by induction that the complexity of solving $t_i^{(m-2)}$, $i = 0, 1, 2, \ldots, m - 1$, is $(r - 1)(m - 1)$ bit operations.

When $l = 0$, from (3) we have

$$
\begin{aligned}
\sum_{i=0}^{m-1} t_i^{(0)} x^i = \sum_{i=0}^{m} s_i x^i &= \sum_{i=0}^{m-1} s_i x^i + s_m x^m \\
&= \sum_{i=0}^{m-1} t_i^{(-1)} x^i + s_m [1 + x^{e_1} + \cdots + x^{e_{r-2}}]
\end{aligned}
$$

Clearly, $t_i^{(0)} = \begin{cases} t_i^{(-1)} + s_m, & \text{if } i = 0, e_1, e_2, \ldots, e_{r-2}, \\ t_i^{(-1)}, & \text{if } 1 \leq i \leq m - 1, \text{ and } i \neq e_1, e_2, \ldots, e_{r-2}. \end{cases}$

It can be seen that $r - 1$ bit additions are required for obtaining $t_i^{(0)}$ from $t_i^{(-1)}$, $i = 0, 1, \ldots, m - 1$.

Assume when $0 \leq l < l'$, $r - 1$ bit-additions are required for obtaining $t_i^{(l)}$ from $t_i^{(l-1)}$, $i = 0, 1, \ldots, m - 1$. Then, when $l = l'$, we have

$$
\begin{aligned}
\sum_{i=0}^{m-1} t_i^{(l')} s^i = \sum_{i=0}^{m+l'} s_i x^i &= \sum_{i=0}^{m+l'-1} s_i x^i + s_{m+l'} x^{m+l'} \\
&= \sum_{i=0}^{m-1} t_i^{(l'-1)} x^i + s_{m+l'} x^{l'} x^m \\
&= \sum_{i=0}^{m-1} t_i^{(l'-1)} x^i + s_{m+l'} x^{l'} [1 + x^{e_1} + \cdots + x^{e_{r-2}}]
\end{aligned}
$$

If $m > l' + e_{r-2}$, then

$$
t_i^{(l')} = \begin{cases} t_i^{(l'-1)} + s_{m+l'}, & \text{if } i = l', l' + e_1, \ldots, l' + e_{r-2}, \\ t_i^{(l'-1)}, & \text{otherwise.} \end{cases}
$$

Obviously, $t_i^{(l')}$ can be computed from $t_i^{(l'-1)}$ using $r-1$ bit additions. Now suppose that $l' + e_{r'-1} < m \leqslant l' + e_{r'}, r' \in \{1, \ldots, r-2\}$, thus it follows

$$\sum_{i=0}^{m+l'} s_i x^i = \{\sum_{i=0}^{m-1} t_i^{(l'-1)} x^i + s_{m+l'} x^{l'} [1 + x^{e_1} + \cdots + x^{e_{r'-1}}]\}$$

$$+ s_{m+l'} x^{l'} [x^{e_{r'}} + x^{e_{r'+1}} + \cdots + x^{e_{r-2}}]$$

$$= \sum_{i=0}^{m-1} t_i^{(l',0)} x^i + s_{m+l'} x^{l'} [x^{e_{r'}} + x^{e_{r'+1}} + \cdots + x^{e_{r-2}}]$$

$$= \sum_{i=0}^{m-1} t_i^{(l',0)} x^i + s_{m+l'} x^{l'+e_{r'}} + \cdots + s_{m+l'} x^{l'+e_{r-2}} \qquad (4)$$

where $\sum_{i=0}^{m-1} t_i^{(l',0)} \triangleq \sum_{i=0}^{m-1} t_i^{(l'-1)} x^i + s_{m+l'} x^{l'} [1 + x^{e_1} + \cdots + x^{e_{r'-1}}]$. Since we have

$$t_i^{(l',0)} = \begin{cases} t_i^{(l'-1)} + s_{m+l'}, & \text{if } i = l', l' + e_1, \ldots, l' + e_{r'-1}, \\ t_i^{(l'-1)}, & \text{otherwise,} \end{cases}$$

it can be seen that r' bit additions are required to obtain $t_i^{(l',0)}$ from $t_i^{(l'-1)}, i = 0, 1, \ldots, m-1$.

In the following we shall prove that $t_i^{(l')}$, $i = 0, 1, \ldots, m-1$, can be obtained from $t_i^{(l',0)}$ with $r - r' - 1$ bit additions. Define

$$\sum_{i=0}^{m-1} t_i^{(l',1)} x^i \triangleq \sum_{i=0}^{m-1} t_i^{(l',0)} x^i + s_{m+l'} x^{l'+e_{r'}} \mod f(x), \ i = 0, 1, \ldots, m-1. \quad (5)$$

Since $0 \leq l' + e_{r'} - m < l'$, we have

$$\sum_{i=0}^{m-1} t_i^{(l'+e_{r'}-m)} x^i = \sum_{i=0}^{m-1} t_i^{(l'+e_{r'}-m-1)} x^i + s_{l'+e_{r'}} x^{l'+e_{r'}} \mod f(x), \ i = 0, 1, \ldots, m-1.$$

$$(6)$$

Since $t_i^{(l'+e_{r'}-m)}$ has been obtained from $t_i^{(l'+e_{r'}-m-1)}$ with $r-1$ bit additions as assumed, comparing (5) to (6), we can see that (5) and (6) can be combined together to save bit operations. That is, when $l = l' + e_{r'} - m$, instead of performing (6), we perform

$$\sum_{i=0}^{m-1} t_i^{(l'+e_{r'}-m-1)} x^i + (s_{l'+e_{r'}} + s_{m+l'}) x^{l'+e_{r'}} \mod f(x) = \sum_{i=0}^{m-1} t_i^{(l'+e_{r'}-m,\star)} x^i$$

$$(7)$$

with r bit additions, while (5) can be saved. In the sense of the count of bit operations, we may equivalently say that (5) requires one bit addition, while

(6) still needs $r - 1$ bit operations. Similar arguments can be applied to the remaining $r - r' - 2$ terms $s_{m+l'} x^{l'+e_j}, j = r' + 1, r' + 2, \ldots, r - 2$ in (4). Thus for $l = l'$, $r - 1$ bit additions are required for obtaining $t_i^{(l')}$ from $t_i^{(l'-1)}$, for $i = 0, 1, \ldots, m - 1$. Therefore, to compute $t_i^{(l)}$ from $t_i^{(l-1)}, i = 0, 1, \ldots, m - 1$, needs $r - 1$ bit additions for any integer l. We conclude that computing $c_i = t_i^{(m-2)}$ from $s_i, i = 0, 1, \ldots, 2m - 2$ requires $(m - 1)(r - 1)$ bit additions. $\qquad\square$

Theorem 1 can be easily extended to $\mathbb{F}_{q^m}$ as it is stated in Theorem 2. A proof for Theorem 2 is analogous to that of Theorem 1.

Theorem 2. If the monic irreducible polynomial $f(x) \in \mathbb{F}_q[x]$ of degree m has the Hamming weight of r, then the modular polynomial reduction in polynomial basis multiplication can be done with $(r - 1)(m - 1)$ multiplications and $(r - 1)(m - 1)$ additions in $\mathbb{F}_q$.

If the conventional method for polynomial multiplication is used, some results of consequent work on finite field multiplier architecture are shown as follows:

If the finite field $\mathbb{F}_{2^m}$ is generated with an irreducible trinomial $f(x) = 1 + x^k + x^m$, $1 \leqslant k \leqslant \left\lfloor \frac{m}{2} \right\rfloor$, then a bit-parallel polynomial basis multiplier can be constructed with $\mathbf{C}_{SA} = m^2$,

(i) $\mathbf{C}_{SX} = m^2 - 1$ and $\mathbf{C}_T = T_A + (\lceil \log_2 m \rceil + 1) T_X$ for $k = 1$;
(ii) $\mathbf{C}_{SX} = m^2 - 1$ and $\mathbf{C}_T = T_A + (\lceil \log_2 m \rceil + 2) T_X$ for $1 < k < \frac{m}{2}$;
(iii) $\mathbf{C}_{SX} = m^2 - \frac{m}{2}$ and $\mathbf{C}_T = T_A + (\lceil \log_2 m \rceil + 1) T_X$ for $k = \frac{m}{2}$.

3 Polynomial Basis Bit-Parallel Squaring

3.1 Complexity of polynomial basis bit-parallel squaring in $\mathbb{F}_{2^m}$

Let $f(x)$ be the irreducible polynomial over $\mathbb{F}_2$ generating the field $\mathbb{F}_{2^m}$. Let $A(x) = \sum_{i=0}^{m-1} a_i x^i$ be the polynomial representation of an arbitrary element of $\mathbb{F}_{2^m}$. The squaring operation of $A(x)$ is $C(x) \triangleq \sum_{i=0}^{m-1} c_i x^i = A^2(x) \bmod f(x) = a_0 + a_1 x^2 + a_2 x^4 + \ldots + a_{\lceil \frac{m}{2} \rceil} x^{2\lceil \frac{m}{2} \rceil} + \ldots + a_{m-1} x^{2m-2} \bmod f(x)$. It can be seen that squaring in $\mathbb{F}_{2^m}$ is actually a case of polynomial modular reduction that has been discussed in the last section, where the degree of each squared terms in $A^2(x)$ is an even integer between 0 and $2m - 2$. From the discussion in the last section, the following corollary is obvious.

Corollary 1. Let the field $\mathbb{F}_{2^m}$ be generated with the irreducible r-term polynomial $f(x)$ of degree m. Then squaring a field element in parallel can be performed with at most $(r - 1)(m - 1)$ addition operations in $\mathbb{F}_2$.

When $f(x)$ is an irreducible trinomial, however, both the size complexity and time complexity can be further reduced.

Theorem 3. Let the field $\mathbb{F}_{2^m}$ be generated with the irreducible trinomial $f(x) = x^m + x^k + 1$, where m is even and k odd. Then squaring a field element in a bit-parallel fashion can be performed with at most $\frac{m+k-1}{2}$ bit operations. □

Proof: Let

$$A^2(x) = \sum_{i=0}^{m-1} a_i x^{2i} = \sum_{i=0}^{2m-2} a_i' x^i,$$

where $a_i' \triangleq a_{\frac{i}{2}}$ if i even, and 0 if i odd. Define

$$\sum_{i=0}^{m+2l} a_i' x^i \bmod f(x) \triangleq \sum_{i=0}^{m-1} t_i^{(l)} x^i, \quad l = -1, 0, 1, \ldots, \frac{m}{2} - 1.$$

The terms $t_i^{(l)}$'s have their initial values $t_i^{(-1)} = a_i'$, and we try to solve the final values $t_i^{(\frac{m}{2}-1)} = c_i, i = 0, 1, \ldots, m-1$.

When $l = 0$,

$$\sum_{i=0}^{m-1} t_i^{(0)} x^i = \sum_{i=0}^{m} a_i' x^i = \sum_{i=0}^{m-1} a_i' x^i + a_m' x^m = \sum_{i=0}^{m-1} a_i' x^i + a_m'(1 + x^k).$$

$$\therefore \ t_i^{(0)} = \begin{cases} a_i' + a_m', & i = 0; \\ a_m', & i = k. \\ a_i', & 0 < i \leqslant m-1, i \text{ even}; \\ 0, & i \text{ odd}, i \neq k; \end{cases}$$

Clearly, one bit addition is needed to compute $t_i^{(0)}$ from $t_i^{(-1)}$, $i = 0, 1, \ldots, m-1$. For $l > 0$, we have

$$\sum_{i=0}^{m-1} t_i^{(l)} x^i = \sum_{i=0}^{m+2l} a_i' x^i$$

$$= \sum_{i=0}^{m+2(l-1)} a_i' x^i + a_{m+2l}' x^{m+2l}$$

$$= \sum_{i=0}^{m-1} t_i^{(l-1)} x^i + a_{m+2l}' x^{2l}(1 + x^k)$$

$$= \sum_{i=0}^{m-1} t_i^{(l-1)} x^i + a_{m+2l}' x^{2l} + a_{m+2l}' x^{k+2l}.$$

If $k + 2l < m$ or $l < \frac{m-k}{2}$, then

$$
t_i^{(l)} = \begin{cases}
t_i^{(l-1)}, & 0 \leqslant i \leqslant m-1, i \neq 2l, i \text{ even; or } i = k, k+2, \ldots, k+2(l-1); \\
0, & i \text{ odd}, i \neq k, k+2, \ldots, k+2l; \\
t_i^{(l-1)} + a'_{m+2l}, & i = 2l; \\
a'_{m+2l}, & i = k+2l.
\end{cases}
$$

(8)

It can be seen that only one bit addition is required to compute $t_i^{(l)}$ from $t_i^{(l'-1)}$ for $0 < l < \frac{m-k}{2}$ and $i = 0, 1, \ldots, m-1$.

In the following we proceed with induction. When $l = \frac{m-k+1}{2}$ ($\because m - k$ is odd) and $l < \frac{m}{2}$, we have

$$
\begin{aligned}
\sum_{i=0}^{m-1} t_i^{(l)} x^i &= \sum_{i=0}^{m-1} t_i^{(l-1)} x^i + a'_{m+2l} x^{m+2l} \\
&= \sum_{i=0}^{m-1} t_i^{(l-1)} x^i + a'_{m+2l} x^{2l} + a'_{m+2l} x^{k+2l} \\
&= \sum_{i=0}^{m-1} t_i^{(l-1)} x^i + a'_{m+2l} x^{2l} + a'_{m+2l} x + a'_{m+2l} x^{k+1}.
\end{aligned}
$$

Then,

$$
t_i^{(l)} = \begin{cases}
t_i^{(l-1)} + a'_{m+2l}, & i = 2l \text{ or } i = k+1; \\
a'_{m+2l}, & i = 1; \\
t_i^{(l-1)}, & i \text{ even}, i \neq 2l, k+1; \\
& \text{or } i = k, k+2, \ldots, k+2(l-1); \\
0, & i \text{ odd}, i \neq k, k+2, \ldots, k+2(l-1) \text{ and } i \neq 1.
\end{cases}
$$

(9)

Obviously, two bit additions are required to compute $t_i^{(l)}$ from $t_i^{(l-1)}, i = 0, 1, \ldots, m-1$.

Assume that for $\frac{m-k+1}{2} \leqslant l < l'$, (9) holds, then for $l = l' < \frac{m}{2}$, we have

$$\sum_{i=0}^{m-1} t_i^{(l')} x^i = \sum_{i=0}^{m-1} t_i^{(l'-1)} x^i + a_{m+2l'} x^{m+2l'}$$

$$= \sum_{i=0}^{m-1} t_i^{(l'-1)} x^i + a'_{m+2l'} x^{2l'} [1 + x^k]$$

$$= \sum_{i=0}^{m-1} t_i^{(l'-1)} x^i + a'_{m+2l'} x^{2l'} + + a'_{m+2l'} x^{2l'+k}$$

$$= \sum_{i=0}^{m-1} t_i^{(l'-1)} x^i + a'_{m+2l'} x^{2l'} + a'_{m+2l'} x^{k+2l'-m} + a'_{m+2l'} x^{2k+2l'-m}$$

$$= \sum_{i=0}^{m-1} t_i^{(l',0)} x^i + a'_{m+2l'} x^{2k+2l'-m}, \tag{10}$$

where $t_i^{(l',0)}$ is defined by $\sum_{i=0}^{m-1} t_i^{(l',0)} x^i \triangleq \sum_{i=0}^{m-1} t_i^{(l'-1)} x^i + a'_{m+2l'} x^{2l'} + a'_{m+2l'} x^{k+2l'-m}$.
Since $2l' < m$, and $k + 2l' - m$ is odd and less than k,

$$t_i^{(l',0)} = \begin{cases} t_i^{(l'-1)} + a'_{m+2l'}, & i = 2l'; \\ a'_{m+2l'}, & i = k + 2l' - m; \\ t_i^{(l'-1)}, & 0 \leqslant i \leqslant m-1, i \neq 2l', i \text{ even}; \text{ or } i = k, k+2, \dots, k+2(l'-1); \\ & \text{or } i = 1, 3, \dots, k+2(l'-1) - m; \\ 0, & i \text{ odd}, i \neq k, k+2, \dots, k+2(l'-1), i \neq 1, 3, \dots, \\ & k + 2(l'-1) - m. \end{cases}$$

$$\tag{11}$$

Thus it requires one bit addition to obtain $t_i^{(l',0)}$ from $t_i^{(l'-1)}$, $i = 0, 1, \dots, m-1$.

When $2k + 2l' - m < m$, we have $t_i^{(l')} = t_i^{(l',0)}$ if $i \neq 2k + 2l' - m$, otherwise $t_i^{(l')} = t_i^{(l',0)} + a'_{m+2l'}$. In this case it is therefore two bit operations are required to compute $t_i^{(l')}$ from $t_i^{(l'-1)}$ for $i = 0, \dots, m-1$.

When $2k + 2l' - m \geqslant m$, consider

$$\sum_{i=0}^{m-1} t_i^{(k+l'-m)} = \sum_{i=0}^{m-1} t_i^{(k+l'-m-1)} + a'_{2k+2l'-m} x^{2k+2l'-m}. \tag{12}$$

It can be seen that the last terms of the right hand side of (10) and (12) are the same except for the coefficient. At the step $l = k + l' - m$, instead of performing (12), if we perform

$$\sum_{i=0}^{m-1} t_i^{(k+l'-m,*)} = \sum_{i=0}^{m-1} t_i^{(k+l'-m-1)} + (a'_{2k+2l'-m} + a'_{m+2l'}) x^{2k+2l'-m} \tag{13}$$

at the cost of one more bit operation, then at step $l = l'$, the term $t_i^{(l')}$ can be computed from $t_i^{(l'-1)}, i = 0, \ldots, m-1$ with only one bit operation. Equivalently, we might say that at step $l = l'$, term $t_i^{(l')}$ can be computed from $t_i^{(l'-1)}$, $i = 0, 1, \ldots, m-1$, at the cost of two bit operations. Thus for $\frac{m-k+1}{2} \leqslant l < \frac{m}{2} - 1$, it requires two bit additions at each step.

We conclude that the total cost for computing $c_i = t_i^{(\frac{m}{2}-1)}$ from $t_i^{(-1)} = a_i', i = 0, 1, \ldots, m-1$ is $\frac{m-k+1}{2} + 2(\frac{m}{2} - 1 - \frac{m-k-1}{2}) = \frac{m+k-1}{2}$ bit operations. $\qquad\square$

Theorem 4. Let the field $\mathbb{F}_{2^m}$ be generated with the irreducible trinomial $f(x) = x^m + x^k + 1$, where m is odd and k even. Then bit-parallel squaring in $\mathbb{F}_{2^m}$ can be performed with at most $\frac{m+k-1}{2}$ bit additions.

Theorem 5. Let the field $\mathbb{F}_{2^m}$ be generated with the irreducible trinomial $f(x) = x^m + x^k + 1$, where both m and k are odd. Then bit-parallel squaring in $\mathbb{F}_{2^m}$ can be performed with at most $\frac{m-1}{2}$ bit additions. $\qquad\square$

Proofs of Theorems 4 and 5 are similar to that of Theorem 3.

Some results of consequent work on implementation of bit-parallel squaring operation done by us is given below.

Theorem 6. Let the field $\mathbb{F}_{2^m}$ be generated with the irreducible trinomial $f(x) = x^m + x^k + 1$, where $m + k$ is odd. Then a bit-parallel squarer can be implemented with at most $\frac{m+k-1}{2}$ XOR gates. For $k = 1$ or 2, the incurred time delay is T_X, and for $2 < k \leqslant \frac{m}{2}$, it is $2T_X$.

Theorem 7. Let the field $\mathbb{F}_{2^m}$ be generated with the irreducible trinomial $f(x) = x^m + x^k + 1$, where both m and k are odd. Then a bit-parallel squarer can be implemented with at most $\frac{m-1}{2}$ XOR gates. The incurred time delay is T_X if $k = 1$, and $2T_X$ if $2 < k < \frac{m}{2}$.

4 Inversion

Inversion operation is required in elliptic curve cryptosystem when computing point multiples. This operation is usually performed with two methods. One is the extended Euclidean algorithm and the other is to exponentiate the element using the following identity

$$x^{-1} = x^{2^m - 2}. \tag{14}$$

The extended Euclid's algorithm usually requires the field element having a polynomial basis representation [5], where the most used operations are field addition, shifting and loading. Efficient algorithms have been proposed for the second method, for example, [8] and [2]. Both algorithms use about $\frac{3}{2}\log(m-1)$

field multiplications on average[1] and $m-1$ squaring operations. It has been generally accepted that normal basis should be used for this method since squaring in normal basis is only a cycle shift of the coefficients [8, 2]. However, with the results presented in this paper, we argue that to solve inverse using (14) polynomial basis representation can be as efficient as normal basis representation.

It has been shown in [7] that a normal basis multiplication can be performed with $2m^2 - 1$ bit operations when a type I optimal normal basis is used. If a type II optimal normal basis or a non-optimal normal basis is used, it takes at least $3m^2$ bit operations to accomplish a field multiplication [16, 6]. On the other hand, If the polynomial basis generated with an irreducible trinomial is used, a bit-parallel multiplication needs at most $2m^2 - 1$ bit operations while a bit-parallel squaring costs not greater than m bit operations. In this case, the complexity to solve the inverse using (14) in terms of bit operations with different bases is given in the following table.

	Multiplications	Squarings
Type I optimal NB	$(2m^2 - 1)(\log_2(m-1) + H(m-1) - 1)$	–
Type II optimal NB	$\geq 3m^2 (\log_2(m-1) + H(m-1) - 1)$	–
Trinomial-generated PB	$(2m^2 - 1)(\log_2(m-1) + H(m-1) - 1)$	$< m(m-1)$

Table 1. The complexity (in bit operations) of inversion using the algorithms in [8].

It can be seen from the table that the complexity using polynomial basis generated with an irreducible trinomial is comparable to that using type I optimal normal basis, while much smaller than that using type II optimal normal basis or non-optimal basis. Moreover, given finite field $\mathbb{F}_{2^m}$ there is more chance that an irreducible trinomial exists than that a type I optimal normal basis does. In fact, for $2 \leq m \leq 1000$, there is an irrducible trinomial in $\mathbb{F}_{2^m}$ for 545 values of m while there exists a type I optimal normal basis for only 67 values of m [4, 9].

5 Concluding Remarks

In this article, we have shown that a bit-parallel multiplication operation in $\mathbb{F}_{2^m}$ using polynomial basis can be performed in $2m^2 + (r-3)m - (r-2)$ bit operations. We have also proven that a bit-parallel squaring operation using polynomial basis costs not more than $\left\lfloor \frac{m+k-1}{2} \right\rfloor$ bit operations if an irreducible trinomial $x^m + x^k + 1$ over $\mathbb{F}_2$ is used. Consequently, it is argued that to solve multiplicative inverse in $\mathbb{F}_{2^m}$ using polynomial basis can be as good as using normal basis.

Consequent work on implementation has shown that the resultant bit-parallel multiplier and bit-parallel squarer also have low time delay.

---

[1] Assume that the Hamming weight of $m-1$ is $\frac{1}{2}\log(m-1)$ on average.

Acknowledgements

This work was done when the author worked for his Ph.D degree with the Dept of ECE, University of Waterloo. The author thanks Professor Hasan and Professor Blake for their encouragement and valuable comments.

References

1. Afanasyev, V.B.: On the complexity of finite field arithmetic. Proc 5th Joint Soviet-Swedish Intern. Workshop on IT, Moscow, USSR, 1991, 9-12
2. Agnew, G.B., Beth, R., Mullin, R.C., Vanstone, S.A.: Arithmetic operations in GF(2^m). J. Cryptology **6** (1993) 3-13
3. Aho, A.V., Hopcroft, J.E., Ullman, J.D.: The Design and Analysis of Computer Algorithms. Addison-Wesley Publ. Co., Reading, MA, 1974
4. Blake, I.F., Gao, S., Lambert, R.: Constructive Problems for Irreducible Polynomials over Finite Fields. Canadian Workshop on IT, Springer-Verlag, 1993
5. Brunner, H., Curiger, A., Hofstetter, M.: On computing multiplicative inverse in GF(2^m). IEEE Trans. Comput. **42** (1993) 1010-1015
6. Gao, S., Vanstone, S.A.: On orders of optimal normal basis generators. Math. Comp. **64** (1995) 1227-1233
7. Hasan, M.A., Wang, M., Bhargava, V.K.: A modified Massey-Omura parallel multiplier for a class of finite fields. IEEE Trans. Comput. **42** (1993) 1278-1280
8. Itoh, T., Tsujii, S.: A fast algorithm for computing multiplicative inverse in GF(2^m) using normal bases. Inform. and Comput. **78** (1988) 171-177
9. Itoh, T., Tsujii, S.: Structure of parallel multipliers for a class of fields GF(2^m). Inform. and Comput. **83** (1989) 21-40
10. Karatsuba, A., Ofman, Y.: Multiplication of multidigit numbers on automata. Sov. Phys.-Dokl. (English translation), **7** (1963) 595-596
11. Knuth, D.E.: The Art of Computer Programming: Seminumerical Algorithms. Addison-Wesley Publishing Company, Reading, MA, 1981
12. Koç, Ç. K., Acar, T.: Montgomery multiplication in GF(2^k). Designs, Codes and Cryptography, **14** (1998) 57-69
13. Massey, J.L., Omura, J.K.: Computational method and apparatus for finite field arithmetic. U.S. Patent No.4587627, 1984.
14. Mastrovito, E.D.: VLSI Architectures for Computations in Galois Fields. Ph.D Thesis, Linköping University, 1991, Linköping, Sweden
15. Montgomery, P.L.: Modular multiplication without trial division. Math. Comp. **44** (1985) 519-521
16. Mullin, R., Onyszchuk, I., Vanstone, S.A., Wilson, R.: Optimal normal bases in GF(p^n). Disc. Appl. Math. **22** (1988) 149-161
17. Paar, C.: Efficient VLSI Architectures for Bit-Parallel Computation in Galois Fields. Ph.D Thesis, VDI-Verlag, Düsseldorf, 1994
18. Schönhage, A.: Schnelle Multiplikation von Polynomen uber Korpern der Charakteristik 2. Acta Inf. **7** (1977) 395-398
19. Wang, C.C.: An algorithm to design finite field multipliers using a self-dual normal basis. IEEE Trans. Comput. **38** (1989) 1457-1459
20. Wu, H.: Efficient Computations in Finite Fields with Cryptographic Significance. Ph.D Thesis, University of Waterloo, Waterloo, Canada, 1998
21. Wu, H., Hasan, M.A., Blake, I.F.: Low complexity weakly dual basis bit-parallel multiplier over finite fields. IEEE Trans. Comput. **47** (1998) 1223-1234

Resistance against Differential Power Analysis for Elliptic Curve Cryptosystems

Jean-Sébastien Coron

Ecole Normale Supérieure
45 rue d'Ulm
Paris, F-75230, France
coron@clipper.ens.fr

Gemplus Card International
34 rue Guynemer
Issy-les-Moulineaux, F-92447, France
coron@gemplus.com

Abstract. Differential Power Analysis, first introduced by Kocher *et al.* in [14], is a powerful technique allowing to recover secret smart card information by monitoring power signals. In [14] a specific DPA attack against smart-cards running the DES algorithm was described. As few as 1000 encryptions were sufficient to recover the secret key. In this paper we generalize DPA attack to elliptic curve (EC) cryptosystems and describe a DPA on EC Diffie-Hellman key exchange and EC El-Gamal type encryption. Those attacks enable to recover the private key stored inside the smart-card. Moreover, we suggest countermeasures that thwart our attack.

Keywords. Elliptic curve, power consumption, Differential Power Analysis.

1 Introduction

The use of elliptic curve in cryptography was first proposed by Miller [17] and Koblitz [12] in 1985. Since that time, a lot of attention has been paid to elliptic curves for cryptographic applications and it has become increasingly common to implement public-key protocols on elliptic curves over large finite field. Elliptic curves (EC) provide a group structure, which can be used to translate existing discrete-logarithm cryptosystems into the context of EC. The discrete logarithm problem in a cyclic group G of order n with generator g refers to the problem of finding x given some element $y = g^x$ of G. The discrete logarithm problem over an EC seems to be much harder than in other groups such as the multiplicative group of a finite field. No subexponential-time algorithm is known for the discrete logarithm problem in the class of *non-supersingular* EC. Consequently, keys can be much smaller in the EC context, typically about 160 bits.

In this paper we consider attacks based on the monitoring of power consumption of smart-card EC implementation. Differential Power Analysis, first described by Kocher *et al.* in [14], is a powerful technique that exploit the leakage of information related to power consumption. The attack was successfully applied to a DES implementation; as few as 1000 encryptions were sufficient to recover the secret key [14]. More recently, the resistance of smart-card implementations of the AES candidates against monitoring power consumption was considered in [1, 3, 5]. The conclusion was that straightforward implementations of AES candidates were highly vulnerable to power analysis. In this paper we show that naive implementations of ECC are also highly vulnerable to power analysis.

The paper is organized as follows. After recalling the principle of EC operations in section 2, we describe in section 3 the principle of our power consumption attack. In section 4, we apply the attack to some common discrete-logarithm based cryptosystems such as Diffie-Hellman key exchange [7] and El-Gamal public-key encryption [8]. Finally we suggest three countermeasures that prevent our attack.

2 Elliptic curve group operation

2.1 Definition of an elliptic curve

An elliptic curve is the set of points (x, y) which are solutions of a bivariate cubic equation over a field K (see [16]). An equation of the form :

$$y^2 + a_1 xy + a_3 y = x^3 + a_2 x^2 + a_4 x + a_6 \tag{1}$$

where $a_i \in K$, defines an elliptic curve over K.

If char $K \neq 2$ and char $K \neq 3$, equation (1) can be transformed to :

$$y^2 = x^3 + ax + b$$

with $a, b \in K$.

In the field $GF(2^n)$ of characteristic 2, equation (1) can be reduced to the form :

$$y^2 + xy = x^3 + ax^2 + b$$

with $a, b \in K$.

The set of points on an elliptic curve, together with a special point $\mathcal{O}$ called the *point at infinity* can be equipped with an Abelian group structure by the following addition operation :

Addition formula [16] for char $K \neq 2, 3$:

Let $P = (x_1, y_1) \neq \mathcal{O}$ be a point, the inverse of P is $-P = (x_1, -y_1)$. Let $Q = (x_2, y_2) \neq \mathcal{O}$ be a second point with $Q \neq -P$, the sum $P + Q = (x_3, y_3)$ can be calculated as :

$$x_3 = \lambda^2 - x_1 - x_2$$
$$y_3 = \lambda(x_1 - x_3) - y_1$$

with

$$\lambda = \begin{cases} \dfrac{y_2 - y_1}{x_2 - x_1}, & \text{if } P \neq Q, \\[2mm] \dfrac{3x_1^2 + a}{2y_1}, & \text{if } P = Q. \end{cases}$$

To subtract the point $P = (x, y)$, one adds the point $-P$.

Addition formula for char $K = 2$:

Let $P = (x_1, y_1) \neq \mathcal{O}$ be a point, the inverse of P is $-P = (x_1, x_1 + y_1)$. Let $Q = (x_2, y_2) \neq \mathcal{O}$ be a second point with $Q \neq -P$, the sum $P + Q = (x_3, y_3)$ can be calculated as :

$$x_3 = \lambda^2 + \lambda + x_1 + x_2 + a$$
$$y_3 = \lambda(x_1 + x_3) + x_3 + y_1$$
$$\lambda = \frac{y_1 + y_2}{x_1 + x_2}$$

if $P \neq Q$ and :

$$x_3 = \lambda^2 + \lambda + a$$
$$y_3 = x_1^2 + (\lambda + 1)x_3$$
$$\lambda = x_1 + \frac{y_1}{x_1}$$

if $P = Q$.

2.2 Computing a multiple of a point

The operation of adding a point P to itself d times is called *scalar multiplication* by d and denoted dP. Scalar multiplication is the basic operation for EC protocols. Scalar multiplication in the group of points of an elliptic curve is the analogous of exponentiation in the multiplicative group of integers modulo a fixed integer m.

Computing dP can be done with the straightforward *double-and-add* approach based on the binary expansion of $d = (d_{\ell-1}, \ldots, d_0)$ where $d_{\ell-1}$ is the most significant bit of d (the method is the analogous of the *square-and-multiply* algorithm for exponentiation) :

```
Algorithm 1 (Double-and-add)
  input P
  Q ← P
  for i from ℓ − 2 to 0 do
      Q ← 2Q
```

```
    if  d_i = 1  then  Q ← Q + P
output  Q
```

Various techniques exist to speed-up scalar multiplication by reducing the number of elementary point operations : see [9] for a good survey. If the point P is known in advance, it may be advantageous to precompute a table of multiples of P [2]. Because elliptic curve subtraction has the same cost as addition, the previous *double-and-add* algorithm can be improved with the *addition-subtraction* algorithm which uses a signed binary expansion of d :

$$d = \sum_{i=0}^{\ell-1} c_i 2^i$$

with $c_i \in \{-1, 0, 1\}$.

The *non-adjacent form* (NAF) of d is a signed binary expansion of d with $c_i c_{i+1} = 0$ for all $i \geq 0$. Each positive integer has a unique NAF. Moreover, the NAF of d has the fewest nonzero coefficients of any signed binary expansion of d [9]. [18] describes an algorithm that generates the NAF of any positive integer.

```
Algorithm 2 (Addition-subtraction method)
    input P
  Q ← P
  for i from ℓ - 2 to 0 do
      Q ← 2Q
      if  c_i = 1  then  Q ← Q + P
      if  c_i = -1  then  Q ← Q - P
  output  Q
```

The double-and-add method and addition-subtraction method can be generalized to the *m-ary method*, the *window method* and the *signed binary window method* [9, 15].

The problem of finding a method to compute dP with the fewest number of elliptic curve group operations for a given d is equivalent to finding the shortest addition-subtraction chain for d [9]. An *addition chain* [11] for d is a sequence of positive integers :

$$a_0 = 1 \rightarrow a_1 \rightarrow a_2 \rightarrow \ldots \rightarrow a_r = d$$

such that $a_i = a_j + a_k$, for some $k \leq j < i$, for all $i = 1, 2, \ldots, r$.

An addition chain can be extended to an *addition-subtraction chain* [11] with $a_i = \pm a_j \pm a_k$ in place of $a_i = a_j + a_k$. The shortest addition-subtraction chain for d gives the fewest number of elliptic group operations for computing dP by computing $a_1 P, a_2 P, \ldots a_r P = dP$.

3 Recovering d in $Q = dP$ from the power consumption

In 1998, Kocher described in a technical draft [14] Simple Power Attacks (SPA) and Differential Power Analysis (DPA) on DES. A SPA consists in observing the power consumption of one single execution of a cryptographic algorithm. A DPA is more sophisticated and powerful. It consists in performing a statistical analysis of many executions of the same algorithm with different inputs.

Here we show that monitoring power consumption during the computation of $Q = dP$ knowing P may enable to recover d. First we show that a naive implementation of scalar multiplication may be vulnerable to SPA. However, it is not difficult to make the implementation resistant against SPA. We then describe a DPA attack of an implementation of scalar multiplication.

3.1 Resistance against SPA

Power consumption attacks are based on the observation that the power consumed at a given time during cryptographic process is related to the instruction being executed and the data being manipulated. Power consumption enables to visually identify large features, for example the main loop in algorithm 1. Power consumption analysis may also enable to distinguish between instruction being executed. For example, it might be possible to distinguish between point doubling and point addition in algorithm 1, thereby revealing the bits of the exponent d.

In order to be resistant against SPA, the instructions performed during a cryptographic algorithm should not depend on the data being processed, *e.g.* there should not be any branch instructions conditioned by the data. It is easy to modify algorithm 1 to achieve this goal :

Algorithm 1' (Double-and-add resistant against SPA)
    input P
    $Q[0] \leftarrow P$
    for i from $\ell - 2$ to 0 do
        $Q[0] \leftarrow 2Q[0]$
        $Q[1] \leftarrow Q[0] + P$
        $Q[0] \leftarrow Q[d_i]$
    output $Q[0]$

3.2 DPA against double-and-add algorithm

In this section we describe a DPA against an implementation of algorithm 1'. We assume that the algorithm is performed in constant time. Otherwise the implementation may be subject to timing attack [13] and Simple Power Attacks [14].

DPA on DES [6] algorithm as described in [14] uses correlation between power consumption and specific key-dependent bits which appear at known steps of the encryption computation. For example, a selected bit b at the output of one SBOX of the first round will depend on the known input message and 6 unknown bits of the key. In [14], the correlation between power consumption and b is computed for the 64 possible values of the 6 unknown bits of the key. The correlation is likely to be maximal for the correct guess of the 6 bits of the key. The attack can be repeated for the remaining SBOXes, thus revealing 48 bits of the key. The remaining 8 bits of the key can be recovered by exhaustive search.

A Differential Power Analysis on algorithm 1' in section 3.1 can be performed by noticing that at step j the processed point Q depends only on the first bits $(d_{\ell-1}, \ldots, d_j)$ of d. Now assume that we know how points are represented in memory during computation and select a particular bit (the same for all points) of this representation. When point Q is processed, power consumption will be correlated to this specific bit of Q. No correlation will be observed with a point not computed inside the card. Thus it is possible to successively recover the bits of the exponent by guessing which points are computed by the card.

The second most significant bit $d_{\ell-2}$ of d can be recovered by computing the correlation between power consumption and any specific bit of the binary representation of $4P$. If $d_{\ell-2} = 0$, $4P$ is computed during algorithm 1', and power consumption is thus correlated with any specific bit of $4P$. Otherwise if $d_{\ell-2} = 1$, $4P$ is never computed, and no correlation will be observed with $4P$. This gives $d_{\ell-2}$. The following bits of d can be recursively recovered in the same way.

Assume that algorithm 1' is performed k times with distinct $P_1, P_2, \ldots, P_k$ to compute $Q_1 = dP_1, Q_2 = dP_2, \ldots, Q_k = dP_k$. Let $C_i(t)$ be the power consumption associated with the i-th execution of the algorithm for $1 \leq i \leq k$. Let s_i be any specific bit of the binary representation of $4P_i$ for $1 \leq i \leq k$. The correlation function $g(t)$ between s_i and $C_i(t)$ can be computed as follows :

$$g(t) = < C_i(t) >_{i=1,2\ldots,k|s_i=1} - < C_i(t) >_{i=1,2,\ldots,k|s_i=0} \qquad (2)$$

Assume that the points $4P_i$ are processed at time $t = t_1$, power consumption $C_i(t_1)$ will then be correlated with the specific bit s_i of the binary representation of $4P_i$. The average of power consumption for those points $4P_i$ for which $s_i = 1$ will be different from the power consumption for the points $4P_i$ for which $s_i = 0$, and function $g(t)$ will present a "peak" at time $t = t_1$. If the points $4P_i$ are never computed, no "peak" will be observed in function $g(t)$. This is illustrated in figure 1 and 2.[1]

3.3 Extending the attack to any scalar multiplication algorithm

In this section we show how to extend the previous attack to any scalar multiplication algorithm executed in constant time with a constant addition-subtraction chain, i.e. for any point P the algorithm computes the sequence of point :

---

[1] Real power consumption curves were *voluntarily* excluded from this paper to avoid straightforward product identification.

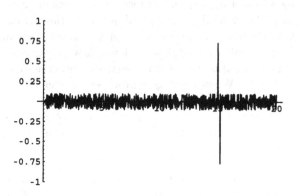

Fig. 1. Simulated correlation function $g(t)$ between the points $4P_i$ and power consumption $C_i(t)$ when $d_{\ell-2} = 0$. A peak is observed corresponding to the computation of $4P_i$ inside the card.

Fig. 2. Simulated correlation function $g(t)$ between the points $4P_i$ and power consumption $C_i(t)$ when $d_{\ell-2} = 1$. No peak is observed since the points $4P_i$ are never computed by the card.

$$a_0 P = P \rightarrow a_1 P \rightarrow a_2 P \rightarrow \ldots \rightarrow a_r P = dP$$

such that $a_i = \pm a_j \pm a_k$, for some $k \leq j < i$, for all $i = 1, 2, \ldots, r$.

The attack consists in successively guessing the a_i starting from $a_0 = 1$ to $a_r = d$. At step $i \geq 1$, one constructs the set A_i of all possible $a_i' = \pm a_j \pm a_k$ for all $0 \leq k \leq j < i$, and for each $a_i' \in A_i$ computes the correlation function $g(t)$ between the point $a_i' P$ and power consumption. If a peak can be observed in $g(t)$, this will indicate that the point $a_i' P$ has been computed by the device and thus $a_i = a_i'$. This enables to recover $d = a_r$ in $\mathcal{O}(r^2)$ time.

4 Attacks on elliptic curve public key protocols

In this section we apply the attack to elliptic curve public key protocols such as El-Gamal encryption and Diffie-Helman key exchange. The attack can not apply to the ECDSA signatures, since in this case scalar multiplication is performed with a random exponent instead of a fixed exponent.

4.1 Elliptic Curve Encryption Scheme

This scheme is analogous to El-Gamal encryption [8].

System parameters :
  An elliptic curve $\mathcal{E}$ over $GF(p)$ or $GF(2^n)$.
  The order of $\mathcal{E}$ denoted $\#\mathcal{E}$ must be divisible by a large prime q.
  $G \in \mathcal{E}$ of order q.

Key generation :
  Secret key : $d \in_R [1, q-1]$.
  Public key : $Q = dP$.

Encryption of a message m :
  Pick $k \in_R [1, q-1]$.
  Compute the points $kP = (x_1, y_1)$ and $kQ = (x_2, y_2)$, and $c = x_2 + m$.
  The ciphertext is (x_1, y_1, c).

Decryption :
  Compute $(x_2', y_2') = d(x_1, y_1)$ and $m = c - x_2'$.

The attack described before enables to recover d when the device decrypts the ciphertext (x_1, y_1, c) for various points (x_1, y_1).

4.2 Elliptic Curve Diffie-Hellman key exchange

The EC Diffie-Hellman protocol derives a common secret value z from one party's private key and another party's public key. The protocol is referenced as ECSVDP-DH (Elliptic Curve Secret Value Derivation Primitive, Diffie-Hellman version) in [10]. If the two parties correctly execute this primitive, they will produce the same output.

System parameter :
  An elliptic curve $\mathcal{E}$ over $GF(p)$ or $GF(2^n)$.
  The order of $\mathcal{E}$ denoted $\#\mathcal{E}$ must be divisible by a large prime q.
  Alice's own private key s.
  Bob's public key W.

Derivation of the shared secret value z :
  Compute the point $P = sW$.
  If $P = \mathcal{O}$ output "error" and stop.
  The shared secret value is $z = x_p$, the x-coordinate of P.

The attack described in the previous section recovers Alice's secret key when she computes the point $P = sW$ for Bob's public key W.

5 Countermeasures against DPA

In this section we describe three countermeasures that prevent from the attack described in section 3. Recall that the attack enables to recover d when $Q_i = dP_i$ are computed inside the card for various P_i for $1 \le i \le k$. These three countermeasures are based on introducing random numbers during the computation of $Q = dP$. We underline that other attacks might of course not be thwarted by our countermeasures.

5.1 First countermeasure : randomization of the private exponent

Let $\#\mathcal{E}$ be the number of points of the curve. The computation of $Q = dP$ is done by the following algorithm :

1. Select a random number k of size n bits. In practice, one can take $n = 20$ bits.
2. Compute $d' = d + k.\#\mathcal{E}$.
3. Compute the point $Q = d'P$. We have $Q = dP$ since $\#\mathcal{E}P = \mathcal{O}$.

This countermeasure makes the previous attack infeasible since the exponent d' in $Q = d'P$ changes at each new execution of the algorithm.

5.2 Second countermeasure : blinding the point P

The method is analogous to Chaum's blind signature scheme for RSA [4]. The point P to be multiplied is "blinded" by adding a secret random point R for which we know $S = dR$. Scalar multiplication is done by computing the point $d(R + P)$ and subtracting $S = dR$ to get $Q = dP$. The points R and $S = dR$ can be initially stored inside the card and refreshed at each new execution by computing $R \leftarrow (-1)^b 2R$ and $S \leftarrow (-1)^b 2S$, where b is a random bit generated at each new execution. This makes the previous attack infeasible since the point $P' = P + R$ to be multiplied by d is not known to the attacker.

5.3 Third countermeasure : randomized projective coordinates

Projective coordinates [16] can be used to avoid the costly field inversion for point addition and doubling. The projective coordinates (X, Y, Z) of a point $P = (x, y)$ are given by :

$$x = \frac{X}{Z} \quad y = \frac{Y}{Z}$$

Another system of projective coordinates may be found in [10]. The projective coordinates of a point are not unique because :

$$(X, Y, Z) = (\lambda X, \lambda Y, \lambda Z) \tag{3}$$

for every $\lambda \neq 0$ in the finite field.

The third countermeasure consists in randomizing the projective coordinate representation of a point $P = (X, Y, Z)$. Before each new execution of the scalar multiplication algorithm for computing $Q = dP$, the projective coordinates of P are randomized according to equation (3) with a random λ. The randomization can also occur after each point addition and doubling.

This makes the attack described above infeasible since it is not possible for the attacker to predict any specific bit of the binary representation of P in projective coordinates.

6 Conclusion

We have shown that unless protected, implementations of elliptic curve cryptosystems such as El-Gamal type encryption or Diffie-Hellman key exchange are vulnerable to Differential Power Analysis. We have introduced three countermeasures that address specifically these attacks. Those countermeasures are easy to implement and do not impact efficiency in a significant way. However, we do not pretend that those countermeasures thwart from all kinds of power attacks, since it may be possible to exploit the information leakage through power consumption in a different way.

Acknowledgments I thank David Naccache and Jean-Marc Robert for their careful reading and valuable suggestions, and the anonymous referees for their helpful comments.

References

1. E. Biham, A. Shamir. Power analysis of the key scheduling of the AES candidates, *Proceedings of the second AES Candidate Conference*, March 1999, pp. 115-121.

2. E. Brickell, D. Gordon, K. McCurley, D. Wilson. Fast Exponentiation with Precomputation (Extended Abstract), *Advances in Cryptology - Eurocrypt '92*, LNCS 658 (1993), Springer-Verlag, pp. 200-207.

3. S. Chari, C. Jutla, J.R. Rao, P. Rohatgi. A cautionary note regarding evaluation of AES candidates on smart-cards, *Proceedings of the second AES Candidate Conference*, March 1999, pp. 133-147.

4. D. Chaum. Security without identification : transaction systems to make Big Brother obsolete, *Communications of the ACM*, vol. 28, n. 10, Oct 1985, pp. 1030-1044.

5. J. Daemen, V. Rijmen. Resistance against implementation attacks A comparative study of the AES proposals, *Proceedings of the second AES Candidate Conference*, March 1999, pp. 122-132.

6. FIPS 46, Data encryption standard, Federal Information Processing Standards Publication 46, U.S. Department of Commerce/National Bureau of Standards, National Technical Information Service, Springfield, Virginia, 1977.

7. W. Diffie and M. Hellman. New directions in cryptography, *IEEE Trans. Info. Theory*, IT-22, 1976, pp 644-654.

8. T. El Gamal. A public key cryptosystem and a signature scheme based on discrete logarithms, *IEEE Trans. Info. Theory*, IT-31, 1985, pp 469-472.

9. D.M. Gordon. A Survey of Fast Exponentiation Methods, *Journal of Algorithms* 27, 129-146 (1998).

10. IEEE P1363/D7. Standard Specifications for Public Key Cryptography. September 11, 1998.

11. D.E. Knuth, *Seminumerical Algorithms*, The Art of Computer Programming, 2 Addison Wesley, 1969.

12. N. Koblitz. Elliptic Curve Cryptosystems, *Mathematics of Computation*, vol. 48, 1987, pp. 203-209.

13. Paul Kocher. Timing attacks on implementations of Diffie-Hellman, RSA, DSS and other systems, *Advances in Cryptology, Proceedings of Crypto' 96, LNCS 1109*, N. Koblitz, Ed., Springer-Verlag, 1996, pp. 104-113.

14. Paul Kocher, Joshua Jaffe, and Benjamin Jun, Introduction to Differential Power Analysis and Related Attacks, http://www.cryptography.com/dpa/technical, 1998.

15. K. Koyama, Y. Tsuruoka, Speeding up elliptic cryptosystems by using a signed binary window method, *Advances in Cryptology - Proceedings of Crypto '92, LNCS* 740, pp. 345-357, Springer-Verlag, Berlin/New-York, 1993.

16. A. J. Menezes, "Elliptic Curve Public Key Cryptosystems", Kluwer Academic Publishers, 1993.

17. V.S. Miller. Use of Elliptic Curves in Cryptography, *Proceedings of Crypto 85*, LNCS 218, Springer, 1986, pp. 417-426.

18. F. Morain, J. Olivos. Speeding up the computation of an elliptic curve using addition-subtraction chains, *Inform. Theory Appl.* 24 (1990), 531-543.

Probing Attacks on Tamper-Resistant Devices

Helena Handschuh[1], Pascal Paillier[1], and Jacques Stern[2]

[1] Gemplus/ENST, France
[2] Ecole Normale Supérieure, France

Abstract. This paper describes a new type of attack on tamper-resistant cryptographic hardware. We show that by locally observing the value of a few RAM or adress bus bits (possibly a single one) during the execution of a cryptographic algorithm, typically by the mean of a probe (needle), an attacker could easily recover information on the secret key being used; our attacks apply to public-key cryptosystems such as RSA or El Gamal, as well as to secret-key encryption schemes including DES and RC5.

1 Introduction

In recent years, many researchers have started investigating the security of tamper-resistant devices such as smart-cards. Along many other cryptanalytic attacks on cryptographic algorithms, new attacks have been suggested. These attacks usually assume the existence of some kind of side-channel and retrieve secret information on the process being executed aboard the device [1].

We distinguish between two types of side-channel attacks, namely passive or intrusive attacks. Typical examples of the first kind are timing attacks [10, 7] and power attacks [11], in which the execution time or the power consumption are monitored while secrets are being handled by the device. Other examples are side-channel attacks described by Schneier *et al.* which include (for instance) carry bit analysis [17].

On the other hand, some authors described agressive scenarios which consist in influencing or perturbating the behavior of the device in order to infer the secret. These attacks include Boneh *et al.*'s induction of transient faults during RSA computations [3] or even cutting wires and forcing given bit values, such as in Differential Fault Analysis [2, 14] of DES.

In this paper we consider a new kind of passive attacks, which appears to be more powerful than the previous ones. No statistical analysis is needed in most cases. We suppose that the attacker simply has access to a probe station, which (for the non-specialist) is a kind of needle that allows to monitor the value of a single bit during the execution of some cryptographic algorithm. Such devices are, of course, not self-sufficient. In practice, the attacker must first prepare the surface of the chip in specific way and overcome a long list of technical problems such as line width, protective layers, the lack of information allowing to match a physical chip location with a given gate, security detectors, and other purely physical phenomena such as the needle's electrical bading or pure signal synchronization problems. More, depending on whether it is known to which register the bit belongs, probing may present a wide range of hardness degrees. Part of the analysis consists of guessing which bit is being recorded and once this is done, infering the secret key or private exponent becomes easy.

The paper is organised as follows. The next two sections investigate probing attacks on RSA and DSA-like cryptosystems. Section 4 and 5 focus on applying specific probing-based cryptanalysis on secret key encryption schemes such as DES and RC5.

2 Probing Attacks in Public-Key Cryptography

ost public-key cryptosystems require modular exponentiations. Unless specifically adressed by a dedicated hardware design, the modular exponentiation is usually available aboard a cryptographic device as a software implementation. Although many variants exist, most real-life devices implement the well-known Square-and-Multiply algorithm in its nominal version.

This section is intended to introduce a new (probing-based) cryptanalytic attack that completely recovers the exponent of a typical Square-and-Multiply implementation, thus providing a tool for breaking RSA, El Gamal, DSA, Schnorr-type signature schemes, and so forth. We first introduce and comment our adversarial model.

2.1 Our Attack Model

We will denote by SM-1 the standard Square-and-Multiply algorithm that outputs $m^d \bmod n$, given a base m, an exponent d and a modulus n. During the modular exponentiation, d is scanned bit by bit from left to right and modular multiplication or squarings are successively applied to an accumulator A depending on the current bit exponent. Denoting by $|d|$ the bitlength of d, we recall the procedure on Fig. 1.

In the sections that follow, we will be considering an attack scenario in which the adversary is given access to some bits of the accumulator

Initialization
$$N \leftarrow n, M \leftarrow m$$
$$A \leftarrow 1, i \leftarrow |d|$$

Scanning Loop
While $(1 \leq i)$
{
$$A \leftarrow A \cdot A \bmod N$$
If $(d[i] == 1)$, $A \leftarrow A \cdot M \bmod N$
$$i \leftarrow i - 1$$
}

Output $A = m^d \bmod n$

Fig. 1. Standard Square-and-Multiply Procedure (SM-1)

throughout the exponentiation and attempts to recover information about the exponent d. Before going further, we state that any attack in the above model could be sophisticated or generalized in various ways to support known variants of SM-1, such as right-to-left or multiple bit scanning.

To be more precise, our model assumes that some "monitoring oracle" provides the adversary with the value of certain accumulator bits suppposedly updated *at each execution* of the internal loop of SM-1. This means that bit-values are collected by the attacker just after the accumulator was squared or squared-then-multiplied. Therefore, we implicitly consider that the monitoring oracle is capable of synchronizing perfectly its observations with the actual execution of SM-1 aboard the device.

2.2   Probing Attack of SM-1

In this section, we show how to infer d by probing a single computation of the form $m^d \bmod n$ for given m and n when it is known that the exponentiation is done by SM-1. We will first consider the case when a few bits (possibly a single one) are probed at known positions $J \subset \{1, \cdots, |A|\}$ in the accumulator. We denote by $A(J)$ the set of bits appearing at positions belonging to J in A. Section 2.3 extends the attack to the case when J is unknown.

Let d_i be the integer formed by the i leading bits of d, for $i = 1, \cdots, |d|$. Clearly, after the i-th step of SM-1 was executed, the accumulator contains the value $A^i = m^{d_i} \bmod n$. Meanwhile, the monitoring oracle has provided the attacker with the sequence

$$T_i = \left(A^1(J), A^2(J), \cdots, A^i(J) \right) . \tag{1}$$

Now if δ is a guess for d_i, the attacker can easily simulate SM-1 given (m, n, δ) and collect the bits at positions J in the simulated accumulator A', i.e. the sequence

$$T_i'(\delta) = \left(A'^1(J), A'^2(J), \cdots, A'^i(J) \right) . \tag{2}$$

It is now clear that δ is a correct guess only if $T_i = T_i'(\delta)$. Then, the procedure can be iterated at step $i+1$ by relying on the surviving guesses at step i. This attack strategy is summarized in Fig. 2.

$$
\begin{array}{l}
\Delta \leftarrow \{0\} \\
\text{For } (i = 1, \cdots, |d|) \\
\{ \\
\quad (1) \quad \Delta_0 \leftarrow \{2\delta, 2\delta + 1 \mid \delta \in \Delta\} \\
\quad (2) \quad \Delta \leftarrow \{\delta \mid \delta \in \Delta_0 \text{ and } T_i = T_i'(\delta)\} \\
\} \\
\\
\text{return } \Delta
\end{array}
$$

Fig. 2. The Adversary's Strategy

Since $T_i = T_i'(d_i)$ for all i, it is clear that when the attack ends Δ contains at least $d = d_{|d|}$. The point here resides in detecting whether the number of guesses to be checked is likely to explode or not while carrying out the attack. Although it seems hard to answer this question in the general case, one can still adress it by the mean of a heuristic reasoning. Let us first consider stage (2) during which wrong guesses are eliminated from Δ_0. At step i, each element δ of Δ_0 entering (2) fulfills $T_{i-1} = T_{i-1}'(\delta_{i-1})$ where $\delta_{i-1} = \lfloor \delta/2 \rfloor$. Furthermore, $\delta \in \Delta_0$ passes the test if (and only if) the equality $A^i(J) = A'^i(J)$ holds. This happens with probability :

$$p(\delta) = P[A^i(J) = A'^i(J) \mid A^{i-1}(J) = A'^{i-1}(J)] .$$

By heuristically assuming that bits $A'^i(J)$ and $A^i(J)$ are decorrelated[1] for any wrong guess, we get that $p(\delta)$ is nearly $\varepsilon = 2^{-|J|}$. Obviously, the

---

[1] in theory, these bits are somehow correlated, but the modular multiplication offers such excellent diffusion properties that our assumption remains highly realistic.

correct exponent yields $p(d_i) = 1$. Since stage (1) just doubles the number of elements in Δ, one has, denoting by u_i the average number of surviving guess after step i, that :

$$u_i = \sum_{\delta \in \Delta_0} p(\delta) = p(d_i) + \sum_{\delta \in \Delta_0 \setminus d_i} p(\delta) = 1 + \varepsilon(2u_{i-1} - 1) . \qquad (3)$$

Consequently, we get :

$$u_i = \frac{1 - \varepsilon - \varepsilon(2\varepsilon)^i}{1 - 2\varepsilon} \leq \frac{1 - \varepsilon}{1 - 2\varepsilon} ,$$

when $2\varepsilon < 1$ and $u_i = 1 + \frac{i}{2}$ if $\varepsilon = 1/2$. This yields :

$$E(|\Delta|) \leq \frac{1 - \varepsilon}{1 - 2\varepsilon}$$

throughout the attack. If J has a single element, that is if a single bit is observed by the monitoring oracle, then $\varepsilon = \frac{1}{2}$ and

$$E(|\Delta|) \leq 1 + \frac{|d|}{2} ,$$

which proves that the attack has a low heuristic complexity.

2.3 Random Hit Attacks

We now address the case when the attacker does not actually know the position of the monitored bits. One could of course address the problem by executing the above strategy with all possible positions simultaneously (there are $|A|^{|J|}$ such possibilities). Instead, we provide a much faster strategy derived from the previously discussed one. The idea exploited here is that the adversary can guess the position of these bits while performing the attack. Again, at step i, δ is a correct guess for d_i if (and only if) T_i coincides with some "simulated trace" that the attacker produces, for instance the complete accumulator history

$$T_i' = \left(A'^1, A'^2, \cdots, A'^i \right) .$$

Let us define $T_i'(j; \delta)$ as the sequence $\left(A'^1(j), \cdots, A'^i(j) \right)$. From now on, if $T_i \not\subset T_i'(j; \delta)$ holds for some $j \in \{1, \cdots, |A|\}$ then either δ is a wrong guess for d_i or $j \notin J$. This motivates the attack strategy depicted in Fig. 3. In this setting, one can show that :

$$E(|\Delta|) \leq \frac{1 - \varepsilon}{1 - 2\varepsilon} + |A|(2\varepsilon)^{|d|} \quad \text{with} \quad \varepsilon = 2^{-|J|} , \qquad (4)$$

which means that the attack would succeed again.

$$\begin{aligned}
&J \leftarrow \{1, \cdots, |A|\} \\
&\forall j \in J \quad \Delta_j \leftarrow \{0\} \\
&\text{For } (i = 1, \cdots, |d|) \\
&\{ \\
&\quad \text{For } (j \in J) \\
&\quad \{ \\
&\qquad \Delta_j \leftarrow \{2\delta, 2\delta + 1 \mid \delta \in \Delta_j\} \\
&\qquad \Delta_j \leftarrow \{\delta \mid \delta \in \Delta_j \text{ and } T_i \subset T_i'(\delta)\} \\
&\quad \} \\
&\quad \text{If } \Delta_j = \emptyset \text{ then } J \leftarrow J - j \\
&\} \\
&\\
&\text{return } \Delta = \cup_J \Delta_j
\end{aligned}$$

Fig. 3. Random Hit Attack

2.4 Discrete-Log Based Signatures

Several Discrete Log (DL)-based cryptosystems have been proposed in the literature. In virtually all of them (El Gamal [5], Schnorr [23], DSA [4]), a fixed known base g has to be raised to a random power k modulo some known prime p. The security of the cryptosystem relies on the randomness and the *secrecy* of the exponent k, which plays in this context a role similar to the one of a secret key.

From the above two attacks, it turns out that probing a single bit during a single signature generation suffices to recover the random exponent k. The device's secret key can thus be easily infered from the knowledge of k and the output signature.

3 Probing Attacks on DES

Following Biham and Shamir's work [2] on Differential Fault Analysis of Secret Key Cryptosystems, as well as Schneier et al.'s ideas on side-channel attacks [17], we take a closer look at probing secret key algorithms, which may be considered as yet another side-channel where information leak. As we saw before, probing is considered as a passive attack, whereas cutting wires would be an agressive attack.

The goal of this section is therefore to show that even a passive attacker may retrieve the secret key of a DES implementation given one single bit of information at each round.

3.1 The Information Leakage Model

The attack we subsequently present works in the following context. Suppose an attacker uses an electronic station to locally observe the value of a given bit during the execution of DES. We require that the attacker have sufficient knowledge of the device to be able to recognize two specific registers which are the R and L data registers on which the round function of an iterated Feistel [6] cipher applies. In the case of DES, any bit of one or the other register is enough to attack the first and the last round subkeys.

3.2 Attacking DES

DES [13] is a 16-round Feistel scheme which can be described as follows :
Let m be a 64-bit message divided into two 32-bit halves : the left half m_L and the right half m_R ; let c be the corresponding ciphertext and k_i the i-th round 48-bit subkey. Finally let IP and IP^{-1} denote the initial and final permutations and F the round function of DES. The algorithm is briefly depicted on Fig. 4.

$$(L_0|R_0) = IP(m_L|m_R)$$
$$\text{for } i = 1 \text{ to } 16 :$$
$$L_i = R_{i-1};$$
$$R_i = F(R_{i-1}, K_i) \oplus L_{i-1};$$
$$\text{end for}$$
$$c = IP^{-1}(R_{16}|L_{16})$$

Fig. 4. Brief Description of DES

In this attack we ignore the initial and final permutations as these are public anyway. Suppose the plaintext is simply $(L_0|R_0)$ and the ciphertext $c = (L_{16}|R_{16})$. We shall now explain how to recover 6 bits of the last round subkey. Assuming that the probe station enables us to record the value of bit number b of register L at each round. The F-function in DES is such that the output bits can be related to a given S-box having a 6-bit input. We refer the reader to [13] for more information on the structure of the F-function. As $L_{16} = R_{15}$, for any ciphertext we can select the 6 bits entering the F-function that produce bit b as an output. These six bits are exored with six bits of the secret subkey k_{16} before entering a specific S-box. Therefore, for each possible value of the 6-bit secret key entering the S-box, the attacker can compute the expected output bit b^* of the F-function. Notice that she also has access to the real value of this bit as she knows bit b from the probe of L_{15} as well as the corresponding bit of the ciphertext R_{16}. Therefore, with one ciphertext, the attacker can eliminate those key guesses where the expected value b^* and the real value of bit b of the output of the round-function are not the same. On average, half of the key candidates will survive. Thus 6 bits of the key are recovered with 6 different ciphertexts.

The same attack can be carried out on six bits of the secret key of the first round. DES happens to be designed such that these 6 bits are different from the 6 bits recovered from the last round subkey. As a matter of fact, the initial permutation on the secret key results in no two consecutive bits entering the same S-box among the first round and the last round subkey. Therefore a total of 12 bits can be recovered by this attack. The remaining 44 bits of the key can be found by exhaustive search.

3.3 Discussion

Note that the attack works on both registers R and L. As a matter of fact, if the attacker probes register R, we can apply exactly the same attack as before because the iterated Feistel structure guarantees that $R_{14} = L_{15}$. Thus we can still compute the input and expected output of the round function and compare the latter to the real value derived from R_{14} as well as from the ciphertext.

This attack uses the same principle as the one described by Biham and Shamir, but does not require the attacker to be able to cut wires or induce faults. In our setting, the prober simply observes the value of a given bit throughout the execution of the block cipher. The complexity is very low : only a handfull plaintext/ciphertext pairs are needed, and the number of offline encryptions is 2^{44}.

Finally, we note that the attack cannot be carried out on two-key triple-DES or triple modes of encryption using DES as a building block. This comes from the fact that only 12 bits (6 from the first encryption component and 6 from the last encryption component) of one of the secret keys can be recovered, therefore the overall exhaustive search on the remaining key bits still amounts to 2^{100} offline encryptions. Additionnally, no output bit of the round-function depends on the input bit at the same bit position due to the bit permutation after the S-box layer. Thus the intermediate values of bit b are of no use to the attacker if she cannot get any other information.

4 Probing RC5

As another example, let us consider RC5. Other iterated algorithms such as RC6 are equally vulnerable to probing. RC5 has been extensively studied from a regular cryptanalysis point of view. See for example [9, 8, 18].

4.1 Description of RC5

RC5 is an iterative secret-key block cipher designed by R. Rivest [15]. It has variable parameters such as the key size, the block size and the number of rounds. A particular (instanced) RC5 algorithm is denoted by RC5-$w/r/b$ where w is the word size (a block is made of two words), r is the number of rounds and b the number of secret key bytes. Our attack works for every choice of these parameters.

RC5 works as follows : the secret key is first extended into a table of $2r + 2$ secret w-bit words S_i. We will assume that RC5's key schedule is one-way and focus on recovering the extended secret key table and not the secret key itself. The detailed description of the key schedule can be found in [15]. By letting (L_0, R_0) denote the left and right halves of the plaintext, the encryption algorithm is depicted on Fig. 5.

$$L_1 = L_0 + S_0$$
$$R_1 = R_0 + S_1$$
for $i = 2$ to $2r + 1$ do
$$L_i = R_{i-1}$$
$$R_i = ((L_{i-1} \oplus R_{i-1}) \lll R_{i-1}) + S_i$$

Fig. 5. Brief Description of RC5

The ciphertext is (L_{2r+1}, R_{2r+1}). The transformation performed for a given i value is called a half-round : there are $2r + 2$ half rounds. Each half-round involves exactly one sub-key S_i. All additions are mod 2^w and the rotations are mod w. As usual, $\oplus$ denotes a bitwise exclusive or.

4.2   Probing Attack on RC5

For our purposes, we suppose that the attacker once again has access to all intermediate values of some bit b of either register L or register R, which is the case when she can probe one of these two in an iterated hardware implementation of the algorithm or a specific RAM buffer. We start by describing an attack where the adversary probes some register R. Our technique uses some of the ideas presented in [7] to derive the subkeys one by one in reverse order starting with S_{2r+1}.

Step 1.

First, collect a few multiples of w plaintext/ciphertext pairs and sort them by the value of the $\log(w)$ least significant bits of the left half of the ciphertext L_{2r+1}. There should be at least a few texts available in each such 'category'. Then consider the texts which belong to category $(w-b)[w]$ (i.e. the value of the $\log(w)$ least significant bits of L_{2r+1} is $(w-b)[w]$). We probe the value of bit b of register $R_{2r-1} = L_{2r}$. Since we know the value of bit b of $R_{2r} = L_{2r+1}$ as well as the value of the last rotation from the left ciphertext half, we can compute the least significant bit of register L_{2r} just before the last subkey addition. Therefore the least significant bit of the last subkey can be found.

Step 2.

Next, consider "category" $(w-b+1)[w]$. After the last rotation, bit b of L_{2r} will be in first position. Applying the same method as in step 1, we can derive the first bit of the last round subkey (taking into account the carry bit created by the addition of the least significant bit, which is by now already known to the attacker and so on for the remaining $(w-2)$ bits of the secret key. They are derived one by one using different ciphertexts, from the low order end to the high order end of the key.

Step 3.

After recovering the last subkey, decrypt one half-round, sort the ciphertexts according to the new value of the $\log(w)$ least significant bits of L_{2r} and derive S_{2r}. Derive all subsequent subkeys up to the very first four.

Step 4. The last four subkeys can be found by cryptanalysing a two-round block cipher, which is straightforward. This concludes successfully the probing attack on RC5.

4.3 Discussion

The attack works in a similar way when we probe register L directly. So the knowledge of a single intermediate bit at each round enables to derive the complete extended secret key and thus to further recover the initial secret key. On the average a few multiples of w known plaintext/ciphertext pairs are needed, in order to be sure that at each round at least one text corresponding to a given rotation value is available. If this should not be the case, the attacker can still query some more pairs of texts while she is working backwards towards the first rounds of the cipher. The complexity of this attack is actually very low and requires less than the exhaustive search of a single 32-bit subkey. Depending on the case, either S_0, S_2 or S_1, S_3 have to be determined otherwise than by the above attack. They can either be recovered by the key schedule, or by guessing a few bits of S_2 or S_3 at a time, and checking for consistency on the corresponding bits of S_0 or S_1.

5 Conclusion

We have shown that probing attacks are a powerful tool to derive information on secret keys in embedded hardware. The interesting feature of these attacks resides in that they are not desctructive, as many previously suggested attacks are. In essence, probing does not require the cutting of wires or inducing faults or even stressing the device to make it behave abnormally, for we just observe (spying) a single bit during execution. We have shown that public key algorithms using exponentiation or the discrete logarithm such as RSA or DSA, as well as secret key algorithms such as DES or RC5 would be vulnerable to such powerful attacks.

Acknowledgements

We are very grateful to David Naccache for motivating this research and would also like to thank the numerous people who contributed to our investigations.

References

1. R. Anderson, M. Kuhn. Low Cost Attacks on Tamper-Resistant Devices. In *Security Protocol Workshop'97, LNCS 1361*, pp. 125-136. Springer-Verlag.1997.

2. E. Biham, A. Shamir. Differential Fault Analysis of Secret Key Cryptosystems. In *Advances in Cryptology - Crypto'97, LNCS 1294*, pages 513-525. Springer-Verlag, 1997.

3. D. Boneh, R. DeMillo and R. Lipton. On the Importance of Checking Cryptographic Protocols for Faults. n *Advances in Cryptology - Eurocrypt'97, LNCS 1233*, pages 37-51. Springer-Verlag, 1997.

4. FIPS PUB 186, February 1, 1993, *Digital Signature Standard*.

5. T. El Gamal. A Public Key Cryptosystem and a Signature Scheme Based on Discrete Logarithms. In *IEEE Transactions on Information Theory*, volume IT-31, no. 4, pages 469-472, July 1985.

6. H. Feistel. Cryptography and computer privacy. In *Scientific american*, 1973.

7. H. Handschuh and H. Heys. A Timing Attack on RC5. In *SAC'98 - Workshop on Selected Areas in Cryptography, LNCS 1556*, pages 306-320. Springer-Verlag, 1999.

8. B. S. Kaliski and Y. L. Yin. On Differential and Linear Cryptanalysis of the RC5 Encryption Algorithm. In *Advances in Cryptology - Crypto'95, LNCS 963*, pages 171-184. Springer-Verlag, 1995.

9. L. R. Knudsen and W. Meier. Improved Differential Attacks on RC5. In *Advances in Cryptology - Crypto'96, LNCS*. Springer-Verlag, 1996.

10. Paul C. Kocher. Timing Attacks on Implementations of Diffie-Hellman, RSA, DSS, and Other Systems. In *Advances in Cryptology - Crypto'96, LNCS*. Springer-Verlag, 1996.

11. P. Kocher, J. Jaffe, B. Jun. Introduction to Differential Power Analysis and Related Attacks. Available from *http://www.cryptography.com/dpa/technical/*.

12. . Matsui. Linear cryptanalysis method for DES Cipher. In *Advances in Cryptology - EUROCRYPT'93, LNCS 765*. Springer-Verlag, 1994.

13. U.S. National Bureau of Standards. *Data Encryption Standard*, Federal Information Processing Standard Publication 46-2, 1977.

14. P. Paillier. Evaluating Differential Fault Analysis of Unknown Cryptosystems. In *Public Key Cryptography - PKC'99, LNCS 1560*. Springer-Verlag, 1999.

15. R. L. Rivest. The RC5 Encryption Algorithm. In *Fast Software Encryption - Second International Workshop, Leuven, Belgium, LNCS 1008*, pages 86-96, Springer-Verlag, 1995.

16. R. L. Rivest, A. Shamir, L. M. Adleman. A method for obtaining digital signatures and public-key cryptosystem. In *Communications of the ACM*, vol. 21, 1978.

17. B. Schneier et al. Side-Channel Attacks. To appear In *Cardis'98 - LNCS*. Springer-Verlag, 1998.

18. A. A. Selçuk. New results in linear cryptanalysis of RC5. In *Fast Software Encryption 5 - LNCS 1372*. pages 1-16, Springer-Verlag, 1998. Springer-Verlag, 1998.

19. J. Kilian, P. Rogaway, "How to protect DES against exhaustive key search, *CRYPTO'96, LNCS 1109*, Springer-Verlag, 1996, pp. 252-267.

20. E. Biham & A. Shamir, *The next stage of differential fault analysis : How to break completely unknown cryptosystems*, Preprint, 1996.

21. R. Anderson, *Robustness principles for public-key protocols*, LNCS, Advances in Cryptology, Proceedings of Crypto'95, Springer-Verlag, pp. 236–247, 1995.

22. R. Anderson & S. Vaudenay, *Minding your p's and q's*, LNCS, Advances in Cryptology, Proceedings of Asiacrypt'96, Springer-Verlag, pp. 26–35, 1996.

23. C. Schnorr, *Efficient Identification and Signatures for Smart-Cards*, Advances in Cryptology: Eurocrypt'89 (G. Brassard ed.), LNCS 435, Springer-Verlag, pp. 239-252, 1990.

Fast Multiplication on Elliptic Curves over $GF(2^m)$ without Precomputation

Julio López[1]* and Ricardo Dahab[2]**

[1] Department of Combinatorics & Optimization
University of Waterloo,
Waterloo, Ontario N2L 3G1, Canada (jclherna@cacr.math.uwaterloo.ca)
[2] Institute of Computing
State University of Campinas,
Campinas, C.P. 6176, 13083-970 SP, Brazil (rdahab@dcc.unicamp.br).

Abstract. This paper describes an algorithm for computing elliptic scalar multiplications on non-supersingular elliptic curves defined over $GF(2^m)$. The algorithm is an optimized version of a method described in [1], which is based on Montgomery's method [8]. Our algorithm is easy to implement in both hardware and software, works for any elliptic curve over $GF(2^m)$, requires no precomputed multiples of a point, and is faster on average than the addition-subtraction method described in draft standard IEEE P1363. In addition, the method requires less memory than projective schemes and the amount of computation needed for a scalar multiplication is fixed for all multipliers of the same binary length. Therefore, the improved method possesses many desirable features for implementing elliptic curves in restricted environments.

Key words. Elliptic Curves over $GF(2^m)$, Point multiplication.

1 Introduction

Elliptic curve cryptography first suggested by Koblitz [5] and Miller [12] is becoming increasingly common for implementing public-key protocols as the Diffie-Hellman key agreement. The security of these cryptosystems relies on the presumed intractability of the discrete logarithm problem on elliptic curves. Since there is no known sub-exponential type algorithm for elliptic curves over finite fields, the sizes of the fields, keys, and other parameters can be considered shorter than other public key cryptosystems such as RSA with the same level of security. This can be especially an advantage for applications where resources such as memory and/or computing power are limited.

Elliptic curves over $GF(2^m)$ are particularly attractive, because the finite field operations can be implemented very efficiently in hardware and software.

* Dept. of Computer Science, University of Valle, A.A. 25130 Cali, Colombia. Research supported by a CAPES-Brasil scholarship
** Partially supported by a PRONEX-FINEP research grant no. 107/97

See for example [1] for a hardware implementation of $GF(2^{155})$, and [19] for a software implementation of $GF(2^{191})$.

Given an elliptic point P and a large integer k of about the size of the underlying field, the operation *elliptic scalar multiplication*, kP, is defined to be the elliptic point resulting from adding P to itself k times. This operation, analogous to exponentiation in multiplicative groups, is the most time consuming operation of the elliptic curve cryptosystems.

In this paper, the calculation of kP for a random integer k and a random point P is considered. An efficient scalar multiplication algorithm, which is an optimized version of an algorithm described in [1], is presented. The proposed algorithm is suitable for hardware and software implementation of random elliptic curves over $GF(2^m)$.

2 Previous Work

The basic method for computing kP is the addition-subtraction method described in draft standard IEEE P1363 [14]. This method is an improved version over the well known "add-and-double" (or binary) method, which requires no precomputations. For a random multiplier k, this algorithm performs on average $\frac{8}{3}\log_2 k$ field multiplications and $\frac{4}{3}\log_2 k$ field inversions in affine coordinates, and $8\frac{1}{3}\log_2 k$ field multiplications in projective coordinates.

Several proposed generalizations of the binary method (for exponentiation in a multiplicative group), such as the k-ary method, the signed window method, can be extended to compute elliptic scalar multiplications over a finite field [11]. These algorithms are based on the use of precomputation and methods for recoding the multiplier. In [3], several algorithms are analyzed under various conditions. However, most of the proposed optimizations may not be worthwhile when memory is at a premium.

Some special classes of elliptic curves defined over $GF(2^m)$ allow efficient implementations. For anomalous curves, the fastest known algorithm to compute kP is given in [17]; for curves defined over small subfields, efficient algorithms are presented in [13].

In [4, 16, 7] some techniques are presented for accelerating methods such as k-ary and window based methods. These methods are suitable for software implementation of random elliptic curves over $GF(2^m)$.

A different approach for computing kP was introduced by Montgomery [8]. This approach is based on the binary method and the observation that the x-coordinate of the sum of two points whose difference is known can be computed in terms of the x-coordinates of the involved points. This method uses the following variant of the binary method:

INPUT: An integer $k > 0$ and a point P.
OUTPUT: $Q = kP$.

1. Set $k \leftarrow (k_{l-1} \ldots k_1 k_0)_2$.
2. Set $P_1 \leftarrow P$, $P_2 \leftarrow 2P$.
3. for i from $l - 2$ downto 0 do
       if $k_i = 1$ then
            Set $P_1 \leftarrow P_1 + P_2$, $P_2 \leftarrow 2P_2$.
       else
            Set $P_2 \leftarrow P_2 + P_1$, $P_1 \leftarrow 2P_1$.
4. return$(Q = P_1)$.

Fig. 1. Algorithm 1: Binary Method

Note that this method maintains the invariant relationship $P_2 - P_1 = P$, and performs an addition and a doubling in each iteration. In [9], Montgomery's method was applied for reducing the number of registers needed to add points in supersingular curves over $GF(2^m)$. However, the authors observed that the benefits in storage provided by Montgomery's method is at a considerable expense of speed.

From the point of view of hardware implementation of elliptic curves over $GF(2^m)$, few papers have discussed efficient methods for computing kP. In [1], Montgomery's method was adapted for non-supersingular elliptic curves over $GF(2^m)$. However, the formulas given for implementing each iteration are not efficient in terms of field multiplications.

In this paper we will present an efficient implementation of Montgomery's method for computing kP on non-supersingular elliptic curves over $GF(2^m)$.

The remainder of the paper is organized as follows. In Section 3 we present a short introduction to elliptic curves over $GF(2^m)$. The proposed algorithm is described and analyzed in Section 4. Some running times of the proposed algorithm based on LiDIA are presented in Section 5. An implementation of the proposed algorithm is given in the appendix.

3   Elliptic Curves over $GF(2^m)$

Here we present a brief introduction to elliptic curves; more information on elliptic curves over finite fields of characteristic two can be found in [10, 14]. Let $GF(2^m)$ be a finite field of characteristic two. A non-supersingular elliptic curve E over $GF(2^m)$ is defined to be the set of solutions $(x, y) \in GF(2^m) \times GF(2^m)$ to the equation,

$$y^2 + xy = x^3 + ax^2 + b ,$$

where a and $b \in GF(2^m), b \neq 0$, together with the point at infinity denoted by $\mathcal{O}$.

It is well known that E forms a commutative finite group, with $\mathcal{O}$ as the group identity, under the addition operation known as the "tangent and chord method". Explicit rational formulas for the addition rule involve several arithmetic operations (adding, squaring, multiplication and inversion) in the underlying finite field. Formulas for adding two points in projective coordinates can be found in [10, 7]. In affine coordinates, the elliptic group operation is given by the following. Let $P = (x_1, y_1) \in E$; then $-P = (x_1, x_1 + y_1)$. For all $P \in E$, $\mathcal{O} + P = P + \mathcal{O} = P$. If $Q = (x_2, y_2) \in E$ and $Q \neq -P$, then $P + Q = (x_3, y_3)$, where

$$
x_3 = \begin{cases} (\frac{y_1 + y_2}{x_1 + x_2})^2 + \frac{y_1 + y_2}{x_1 + x_2} + x_1 + x_2 + a \;, & P \neq Q \\ x_1^2 + \frac{b}{x_1^2} \;, & P = Q. \end{cases} \tag{1}
$$

and

$$
y_3 = \begin{cases} (\frac{y_1 + y_2}{x_1 + x_2})(x_1 + x_3) + x_3 + y_1 \;, & P \neq Q \\ x_1^2 + (x_1 + \frac{y_1}{x_1})x_3 + x_3 \;, & P = Q. \end{cases} \tag{2}
$$

Notice that the x-coordinate of $2P$ does not involve the y-coordinate of P. This observation will be used in the derivation of the improved method.

4 Improved Method

This section describes the improved method for computing kP. We first develop an algorithm in affine coordinates which requires two field inversions in each iteration. Next a "projective" version is presented with more field multiplications, but with only one field inversion at the end of the computation.

4.1 Affine version

The extension of Montgomery's method [8] to elliptic curves over $GF(2^m)$ requires formulas for implementing Step 3 of Algorithm 1. In what follows we give efficient formulas that use only the x-coordinates of P_1, P_2 and P for performing the arithmetic operations needed in Algorithm 1. At the end of the lth iteration of Algorithm 1, we obtain the x-coordinates of kP and $(k+1)P$. We also provide a simple formula for recovering the y-coordinate of kP.

The following lemma gives another formula for computing the x-coordinate of the addition of two different points.

Lemma 1 *Let* $P_1 = (x_1, y_1)$, *and* $P_2 = (x_2, y_2)$ *be elliptic points. Then the* x-*coordinate of* $P_1 + P_2$, x_3, *can be computed as follows.*

$$
x_3 = \frac{x_1 y_2 + x_2 y_1 + x_1 x_2^2 + x_2 x_1^2}{(x_1 + x_2)^2} . \tag{3}
$$

Proof. Since P_1 and P_2 are elliptic points, it follows that $y_1^2 + y_2^2 + x_1 y_1 + x_2 y_2 + x_1^3 + x_2^3 = 0$. The result then follows easily from formula (1).

The following lemma shows how to compute the x-coordinate for the addition of two points whose difference is known.

Lemma 2 *Let* $P = (x, y), P_1 = (x_1, y_1),$ *and* $P_2 = (x_2, y_2)$ *be elliptic points. Assume that* $P_2 = P_1 + P.$ *Then the x-coordinate of* $P_1 + P_2,$ $x_3,$ *can be computed in terms of the x-coordinates of* P, P_1 *and* P_2 *as follows.*

$$x_3 = \begin{cases} x + (\dfrac{x_1}{x_1 + x_2})^2 + \dfrac{x_1}{x_1 + x_2} & , P_1 \neq P_2 \\ x_1^2 + \dfrac{b}{x_1^2} , & P_1 = P_2. \end{cases} \tag{4}$$

Proof. The case $P = \mathcal{O}$ follows directly from (1). Applying formula (3), we obtain that the x-coordinate of $P_2 + P_1$ can be rewritten as

$$x_3 = \frac{x_1 y_2 + x_2 y_1 + x_1 x_2^2 + x_2 x_1^2}{(x_1 + x_2)^2} . \tag{5}$$

Similarly, the x-coordinate of $P_2 - P_1$ satisfies

$$x = \frac{x_1 y_2 + x_2(x_1 + y_1) + x_1 x_2^2 + x_2 x_1^2}{(x_1 + x_2)^2} . \tag{6}$$

The result follows from adding (5) and (6).

The next lemma allows one to compute the y-coordinate of P_1 when P and the x-coordinates of P_1 and $P_1 + P$ are known.

Lemma 3 *Let* $P = (x, y), P_1 = (x_1, y_1),$ *and* $P_2 = (x_2, y_2)$ *be elliptic points. Assume that* $P_2 = P_1 + P$ *and* $x \neq 0.$ *Then the y-coordinate of* P_1 *can be expressed in terms of* $P,$ *and the x-coordinates of* P_1 *and* P_2 *as follows.*

$$y_1 = (x_1 + x)\{(x_1 + x)(x_2 + x) + x^2 + y\}/x + y . \tag{7}$$

Proof. Since $P_2 = P_1 + P,$ we obtain from (3) that y_1 satisfies the following equation:

$$x_2(x_1 + x)^2 = x_1 y + x y_1 + x_1 x^2 + x x_1^2 .$$

Therefore,

$$\begin{aligned} x y_1 &= x_2 x_1^2 + x_2 x^2 + x_1 y + x_1 x^2 + x x_1^2 \\ &= x_1\{x_1 x_2 + x_1 x + x^2 + y\} + x\{x x_2\} \\ &= x_1\{x_1 x_2 + x_1 x + x^2 + x x_2 + x^2 + y\} \\ &\quad + x\{x_1 x_2 + x_1 x + x x_2 + y\} + x y \\ &= (x_1 + x)\{(x_1 + x)(x_2 + x) + x^2 + y\} + x y. \end{aligned}$$

The following algorithm, based on Lemmas 2 and 3, implements Montgomery's method in affine coordinates.

Fig. 2. Algorithm 2A: Montgomery Scalar Multiplication

INPUT: An integer $k \geq 0$ and a point $P = (x, y) \in E$.
OUTPUT: $Q = kP$.

1. if $k = 0$ or $x = 0$ then output$(0,0)$ and stop.
2. Set $k \leftarrow (k_{l-1} \ldots k_1 k_0)_2$.
3. Set $x_1 \leftarrow x, \quad x_2 \leftarrow x^2 + b/x^2$.
4. for i from $l - 2$ downto 0 do
   Set $t \leftarrow \dfrac{x_1}{x_1 + x_2}$.
   if $k_i = 1$ then
       Set $x_1 \leftarrow x + t^2 + t, \quad x_2 \leftarrow x_2^2 + b/x_2^2$.
   else
       Set $x_1 \leftarrow x_1^2 + b/x_1^2, \quad x_2 \leftarrow x + t^2 + t$.
5. Set $r_1 \leftarrow x_1 + x, \quad r_2 \leftarrow x_2 + x$.
6. Set $y_1 \leftarrow r_1(r_1 r_2 + x^2 + y)/x + y$
7. return$(Q = (x_1, y_1))$.

Observe that Algorithm 2A, in each iteration of Step 4, performs two field inversions, one general field multiplication, one multiplication by the constant b, two squarings, and four additions; it follows that the total number of field operations to compute kP is given in the following lemma:

Lemma 4 *For computing kP, Algorithm 2A takes exactly the following number of field operations in $GF(2^m)$:*

$$\#INV. = 2\lfloor \log_2 k \rfloor + 1 , \quad \#MULT. = 2\lfloor \log_2 k \rfloor + 4 ,$$
$$\#ADD. = 4\lfloor \log_2 k \rfloor + 6 , \quad \#SQR. = 2\lfloor \log_2 k \rfloor + 2.$$

Remark. A further improvement to Algorithm 2A is to use an optimized routine to multiply by the constant b. Another potential improvement is to compute in parallel x_1 and x_2 from Step 4, since these calculations are independent of each other.

4.2 Projective Version

When field inversion in $GF(2^m)$ is relatively expensive (e.g., inversion based on Fermat's theorem requires at least 7 multiplications in $GF(2^m)$ if $m \geq 128$), then it may be of computational advantage to use fractional field arithmetic to perform elliptic curve calculations.

Let P, P_1 and P_2 be points on the curve E such that $P_2 = P_1 + P$. Let the x-coordinate of P_i be represented by X_i/Z_i, for $i \in \{1, 2\}$. From Lemma 2, when the x-coordinate of $2P_i$ is converted to projective coordinates it becomes

$$\begin{cases} x(2P_i) = X_i^4 + b \cdot Z_i^4 \ , \\ z(2P_i) = Z_i^2 \cdot X_i^2. \end{cases} \tag{8}$$

Similarly, the x-coordinate of $P_1 + P_2$ in projective coordinates can be computed as the fraction X_3/Z_3, where

$$\begin{cases} Z_3 = (X_1 \cdot Z_2 + X_2 \cdot Z_1)^2 \ , \\ X_3 = x \cdot Z_3 + (X_1 \cdot Z_2) \cdot (X_2 \cdot Z_1). \end{cases} \tag{9}$$

The addition formula requires three general field multiplications, one multiplication by x (i.e., the x-coordinate of P, which is fixed during the computation of kP), one squaring and two additions; doubling requires one general field multiplication, one multiplication by the constant b, five squarings, and one addition. A method based on these formulas is described in the next algorithm.

Fig. 3. Algorithm 2P: Montgomery Scalar Multiplication

INPUT: An integer $k \geq 0$ and a point $P = (x, y) \in E$.
OUTPUT: $Q = kP$.

1. if $k = 0$ or $x = 0$ then output$(0,0)$ and stop.
2. Set $k \leftarrow (k_{l-1} \ldots k_1 k_0)_2$.
3. Set $X_1 \leftarrow x, \quad Z_1 \leftarrow 1, \quad X_2 \leftarrow x^4 + b, \quad Z_2 \leftarrow x^2$.
4. for i from $l - 2$ downto 0 do
   if $k_i = 1$ then
       Madd(X_1, Z_1, X_2, Z_2),   Mdouble(X_2, Z_2).
   else
       Madd(X_2, Z_2, X_1, Z_1),   Mdouble(X_1, Z_1).
5. return$(Q = Mxy(X_1, Z_1, X_2, Z_2))$.

An implementation of the procedures Madd, Mdouble and Mxy is given in the appendix.

Lemma 5 *Algorithm 2P performs exactly the following number of field operations in* $GF(2^m)$:

$$\#INV. = 1 \ , \qquad\qquad \#MULT. = 6\lfloor \log_2 k \rfloor + 10 \ ,$$
$$\#ADD. = 3\lfloor \log_2 k \rfloor + 7 \ , \quad \#SQR. = 5\lfloor \log_2 k \rfloor + 3.$$

Remark. Since the complexity of both versions of Algorithm 2 does not depend on the number of 1's (or 0's) in the binary representation of k, this may help to prevent timing attacks. On the other hand, the use of restricted multipliers (e.g., with small Hamming weight) does not speedup directly Algorithms 2A and 2P, and this is a disadvantage compared to methods such as the binary method. However, from a practical point of view, most protocols in cryptographic applications use random multipliers.

4.3 Complexity Comparison

In the sequel, we assume that adding and squaring in $GF(2^m)$ is relatively fast. Now we compare the complexities of the addition-subtraction method to the complexity of the proposed method. This is a fair comparison since both methods do not use precomputation. For a random multiplier k, the addition-subtraction method in projective coordinates, given in [14], performs $8.3\log_2 k$ field multiplications; it follows we expect Algorithm 2P to be about 28% faster on average. However, if we use the formulas given in [7] for implementing the group operation in projective schemes, Algorithm 2P is about 14% faster than the addition-subtraction method. In the following table we summarize the complexities of these methods.

Table 1. Complexity Comparison of Algorithm 2P with other algorithms ($a = 0, 1$).

Method	Projective Coordinates
Binary [10]	$13\log_2 k$
Add-Sub [14]	$8.3\log_2 k$
Add-sub[7]	$7\log_2 k$
Algorithm 2P	$6\log_2 k$

Now we derive the cost of the addition-subtraction method (using affine coordinates) in terms of field multiplications. As mentioned in Section 2, this method performs on average $\frac{8}{3}\log_2 k$ field multiplications and $\frac{4}{3}\log_2 k$ field inversions. Thus, the total cost is $\frac{1}{3}(4r + 8)$ multiplications, where r is the cost-ratio of inversion to multiplication. This shows that for implementations of the finite field $GF(2^m)$ where $r > 2.5$ (see for example [1, 19, 4]), Algorithm 2P gives a computational advantage over the addition-subtraction method.

5 Running Times

In this section we present some running times we obtained in our software implementation of the proposed algorithm over the finite fields $GF(2^m)$, where $m = 163, 191$ and 239. To represent the finite fields we used LiDIA [6], a C++ based library. This finite field implementation uses a polynomial basis representation and the irreducible modulus is chosen as sparse as possible. We used a Sun UltraSPARC 300MHz machine. For comparison, we list in Table 2 the timings for the basic arithmetic operations in $GF(2^m)$.

Notice that one field inverse costs more than 9 field multiplications; therefore, the use of LiDIA may illustrate the performance of the proposed algorithm in situations where a field inverse is relatively expensive compared to field multiplication.

In Table 3 we present average running times for computing a scalar multiplication using several methods. These values were obtained using the following

Table 2. Average running times (in microseconds) for $GF(2^m)$ using LiDIA.

Extension m	Add.	Sqr.	Mult.	Inv.
163	0.6	2.3	10.5	96.2
191	0.7	2.0	10.9	118.1
239	0.8	2.6	14.6	162.8

Table 3. Average running times (in milliseconds) for computing mP.

Extension m	Binary[10]	Add-Sub.[14]	Algorithm 2P
163	27.5	19.1	13.5
191	33.1	22.4	16.0
239	52.3	35.1	25.6

test: we select 10 random elliptic curves ($a = 0$) over $GF(2^m)$, then we multiply a random point P in each curve with 100 randomly chosen integers of size $< 2^m$. We implemented the binary method in projective coordinates (see [10]), the addition-subtraction method [14] and Algorithm 2P. From Table 3 we conclude that the proposed method on average is 27-29% faster than the addition-subtraction method and 51% faster than the binary method. These timings show that the theoretical improvement of Algorithm 2P, given in Table 1, is observed in a actual implementation.

6 Conclusion

In this paper, we have presented an efficient method for computing elliptic scalar multiplications, which is an optimized version of an algorithm presented in [1]. The method performs exactly $6\lfloor \log_2 k \rfloor + 10$ field multiplication for computing kP on elliptic curves selected at random, is easy to implement in both hardware and software, requires no precomputations, works for any implementation of $GF(2^n)$, is faster than the addition-subtraction method on average, and uses fewer registers than methods based on projective schemes. Therefore, the method appears useful for applications of elliptic curves in constraint environments such as mobile devices and smart cards.

7 Acknowledgments

The first author would like to thank Alfred Menezes for making him possible to visit the center CACR and for his encouragement and support during this work. We also thanks the referees for their helpful comments.

References

1. G. B. Agnew, R. C. Mullin and S. A. Vanstone, "An Implementation of Elliptic Curve Cryptosystems Over $F_{2^{155}}$", *IEEE journal on selected areas in communications, Vol 11. No. 5*, June 1993.
2. ANSI X9.62: "The Elliptic Curve Digital Signature Algorithm (ECDSA)", draft, July 1997.
3. D. M. Gordon, "A survey of Fast Exponentiation Methods", *Journal of Algorithms*, 27, pp. 129-146, 1998.
4. J. Guajardo and C. Paar, "Efficient Algorithms for Elliptic Curve Cryptosystems", *Advances in Cryptology, Proc. Crypto'97, LNCS 1294*, B. Kaliski, Ed., Springer-Verlag,1997,pp. 342-356.
5. N. Koblitz, "Elliptic Curve Cryptosystems", *Mathematics of Computation*, 48, pp. 203-209, 1987.
6. LiDIA Group **LiDIA v1.3**- A library for computational number theory. TH-Darmstadt, 1998.
7. J. Lopez and R. Dahab, "Improved Algorithms for Elliptic Curve Arithmetic in $GF(2^n)$", *SAC'98, LNCS* Springer Verlag, 1998.
8. P. Montgomery, "Speeding the Pollard and elliptic curve methods of factorization", *Mathematics of Computation*, vol 48, pp. 243-264, 1987.
9. A. Menezes and S. Vanstone, "Elliptic curve cryptosystems and their implementation", *Journal of Cryptology*, 6, 1993, pp. 209-224.
10. A. Menezes, *Elliptic curve public key cryptosystems*, Kluwer Academic Publishers, 1993.
11. A. Menezes, P. van Oorschot and S. Vanstone, *Handbook of applied cryptography*, CRC Press, 1997.
12. V. Miller, "Uses of elliptic curves in cryptography", *Advances in Cryptology: proceedings of Crypto'85, Lecture Notes in Computer Science*, vol. 218. New York: Springer-Verlag, 19986, pp. 417-426.
13. V. Müller, "Fast Multiplication on Elliptic Curves over Small Fields of Characteristic Two", *Journal of Cryptology*, 11, 1998, pp. 219-234.
14. IEEE P1363: "Editorial Contribution to Standard for Public Key Cryptography", *draft*, 1998.
15. R. Schroeppel, H. Orman, S. O'Malley and O. Spatscheck, "Fast key exchange with elliptic curve systems," *Advances in Cryptology, Proc. Crypto'95, LNCS 963*, D. Coppersmith, Ed., Springer-Verlag, 1995, pp. 43-56.
16. R. Schroeppel, "Faster Elliptic Calculations in $GF(2^n)$," *preprint*, March 6, 1998.
17. J. Solinas, "An improved algorithm for arithmetic on a family of elliptic curves," *Advances in Cryptology, Proc. Crypto'97, LNCS 1294*, B. Kaliski, Ed., Spring-Verlag, 1997, pp. 357-371.
18. E. De Win, A. Bosselaers, S. Vanderberghe, P. De Gersem and J. Vandewalle, "A fast software implementation for arithmetic operations in $GF(2^n)$," *Advances in Cryptology, Proc. Asiacrypt'96, LNCS 1163*, K. Kim and T. Matsumoto, Eds., Springer-Verlag, 1996, pp. 65-76.
19. E. De Win, S. Mister, B. Prennel and M. Wiener, "On the Performance of Signature based on Elliptic Curves", *LNCS*, 1998.

8 Appendix

Mdouble (Doubling algorithm)

Input: the finite field $GF(2^m)$; the field elements a and $c = b^{2^{m-1}} (c^2 = b)$ defining a curve E over $GF(2^m)$; the x-coordinate X/Z for a point P.
Output: the x-coordinate X/Z for the point $2P$.

1. $T_1 \leftarrow c$
2. $X \leftarrow X^2$
3. $Z \leftarrow Z^2$
4. $T_1 \leftarrow Z \times T_1$
5. $Z \leftarrow Z \times X$
6. $T_1 \leftarrow T_1^2$
7. $X \leftarrow X^2$
8. $X \leftarrow X + T_1$

This algorithm requires one general field multiplication, one field multiplication by the constant c, four field squarings and one temporary variable.

Madd (Adding algorithm)

Input: the finite field $GF(2^m)$; the field elements a and b defining a curve E over $GF(2^m)$; the x-coordinate of the point P; the x-coordinates X_1/Z_1 and X_2/Z_2 for the points P_1 and P_2 on E.
Output: The x-coordinate X_1/Z_1 for the point $P_1 + P_2$.

1. $T_1 \leftarrow x$
2. $X_1 \leftarrow X_1 \times Z_2$
3. $Z_1 \leftarrow Z_1 \times X_2$
4. $T_2 \leftarrow X_1 \times Z_1$
5. $Z_1 \leftarrow Z_1 + X_1$
6. $Z_1 \leftarrow Z_1^2$
7. $X_1 \leftarrow Z_1 \times T_1$
8. $X_1 \leftarrow X_1 + T_2$

This algorithm requires three general field multiplications, one field multiplication by x, one field squaring and two temporary variables.

Mxy (Affine coordinates)

Input: the finite field $GF(2^m)$; the affine coordinates of the point $P = (x, y)$; the x-coordinates X_1/Z_1 and X_2/Z_2 for the points P_1 and P_2.

Output: The affine coordinates $(x_k, y_k) = (X_2, Z_2)$ for the point P_1.

1.   if $Z_1 = 0$ then output $(0,0)$ and stop.
2.   if $Z_2 = 0$ then output $(x, x + y)$ and stop.
3. $T_1 \leftarrow x$
4. $T_2 \leftarrow y$
5. $T_3 \leftarrow Z_1 \times Z_2$
6. $Z_1 \leftarrow Z_1 \times T_1$
7. $Z_1 \leftarrow Z_1 + X_1$
8. $Z_2 \leftarrow Z_2 \times T_1$
9. $X_1 \leftarrow Z_2 \times X_1$
10. $Z_2 \leftarrow Z_2 + X_2$
11. $Z_2 \leftarrow Z_2 \times Z_1$
12. $T_4 \leftarrow T_1^2$
13. $T_4 \leftarrow T_4 + T_2$
14. $T_4 \leftarrow T_4 \times T_3$
15. $T_4 \leftarrow T_4 + Z_2$
16. $T_3 \leftarrow T_3 \times T_1$
17. $T_3 \leftarrow inverse(T_3)$
18. $T_4 \leftarrow T_3 \times T_4$
19. $X_2 \leftarrow X_1 \times T_3$
20. $Z_2 \leftarrow X_2 + T_1$
21. $Z_2 \leftarrow Z_2 \times T_4$
22. $Z_2 \leftarrow Z_2 + T_2$

This algorithm requires one field inversion, ten general field multiplications, one field squaring and four temporary variables.

NICE - New Ideal Coset Encryption -

Michael Hartmann[1], Sachar Paulus[2], and Tsuyoshi Takagi[3]

[1] Darmstadt University of Technology,
Alexanderstr. 10, 64283 Darmstadt, Germany,
mhartman@cdc.informatik.tu-darmstadt.de,
[2] SECUDE Sicherheitstechnologie Informationssysteme GmbH,
Landwehrstr. 50a, 64283 Darmstadt, Germany,
paulus@secude.com,
[3] NTT Information Sharing Platform Laboratories,
Immermannstr. 40, D-40210 Düsseldorf, Germany,
ttakagi@ntt.de

Abstract. Recently, a novel public-key cryptosystem constructed on number fields is presented. The prominent theoretical property of the public-key cryptosystem is a quadratic decryption bit complexity of the public key, which consists of only simple fast arithmetical operations. We call the cryptosystem NICE (New Ideal Coset Encryption). In this paper, we consider practical aspects of the NICE cryptosystem. Our implementation in software shows that the decryption time of NICE is comparably as fast as the encryption time of the RSA cryptosystem with $e = 2^{16} + 1$. To show if existing smart cards can be used, we implemented the NICE cryptosystem using a smart card designed for the RSA cryptosystem. Our result shows that the decryption time of NICE is comparably as fast as the decryption time of RSA cryptosystem but not so fast as in software implementation. We discuss the reasons for this and indicate requirements for smartcard designers to achieve fast implementation on smartcards.

Key words: public-key cryptosystem, fast decryption, quadratic order, smart card implementation.

1 Why NICE?

Plenty of public-key cryptosystem not relying only on the RSA cryptosystem have been proposed. They are stemming from deep number theory (hyper- and superelliptic curves) to geometry and combinatorics (LLL-based systems). One major advantage of RSA is its simplicity: it can be easily implemented, one only needs a moderate background in mathematics to understand it. Moreover, RSA is quite fast - there exist public key cryptosystems which are much faster, but the combination of simplicity, speed and confidence in its security makes it the most practical cryptosystem. Until now, only one other system may be considered as equally interesting: ElGamal type systems on elliptic curves. Elliptic curves are somewhat more complicated, and the best known algorithms to break elliptic curve cryptosystems are much slower, in the order of exponential complexity.

So, both speed and security are worth being paid by the higher mathematical complexity.

But: both systems have one drawback in common: the decryption/signing time is of cubic complexity in the bit length of the public key for both systems because these steps consist of modular multiplication(s). This becomes even more important when thinking of smart cards. The smart card is considered to be the personal security computing device of tomorrow. It contains all personal secret information, especially the private keys for public key systems like decryption and signing. The complexity of PC operating systems is too high to be reliably secure; every relevant security operation should be effected by the smart card. Non-relevant operations like public key encryption and signature verification could be done by the PC. Operations which cannot be transferred to the PC at all due to security reasons are decryption and signing. Considering RSA, the task which has to be effected by a low power computing device is precisely the most complex task. Moreover, we can expect that the key length of the public key will increase with the progress of hardware technology. In addition, there are no guarantees that new sub-exponential attacks for the basic number theoretic problem will not be suddenly proposed. Therefore it would be better to have a more efficient public key cryptosystem.

As an alternative, we might use a new public-key cryptosystem constructed over number fields [18]. The cryptosystem has a theoretically fast decryption process such as a quadratic decryption complexity of bit-length of a public-key, which consists of only simple fast arithmetical operations. So even if the key length gets bigger in the future, there will be no great increase of the computational complexity. This becomes even more important when thinking of smart cards. In this paper, we call the new cryptosystem NICE, (New Ideal Coset Encryption). We focus on the practical aspects of NICE cryptosystem. We implement the NICE cryptosystem over different architectures, namely software on a standard PC and on a smartcard designed for the RSA cryptosystem. Our implementation in software shows NICE is as fast as the encryption time of the RSA cryptosystem with $e = 2^{16} + 1$. Implementation on a smartcard designed for the RSA cryptosystem is comparably as fast as the decryption time of the RSA cryptosystem but not so fast as in software implementation. We discuss the reasons for this and indicate requirements to achieve fast implementation on smartcards.

This paper is organized as follows: In section 2, we give several applications based on NICE cryptosystem. In section 3, we explain the details of the algorithms of the NICE cryptosystem. In section 4, we show timings of the implementation in software. In section 5, we discuss a smart card implementation and its problems.

2 Applications

In section 3, we will present NICE in the formulation of an encryption scheme. An immediate application is therefore session key distribution from a powerful server to a device which has a limited computing power or where time is important. An example for such a device is e.g. a mobile phone.

Another application of NICE is the use as an authentication scheme. The usual protocols (2-way and 3-way) can be adapted to use NICE as encryption component. Of course, it could be combined with RSA, where the modulus is the absolute value of the discriminant of the non-maximal order, i.e. Δ_q. In that case, the 3-way protocol can be realized in such a way that the client (= the low computing power device) only effects the fast components of both algorithms.

As a last application, we propose an undeniable signature scheme. NICE itself cannot be used as a classical signature scheme; undeniable signature schemes have their own use e.g. in online transactions. A signature of this kind cannot be verified without the interaction of the signer. The standard example for its application is its use by a software development company: the distributed software is signed by means of an undeniable signature of the company to allow legal users to ensure themselves that they use unmodified software. Since interaction with the seller is needed to check the signature, illegal users either cannot check and risk to use some virus-infected software or will be traced by the software-seller as soon as they ask for interactive verification. Details can be found in [4]. Again, the low computing power device only effectuates the NICE decryption steps, so smart cards can be used for assuring the security of online transactions.

3   The NICE cryptosystem

In this section, we present an overview of the NICE cryptosystem. Details can be found in [18]. The idea of NICE is roughly as follows: consider two finite abelian groups G and H which are related by a surjective map $\pi : G \to H$. Moreover, there exists a well-defined bijective mapping of sets $\phi : H \to U$ of H onto a subset of U of G such that $\pi(\phi(M))) = M$ for all $M \in H$. The representation of elements of G and the group operation algorithm of G are publicly known, as well as an element h of the kernel of π. U is chosen such that a consecutive set of representations of elements of G are representations of elements of U. This information is publicly known. Assume that you know the group H (i.e. representation of group elements and group operation) and how to compute π, but no one else does. The message space consists of the publicly known elements of U. Now, a message m is probabilistically encrypted by randomly multiplying an element h^r of Kerπ onto it: the ciphertext is $c = m * h^r$. Decryption simply works as follows: compute $\phi(\pi(m * h^r))$.

This is a secure cryptosystem if the computation of the map π cannot be deduced from the given information, namely the group G, the kernel element h and the test for U. There exist some constructions of this scheme using number theoretic problems, e.g. [17]. An overview can be found in [19].

The following implementation of this scheme is especially interesting: Generate two random primes $p, q > 4$ such that $p \equiv 3 \pmod 4$ and let $\Delta_1 = -p$. Let $H = Cl(\Delta_1)$ be the ideal class group of the maximal order with discriminant Δ_1 and $G = Cl(\Delta_q)$ be the ideal class group of the non-maximal order with conductor q. Δ_q will be public, whilst its factorization into Δ_1 and q will be kept private.

Any element of G is given by a pair of numbers (a, b) such that $0 < a \leq \sqrt{|\Delta_q|/3}$, $-a < b \leq a$ and $b^2 \equiv \Delta_q \bmod 4a$ (and some other minor requirements). Elements of H are represented in the same way with Δ_1 instead of Δ_q. The group operation works as follows:

Composition in $\mathrm{Cl}(\Delta_q)$
Input: $(a_1, b_1), (a_2, b_2) \in \mathrm{Cl}(\Delta_q)$, the discriminant Δ_q
Output: $(a, b) = (a_1, b_1) * (a_2, b_2)$.

1. /* Multiplication step */
   1.1. Solve $d = ua_1 + va_2 + w(b_1 + b_2)/2$ for $d, u, v, w \in \mathbb{Z}$ using the extended Euclidean algorithm
   1.2. $a \leftarrow a_1 a_2 / d^2$
   1.3. $b \leftarrow b_2 + (va_2(b_1 - b_2) + w(\Delta_q - b_2^2)/2)/d \bmod 2a$
2. /* Reduction step */
   2.1. $c \leftarrow (\Delta_q - b^2)/4a$
   2.2. WHILE $\{-a < b \leq a < c\}$ or $\{0 \leq b \leq a = c\}$ DO
      2.2.1. Find $\lambda, \mu \in \mathbb{Z}$ s.t. $-a \leq \mu = b + 2\lambda a < a$ using division with remainder
      2.2.2. $(a, b, c) \leftarrow (c - \lambda\frac{b+\mu}{2}, -\mu, a)$
   2.3. IF $a = c$ AND $b < 0$ THEN $b \leftarrow -b$
   2.4. RETURN (a, b)

This algorithm has quadratic bit complexity $O((\log_2 \Delta_q)^2)$ and is only needed for the encryption step. For the decryption step, we need only to compute π. The computation of the map π works as follows:

Computation of π
Input: $(a, b) \in Cl_{\Delta_q}$, the fundamental discriminant Δ_1, the discriminant Δ_q and the conductor q
Output: $(A, B) = \pi((a, b))$.

1. $b_O \leftarrow \Delta_q \bmod 2$
2. Solve $1 = uq + va$ for $u, v \in \mathbb{Z}$ using the extended Euclidean algorithm
3. $B \leftarrow bu + ab_O v \bmod 2a$
4. $C \leftarrow (\Delta_1 - B^2)/4A$
5. WHILE NOT $(\{-A < B \leq A < C\}$ or $\{0 \leq B \leq A = C\})$ DO
   5.1 Find $\lambda, \mu \in \mathbb{Z}$ s.t. $-A \leq \mu = B + 2\lambda A < A$ using division with remainder
   5.2 $(A, B, C) \leftarrow (C - \lambda\frac{B+\mu}{2}, -\mu, A)$
6. IF $A = C$ AND $B < 0$ THEN $B \leftarrow -B$
7. RETURN (A, B)

This algorithm π has quadratic bit complexity $O((\log_2 \Delta_q)^2)$ ([18]). Moreover, only simple well-known operations are needed, thus this algorithm can easily be implemented on an existing smart card.

Finally, the test belonging to U is simply whether for an element (a, b) a is smaller than $[\sqrt{|\Delta_1|/4}]$ (or a lower bound thereof of the form 2^k). The computation of the map ϕ will not be needed in that implementation.

We explain how the NICE encrytion scheme works: In the key generation, we choose an element (h, b_h) from the kernel Kerπ and make (h, b_h) public. The message m is embedded in an element (m, b_m) of $Cl(\Delta_q)$ with m smaller than $[\sqrt{|\Delta_1|/4}]$. Encryption is done in the class group $Cl(\Delta_q)$ by computing $(c, b_c) = ((m, b_m) * (h, b_h)^r)$, where r is a random integer smaller than 2^l with $l = \log_2\left(q - \left(\frac{\Delta_1}{q}\right)\right)$. Then, having the secret information, namely the knowledge of the conductor q, one can go to the maximal order and the image of the message (m, b_m) in the maximal order is revealed, since $\pi(c, b_c) = \pi(m, b_m)$ and m can be recovered without computing ϕ.

The NICE encryption protocol

1. **Key generation:** Generate two random primes $p, q > 4$ with $p \equiv 3 \pmod 4$ and $\sqrt{p/4} < q$. Let $\Delta_1 = -p$ and $\Delta_q = \Delta_1 q^2$. Let k and l be the bit lengths of $[\sqrt{|\Delta_1|/4}]$ and $q - \left(\frac{\Delta_1}{q}\right)$ respectively. Choose an element (h, b_h) in $Cl(\Delta_q)$, where

$$\pi((h, b_h)) = (1, 0) \tag{1}$$

   Then $((h, b_h), \Delta_q, k, l)$ are the system parameters, and q is the secret key.

2. **Encryption:** Let (m, b_m) be the plaintext, in $Cl(\Delta_q)$ with $\log_2 m < k$. Pick up a random $l - 1$ bit integer and we encrypt the plaintext as follows using binary exponentiation and precomputation techniques:

$$(c, b_c) = (m, b_m) * (h, b_h)^r \tag{2}$$

   Then (c, b_c) is the ciphertext.

3. **Decryption:** Using the secret key q, we compute $(d, b_d) = \pi((c, b_c))$. The plaintext is then $m = d$.

A message embedding technique and security aspects of this cryptosystem, we refer to [18]. Again, please note that this cryptosystem can easily be implemented using well-known techniques and existing smart cards. This will be shown in the next section.

4 NICE running times in software

The prominent property of the proposed cryptosystem is the running time of the decryption. Most prominent cryptosystems require decryption time $O((\log_2 n)^3)$, where n is the size of the public key. The total running time of the decryption process of our cryptosystem is $O((\log_2 \Delta_q)^2)$ bit operations. In order to demonstrate the improved efficiency of our decryption, we implemented our scheme using the LiDIA library [2]. It should be emphasized here that our implementation was not optimized for cryptographic purposes — it is only intended to provide a comparison between RSA and NICE. The results are shown in table 1. In these tests, we did choose $p \approx q \approx n^{1/3}$, so that breaking RSA and NICE by factoring is approximately equally hard. Other variations and a discussion of the variants can be found in [18].

$\log_2(\Delta_q)$	1024	1536	2048	3072
RSA encryption	2.2 ms	4.8 ms	7.5 ms	15.2 ms
RSA classical decryption	259.2 ms	751.3 ms	1643.9 ms	4975.6 ms
RSA decryption with CRT	110.5 ms	291.7 ms	629.7 ms	1855.5 ms
NICE encryption	602.9 ms	1180.1 ms	1902.0 ms	3933.5 ms
NICE decryption	3.8 ms	6.2 ms	10.0 ms	19.3 ms

Table 1. Average timings for the new cryptosystem with 80-bit encryption exponent r compared to RSA with encryption exponent $2^{16}+1$ over 100 randomly chosen pairs of primes of the specified size on a Pentium 266 Mhz using the LiDIA library

Observe that one can separate the fast exponentiation step of the encryption as a "precomputation" stage. Indeed, if we can securely store the values $(p, b_p)^r$, then the actual encryption can be effected very rapidly, since it requires only one ideal multiplication and one ideal reduction. Moreover, using well-known techniques for randomized encryption, we can even reduce the encryption time much more. Note that no square root technique like the Pollard-rho method or Shanks' algorithm are directly applicable to the ciphertext (c, b_c), because the encryption consists of $(c, b_c) = (m, b_m)(p, b_p)^r$ where r is a random exponent and (m, b_m) is the secret plaintext. This means that we can use a very short random exponent r having e.g. about 80 bits.

It should be mentioned that the size of a message for our cryptosystem is significantly smaller than the size of a message for the RSA encryption (e.g. 256 bit vs. 768 bit, or 341 bit vs. 1024 bit). In connection with the very fast decryption time, an excellent purpose for our cryptosystem could be (symmetric) key distribution. In that setting, the short message length is not a real drawback. On the other hand, the message length is longer than for ElGamal encryption on "comparably" secure elliptic curves (e.g. 341 bit vs. 180 bit).

$\log_2(n)$	1024	1536	2048	3072
RSA encryption	1	2.18	3.41	6.91
RSA classical decryption	1	2.90	6.34	19.20
RSA decryption with CRT	1	2.64	5.70	16.79
NICE encryption	1	1.96	3.15	6.52
NICE decryption	1	1.63	2.63	5.08

Table 2. Rate of the speed increasing when the bit-length of a public-key becomes larger

Note that even if a public-key becomes large, the rate at which the speed of decryption of NICE increases is not so large as that of the RSA cryptosystem. This shows the effectiveness of a quadratic decryption time of NICE cryptosystem. The ratio is given in table 2.

5 A smart card implementation and its problems

Moreover, we implemented the NICE decryption on a smart card. More precisely, we implemented the NICE decryption algorithm using the Siemens development kit for chip card controller ICs based on Keil PK51 for Windows. As assembler we used A51, as linker L51 to generate code for the 8051 microcontroller family. The software simulation were made using dScope-51 for Windows and the drivers for SLE 66CX160S. Thus, we realized a software emulation of an assembler implementation of the NICE decryption algorithm to be run on the existing Siemens SLE 66CX160S. Unfortunately, the timings of this software simulation were unrealistic. So, we did run some timings on a hardware simulator for SLE 66CX160S. Thanks to Deutsche Telekom AG, Produktzentrum Telesec in Siegen and Infineon/Siemens in Munich for letting us use their hardware simulator. See the timings for decryption in table 3 for a smart card running at 4.915 MHz.

$\log_2(n)$	1024
RSA decryption with CRT	490 ms
New CS decryption for $p \approx q \approx \Delta_q^{1/3}$	1242 ms
Improved version	1035 ms

Table 3. Timings for the decryption of the new cryptosystem compared to RSA using the hardware simulator of Siemens 66CX160S at 4.915 MHz

The very first implementation was very inefficient; the straightforward algorithms used in the software comparison proved to be much slower on the smart card than the existing RSA on the card. This was surprising, but after a while this could be easily explained: the cryptographic coprocessor has been optimized for modular exponentiation. On the other side, NICE uses mostly divisions with remainder and comparisons. These operations are slow on the coprocessor, so we had to modify the decryption algorithm to speed it up in hardware. We describe here two significant changes:

The computation of the inverse of a modulo q (step 3 in the computation of π) using the extended Euclidean algorithm took (with $q \approx p \approx 341$ bit) about 9 seconds (!), whereas computing the inverse using Fermat's little theorem - by using fast exponentiation mod q - took less than 1 second. Note that the decryption time using this method is no longer of quadratic complexity.

In the reduction process the quotient in the division with remainder step is most of the time very small (say ≤ 10, see Appendix A); to effect a division is this case is much more time consuming than subsequent subtractions. We did replace reduction step 5

5.1 Find $\lambda, \mu \in \mathbb{Z}$ s.t. $-A \leq \mu = B + 2\lambda A < A$ using division with remainder
5.2 $(A, B, C) \leftarrow (C - \lambda\frac{B+\mu}{2}, -\mu, A)$

by the following algorithm.

5.1 WHILE $B \leq -A$ OR $B > A$ DO
   5.1.1 IF $B < 0$ THEN $B \leftarrow B + 2A$; $C \leftarrow C - (B + A)$;
   5.1.2 ELSE $B \leftarrow B - 2A$; $C \leftarrow C - (B - A)$;

Every time that the bitlength of B was exceeding the bitlength of variant B by at least 3. Using this improvement, we could decrease the running time from about 1.8 s to 1.2 s. This is already faster than a 1024 bit RSA decryption without Chinese remainder theorem (approx. 1.6 ms).

The timings in table 3 were made including these two improvements. Moreover, a detailed timing analysis in Siegen showed that both our static memory management and as well as the cryptographic coprocessor are not optimal for this algorithm. We discuss this in the sequel. An average overview of the most time consuming parts is given in table 4.

Function	Average time over the whole computation
mul, multiplications on the coprocessor	170 ms
div, divisions on the coprocessor	231 ms
left_adjust, length correction of the variables	376 ms
C2XL, moving numbers into the coprocessor	118 ms
XL2C, moving numbers out of the coprocessor	223 ms
others (comparisons, small operations)	114 ms
Overall time	1242 ms

Table 4. Detailed timings for different functions in the decryption of the new cryptosystem

One major difference between RSA and NICE is the number of variables needed during the computation of the decryption algorithm. In our implementation, we need to store 11 variables of length at most 2048 bit. Computations of the cryptographic coprocessor are shortening these variables. Thus, we had to adjust the length of the variables after each important operation. This was done by moving the top nonzero bytes of the number to the fixed address of the number and so "erasing" leading zero bytes. To do this, we used the cryptographic coprocessor. Now the exact timings showed that about 33 % of the running time is spent by the function left_adjust, which effects this correction.

Now changing the memory management from static to dynamic (i.e. in the XL2C and C2XL functions making the appropriate changes and having additionally some registers holding the starting address of the numbers), we got an improvement to **637 ms** using the software simulator. Both Infineon/Siemens and Deutsche Telekom reported the overall time of the hardware simulator now to be **1035 ms**. Note that the amount of memory required is **965 Bytes** and thus fits into a real SLE 66CX160S.

As one can see from table 4, another important time consuming operation is to move numbers into and out of the coprocessor. At this point, we would get a speedup of about 350 ms if we could leave the numbers in registers inside the coprocessor. It is clear that the currently used processor is not prepared for such operations, since it is optimized for RSA, thus operations with very few variables. At this point we ask the hardware community to present solutions to this problem.

6 Conclusion and acknowledgements

The NICE cryptosystem is fast and well suited for software implementation. To get an equally fast speedup compared to RSA on a smart card, we think that the underlying hardware must be developed adequately. Nevertheless, if this is done, the think that NICE can be a competitor to RSA whenever fast decryption is needed.

We thank Deutsche Telekom AG, Produktzentrum Telesec for letting us testing NICE on the hardware simulator and Siemens AG/Infineon GmbH for their valuable help concerning the use of the development kit as well as running our code on their hardware simulator.

References

1. L. M. Adleman and K. S. McCurley, "Open problems in number theoretic complexity, II" proceedings of ANTS-I, LNCS 877, (1994), pp.291-322.
2. I. Biehl, J. Buchmann, and T. Papanikolaou. *LiDIA - A library for computational number theory*. The LiDIA Group, Universität des Saarlandes, Saarbrücken, Germany, 1995.
3. I. Biehl and J. Buchmann; "An analysis of the reduction algorithms for binary quadratic forms," Technical Report No. TI-26/97, Technische Universität Darmstadt, (1997).
4. I. Biehl and S. Paulus and T. Takagi, "Efficient Undeniable Signature Schemes based on Ideal Arithmetic in Quadratic Orders," in preparation.
5. J. Buchmann and H. C. Williams; "A key-exchange system based on imaginary quadratic fields," Journal of Cryptology, 1, (1988), pp.107-118.
6. J. Buchmann and H. C. Williams; "Quadratic fields and cryptography," London Math. Soc. Lecture Note Series 154, (1990), pp.9-26.
7. J. Buchmann, S. Düllmann, and H. C. Williams. On the complexity and efficiency of a new key exchange system. In *Advances in Cryptology - EUROCRYPT '89*, volume 434 of *Lecture Notes in Computer Science*, pages 597-616, 1990.
8. J. Cowie, B. Dodson, R. Elkenbracht-Huizing, A. K. Lenstra, P. L. Montgomery, J. Zayer; "A world wide number field sieve factoring record: on to 512 bits," Advances in Cryptology - ASIACRYPT '96, LNCS 1163, (1996), pp.382-394.

9. D. A. Cox: *Primes of the form $x^2 + ny^2$*, John Wiley & Sons, New York, 1989

10. W. Diffie and M. Hellman, "New direction in cryptography," IEEE Transactions on Information Theory, 22, (1976), pp.472-492.

11. ECMNET Project; http://www.loria.fr/~zimmerma/records/ecmnet.html

12. T. ElGamal, "A public key cryptosystem and a signature scheme based on discrete logarithm in $GF(p)$," IEEE Transactions on Information Theory, 31, (1985), pp.469-472.

13. D. Hühnlein, M. J. Jacobson, Jr., S. Paulus, and T. Takagi; "A cryptosystem based on non-maximal imaginary quadratic orders with fast decryption," Advances in Cryptology – EUROCRYPT '98, LNCS 1403, (1998), pp.294-307.

14. H. W. Lenstra, Jr., "Factoring integers with elliptic curves", Annals of Mathematics, 126, (1987), pp.649-673.

15. A. K. Lenstra and H. W. Lenstra, Jr. (Eds.), *The development of the number field sieve*. Lecture Notes in Mathematics, 1554, Springer, (1991).

16. A. J. Menezes, P. van Oorschot, S. Vastone: *Handbook of applied crytpography*. CRC Press, 1996.

17. T. Okamoto and S. Uchiyama, "A new public-key cryptosystem as secure as factoring," Advances in Cryptology – EUROCRYPT '98, LNCS 1403, (1998), pp.308-318.

18. S. Paulus and T. Takagi, "A new public-key cryptosystem over the quadratic order with quadratic decryption time," to appear in Journal of Cryptology.

19. S. Paulus and T. Takagi, "A generalization of the Diffie-Hellman problem based on the coset problem allowing fast decryption ," Proceedings of ICISC'98, Seoul, Korea, 1998.

20. R. Peralta and E. Okamoto, "Faster factoring of integers of a special form," IEICE Trans. Fundamentals, Vol.E79-A, No.4, (1996), pp.489-493.

21. R. J. Schoof: *Quadratic Fields and Factorization*. In: H.W. Lenstra, R. Tijdeman, (eds.): *Computational Methods in Number Theory*. Math. Centrum Tracts **155**. Part II. Amsterdam, 1983. pp. 235-286.

A   An example of the reduction step

Input: discriminant Δ_1, an ideal $\mathfrak{A} = (A, B)$ before the reduction step

```
D1 = -39195532988111573684755239909947774867128796201606366860749809266866357745652816544165356132253122452663
A  = 14709196987236217470310344790336554582213450995580439354777614852261315535437072993917565044745312551892647190947572
     659808191301659419506155735441417414491
B  = -16120786611173332936175433825318881027235314017722179961013037966005831645873381632911088594573172128409548204320324227
     6612069695877785023298037626557211503213
```

Reduction step 5.1.: quotient λ, remainder μ such that $B = 2A\lambda + \mu$

```
(quotient, remainder) =
(-1,   13297607363299102004445255755354228137191587973438698748542191738516799425000764354924041494917452975375746178653
       76584094130155403366825077082271136787693)
(-2,  -12761243590327174728420394096256304728831977979402420516873134568804685699281859725103664466875812641319914710949015
       92835564639348867895038964130696321056)
(5,   -78528290011262457388456843941329248503454444074743493793783111760006835153040692321204584715201859717103940903082618
       68661757907073551937411599271404065)
(3,   10244818933510891787370810612951117737019488809536816914420866112436574171352321538119941956085359074784832443404744
       03297520916097275928966085178754669)
(-2,  -10224129845590147501382365118494266252021893092244629712061058090686680852197387614386894226954222864793077231352357
       416590511833810655070898694329205)
(5,   -111052964327862543248014454108681029972621611840532971461250154942477651873609548116068054690931838077804710928090934
       7356285560186199994948199065031 5)
(2,   22472176875075749176609304144458704509990853674902869504729305667589920231890466102920134589953378731011273003071265369
       6685205138129287433281611151)
(-24, -1939976826223861607475092562034594570013024007436439401577693918681279881275209886159965930558543868196481590174064685
       36902929528949871546723550 3)
(5,   -87214024755816152729750202158525445031303615368933392297827179212874361504535183941307866086681685052251673544184 91)
```

```
                 910488305041328185564443763927)
(5,    -6589256154220428348763484804840290820585070540526813862090115099621050606775639083942326160598304048110107273013 59
        351206968420846678418077 44933)
(3,    -35372540965036205934141340575440504523333071659818592716658193807183159467823011633885293333993006718946556400076 9
        28723389657964699502098981 5)
(7,    -24722228429459372714248962154023865024293254144916063323433706888146107111333799338478440615432732354638347537820 3
        76474826892936909497877111)
(2,    21123896082367134246190321628964774052168904073609642897109183311547987453027704896019427995726765091708200095299 2
        7566081981468920686945135)
(-5,   -13882849961083763068299825465068008772978539934410435686028126703692746197995757231193841008190833004541581261177 9
        6117968165387193074055 05)
(3,    -7658114509885474231440468833038946684623106412514491556907595574141977825664772294590168889402849912338147698286 08
        31385805486639937 32859)
(6,    45171597597333763623992365562270451301836956090008641386920496283515261556967185885324296557967735725107983939203 53
        84463045341798439 22931)
(-3,   57949399291153659951735032186608343843568674256297346337247982645939248022677985693922616046011747896784792467175 3
        9296859648971048121 3)
(-3,   13506775252588510548399455966472193855917717568054168943345013045530491969273737406572000625840454580937146559331 8
        90497387788289 83289)
(-15,  33719331931229543174596901666464315138791547976559206823565738527321556044311098226988811614870540130943134313822 5
        969137699779 80851)
(-3,   32387847383134384596896074551207436375450678276080867668177408072318238232294269933210033680515064228455159232690 4
        42961002663 73941)
(-4,   16714419738146443196752822573871854231405166017209952769083137312359031623980857178233078983102193989180745595809 6
        7325428762 00243)
(-5,   -30939001159755959315976107374558197066820019463549929232211643659606767979683020394240493706340093182264070243126 5
        390693908 2883)
(11,   -11531066580582167360450543647526376555451729341586832024627696861235896647014417982287287152417816575504218178089 20
        1218624 24193)
(3,    -21416477884929434181388631924023741361264598809816873076331133830318602596332088069733933047463884290696572783831 21
        5217010 9453)
(2,    12888523591160334106137912439941642415919053550737356342471653928056183779460149050157015086812555259924781853101 4
        6905542 061)
(-8,   31633578159439213064953799152697863204115922676726434102233580962225494334700961557207697651285649976347162685798 6
        37456 771)
(-5,   -13210018897051880813498927374101995339595816992508346895849609154802928220757881013812918935087788634639351547161 9
        944989 1)
(5,    -11820767885517306027346197667704751166610749535623823332585074677744843395288704196222344446875121774461329878842
        257829)
(2,    21119279461713277596207017893502266538313143311963030857504415107811710131828253822345631758424093515908565819438 9
        75697)
(-2,   -27432853051141724255411105408025611057200201419389184266003977457557840592193796232158014760553605820295547745472 2
        6713)
(3,    28607481588919854014744297131365927822747279893781042870977736270566347391699980867676818174655427818291322493595 17
        975)
(-3,   13675329979120524654436448689915415134775488207798519854672704225197836768913822075713964399426344261201102817403 3
        09)
(-7,   -57339256581978408901576164776560926547521320212994406708963853136151783979312889475967541162999938676928609715757 3)
(3,    46420141247202899632758735492428308491655665018737859648769488246653936727190431310329578844089903814145126014811)
(-4,   26487529779478453968178323538670149346465038648366123949432443617527045265166163949185284933213945742464607485573)
(-5,   21085592458638552270229199687537556443283490914878826331361646499582326571191045770185801157558886696640500937)
(-3,   22358943854775137747028532971526452088757418214703795026514539955093835452771449594075431944337959109115485481 1)
(-3,   -30664289187300506323250976607341637378515722302236146563153692657856530663189799924090781836211749637768 78039)
(2,    14208659898740727556652822302984768311062999079336719320748796162184523124947445076271687833345276817037 1307)
(-10,  -40138331862750510542615290024482660018544628657362830007302092623863113001768197144805848997953649188768 987)
(3,    51331878277981454009993341485265726497471349346937458545337954364783518545662156086252390856287655321 7709)
(-2,   -61677436882563203815369113178313444194666063177138046752802519079030109397743003273918060562207208 817421)
(4,    -57053438120635304110047793680658649566299800066386525784794374869156847276046884586275056248195423 03243)
(3,    -91558187817393163469667420478915357008167385723230297957370535226883331905337014079460712753247612 7053)
(21,   45386853368506971878424996536708887813578856708661618056040478300528997110425684029899742727201843 537)
(-10,  20964009631208956452091856125365379202225167263769646847877535223464862248383150866105858045876 6883)
(-2,   -24526541647962635129591655653775971471231657933925013542908668110590277226101198472717238287 96771)
(4,    -14705279367346153808650653066838576510942226478246315871086537382991040364161674649916472973 8165)
(4,    13963798812911120130279709463801187300299288356253613041701058365238079493066888731128639700 0735637)
(-2,   -22430594690298603417335824980938203477805571947687436652981292220078494230642566854465655 258585)
(3,    -33253698473707029335678153634141091828891003805965525320260309200314966382689306783552 27979)
(3,    -53869672449708954071201037343980172216194792353383517263934396934658455469016876949250 112145)
(2,    88092436625898997244418623024247489374063110471015836523807286828879054858212546 51162082773)
(-3,   15239168503962249296834111165208224834733656939678806458689394255853391799372576 3450387843)
(-2,   -17544893643921935147494319040814098488015209648816507428601473576813341763276791 632582051)
(4,    -71721195253926047585869127579004116736732795608823000735576460232778526734332 0520772285)
(6,    41198358621779132573379802562967284951378028678577848604847304481941247839971 1749331293)
(-3,   39952378254046331640985245359853148480439674672046852693477940886873711190791 503779555)
(-4,   24304189642226944806712129161363218927714261648078945847751835949058682645713 97226917)
(-4,   -20048163895067921265137069623754996983678700353077266808465268084655280987822 9675431254661)
(3,    -15809036670606058719969662268090458772980604046280756287155668401840558844802 721257)
(5,    -15235857183157173338176030634817789382223536284729626135531586803908322657267 143003)
(2,    18431149369561694777795610580829154959771200085283101706038290472860989663874 931)
(-4,   18608721351129642487382805128934749545584400507726296787447608402705855634753 005)
(-3,   18693474451732730852080232290147818780501285481113119601447916323894501100859 91)
(-4,   17838073009688435403350490917190161689420852951826860570730599938052892637 777)
(-3,   80874852893664444631036106630666728692803032028149868435266439607722241 722651)
(-8,   33098987051860433032150149826166042111642048557203333097938938487614 3961861)
(-3,   -18664536615969020483601618613669413560772978432066084296155744931 355960393)
(6,    -14125048754972600933763966266435300472874357880857601784036629824 70742019)
(2,    21999649273484884414578537229346505053489427558714791359395530729 64463795)
(-3,   23516043605665539210119254241067982862993090461007215280845515 35870327)
(-3,   -30775116246568297851944351874963723288005308338329375129763801 87853603)
(2,    29715211291742818647348740253017800449838161047883641603879225 7267944435)
(-5,   20430479687841138388651786490336214660319513581933064107387 538439725)
```

```
(-3,   -715710522544461579764349638671583037687849172262961920648424641623)
(9,      14179513470875975149484884956525606261632898759917041924304692531)
(-5,   -128252533327126479297702878036187960221874533930004646524692531)
(2,      71335806135296587373319903350868126871255530166512866750439935)
(-89, -3140274313351805370923868077412836021349838155479525780492491)
(3,     -56700143472800978277991245437962976894577596798463184088001)
(2,      46688435917208741704878430078453065590251036868441561054)
(-6,     35846124997992718851920200094845005686531518229928579835)
(-2,    -51749469791186864639491653593057747230284128521870150151)
(3,     -63863243252949591186892355477373614132946242509565125)
(27,     74259357819904732203673579824542737377325192815149)
(0,     -742593578199047322036735798245427373773251928151497)
```

Output: the reduced ideal $(A1, B1)$ equivalent to $\mathfrak{A}$

```
A1 =  9562398414327226521335535763343291861775258207781149
B1 = -74259357819904732203673579824542737377325192815149
```

Encryption with Statistical Self-Synchronization in Synchronous Broadband Networks

Oliver Jung and Christoph Ruland

University of Siegen, Institute for Data Communications Systems,
Hölderlinstraße 3, D-57068 Siegen, Germany
{jung, ruland}@nue.et-inf.uni-siegen.de
WWW home page: http://www.nue.et-inf.uni-siegen.de

Abstract. ost of the data transmission networks used today are based on the technology of the Synchronous Digital Hierarchy (SDH) or Synchronous Optical Networks (SONET) respectively. However rarely, they support any security services for confidentiality, data integrity, authentication or any protection against unauthorized access to the transmitted information. It is the subscriber's responsibility to apply security measures to the data before the information is passed on to the network. The use of encryption provides data confidentiality. This, however, requires consideration of the underlying network technology. The method described in this paper allows the use of encryption in broadband networks. The advantages of this method are the transparency of the encryption applied to the signal structure and signal format, and the automatic re-synchronization after transmission errors. The used mode of operation, is called "statistical self-synchronization", because the synchronization between encryption and decryption is initiated by the presence of a certain bit pattern in the ciphertext, which occurs statistically. An encryption device, designed for SDH/SONET-networks with transmission rates of 622 Mbit/s, is to be presented.

Keywords: Broadband Networks, SDH/SONET, Confidentiality, Cryptography, Encryption, Modes of Operation, Self-Synchronization

1 Introduction

The increased requirement of bandwidth capacity over the last years has prompted for the development of efficient, digital transmission systems. Modern transmission systems provide higher transmission rates with a better bandwidth/cost ratio. SDH or SONET are the technologies that do not only fulfill this demand, but also offer the possibility of enhanced network management, controllable quality of service and simple multiplex structure. They are based on common international standards. One specific property of these types of networks is the supply of one central clock for all of the network components, like multiplexers or cross-connects. These types of networks are thus also called synchronous networks.

The whole range of information transmission, e.g. from phone call to video transmission, is handled via this type of networks. SDH/SONET is also largely used for Corporate Networks, which span long distances that cannot be controlled. The necessity for encryption technology is therefore especially important in this particular instance. There are four main reasons that cause difficulties in realizing an encryption technique for synchronous broadband networks:

- The maximum data transmission rate of 622 Mbit/s, and minimum of 155 Mbit/s respectively, and forecast 2.4 Gigabit/s or 10 Gigabit/s in the future
- Synchronous processing
- Bit slipping
- Complex management information

The synchronous transmission in these networks requires, that the encipherment is done with the same rate as network transmission rate. In Europe, there are no VLSI encryption components available today, supporting such high throughput rates. Therefore, it is necessary to use multiple encryption chips in parallel. The standardized and fixed frame structure does not allow for additional synchronization information of the crypto algorithm. We use the technique of statistical self-synchronization, which allows for the synchronization of the decryption algorithm even in the case of bit slipping. This mode of operation guarantees that the correct plaintext is computed at the receiver's side after an error-propagation has occurred. Bit slipping can occur if bits are deleted or additional bits are added by transmission components due to small differences of transmission lines data rates (jitter).

Bit slipping up to a certain extent is however covered by the synchronous network components. Even up to three bytes can positively or negatively be stuffed into one transmission frame. Therefore, bit- or byte slipping happens only if these thresholds are exceeded. This error needs to be taken into consideration even if the bit- or byte slipping probability is extremely small as also automatic switching of routing due to line drops can cause this.

The management information contained in the transmission frame requires, that specific parts of the frame are not to be encrypted as information which is important for network management and processed by the network components has to stay in plaintext. It is therefore required that management information bypasses encryption.

Chapter 2 shortly describes the structure of the STM-1-frame, whose payload is to be encrypted. Chapter 3 focuses on the modes of operation of block ciphers and shows that the standardized CFB- and OFB-mode are not sufficient for use in synchronous broadband networks. A new mode of operation, which we call statistical self-synchronization, is presented in chapter 4. Chapter 5 gives a layout of the realization of an encryption device for 622 Mbit/s with STM-4 interface or STS-12c respectively. This chapter furthermore gives additional information on selected implementation aspects. The presentation concludes with a summary and an outlook in chapter 6.

2 SDH and SONET

SDH and SONET are transmission systems with unlimited increasing rates that originally have been developed for the use in Wide Area Networks (WANs). Nowadays, they are however also used for Asynchronous Transfer Mode (ATM) in Local Area Networks (LANs). The standards for SDH and SONET contain not only the definitions of interfaces, e.g. transmission rates, formats and multiplexing techniques, but also recommendations for network management. SDH and SONET have similar characteristics. SONET is however mainly used in North America and is based on a standard frame, called STS-1, with a transmission rate of 51,84 Mbit/s whereas SDH, based on a standard rate of 155,52 Mbit/s is widely spread in Europe and Asia. The standard SDH frame, called STM-1, contains three concatenated STS-1 frames. Both transmission methods offer the basis for international data transmission and both support interfaces for existing as well as future techniques. The SDH-network interface is specified in ITU-T G.707 [5].

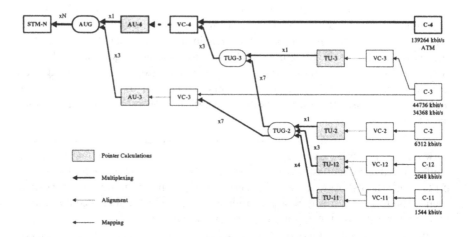

Fig. 1. SDH-Multiplex Hierarchy

Figure 1 shows, how the Synchronous Transport Module 1 (STM-1) can be constructed. The signals of the tributary systems that use SDH as a transport network are mapped into a standardized container (C-x). Through the adding of stuffing bits and control information, the so-called Path OverHead (POH), we get a Virtual Container (VC-x). The STM-1 module is constructed by adding a pointer which directs to the first byte of the Virtual Container and adding the Section OverHead (SOH) which consists of the Regenerator Section OverHead (RSOH) and the Multiplexer Section OverHead (MSOH) (see Figure 2).

The SDH is built in a modular way, whereby the STM-1 builds the basis for all higher transmission rates. Higher transmission rates are gained by bytewise multiplexing 4·n STM-1 frames (n = 1,2, etc) to one STM-4·n-frame. In this way, the next level of the hierarchy is the STM-4, which offers a capacity of 622 Mbit/s. A same structure is also defined in SONET, however, is called differently. The STM-1 corresponds to a SONET STS-3c and the STM-4 to an STS-12c frame.

The encryption device described in this paper processes STM-4 frames. The 622 Mbit/s data stream is internally split into four 155 Mbit/s streams, which are passed over to the encryption modules byte by byte.

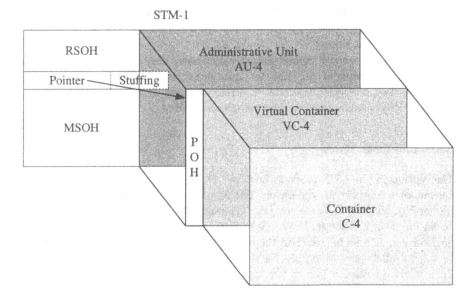

Fig. 2. SDH-Frame Structure

SOH and POH contain bytes for frame synchronization, signaling of the frame structure, service quality monitoring, path identification, alerts and alert responses. Some overhead bytes bypass the encryption and enter the next SOH or POH respectively, to be transmitted in plaintext. Others have to be re-calculated, e.g. parity bytes, which check the integrity of the payload. Again others have to be encrypted as they contain stuffed user information (stuffing). Consequently, it is required that the encryption modules are indicated which bytes are not to be enciphered.

3 Encryption

Encryption algorithms can be split into two groups. They either belong to the category of stream ciphers or block ciphers. Block ciphers encrypt blocks of bits in one step; a block length of 64 bit is usual. In contrast to this, only single bits or bytes are encrypted by stream ciphers. It would be beneficial to use stream ciphers for encryption of data units transmitted in broadband networks, as they reduce the delays in encryption devices. Cryptographically secure stream ciphers are however rarely known. The correspondingly required VLSI-chips, which in any case are needed for high-speed data rates, are not available. On the other hand, each block cipher can be turned into a stream cipher if it is used in an appropriate mode of operation. This approach is chosen here.

The four modes of operation, defined so far in ISO 10116 [4], are quite different in their properties regarding security, synchronization, error propagation, delay and throughput. (Note: We expect that this standard will be extended by new modes appropriate for high speed applications, e.g. the ATM Counter Mode, at the next release).

In order to turn block ciphers into stream ciphers they are used as key stream generators. Two modes do exist for this: either the Cipher FeedBack Mode (CFB) is used for self-synchronizing stream ciphers or the Output FeedBack Mode (OFB) is taken for synchronous stream ciphers.

3.1 The CFB-Mode

Encryption in the CFB-mode is achieved by XOR-ing the plaintext with the output of a key stream generator. The key stream is generated by the block cipher E_K, whereby K is a secret key. The input data of the algorithm is buffered in an input shift register. Since the last revision of ISO 10116, this input shift register can also be bigger than the block length of the block cipher in order to enable parallel encryption units for the high-speed generation of the key stream. In a standard case, n bits of the ciphertext are fed back into the input shift register, i.e. if n bits of the generated key stream are used for the encryption of n bits of the plaintext. Adjustments of the word formats could require the stuffing of the fed back ciphertext. This stuffing is explained in ISO 10116 in detail. Therefore, the definition of the CFB-mode in ISO 10116 is a bit more complicated than shown in Figure 3.

The CFB-mode offers the huge benefit of self-synchronizing. If a synchronization error occurs by erasing or adding a ciphertext unit of n bits length (corresponding to an n-bit slipping), the de-crypting side only generates wrong plaintext until the defect ciphertext units are shifted out of the input shift register. The same behavior occurs, if bits have been modified during transmission.

It however turns out that the implementation of the CFB-mode requires a very high encryption throughput rate. Assuming that n-bit by n-bit of plaintext are to be enciphered, then a complete input block needs to be encrypted in order to gain n-bit of cipher text. If V is the throughput rate of the block cipher implementation, the effective encryption rate, with which plain text in CFB-mode can be encrypted, applies as follows:

$$\nu = V \cdot \frac{n}{64}$$

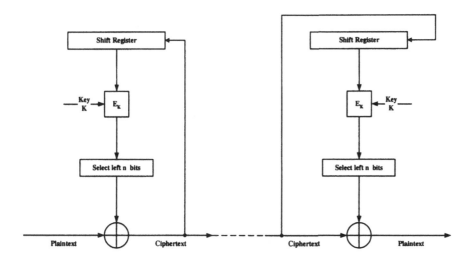

Fig. 3. CFB-Mode

If n = 1 is selected to receive a self-synchronization even in the case of bit slipping, then an encryption capacity of approximately 40 Gigabit/s per transmission direction is required to encrypt a STM-4-interface of 622 Mbit/s with a payload of approximately 600 bit/s. The CFB-mode therefore cannot be used in this way for broadband networks.

3.2 The OFB-Mode

The OFB-mode, in contrast to the CFB-mode, does not feed back the ciphertext into the input shift register, but the generated key stream. In this way, a complete output block of the key stream can be XORed with the plaintext for encryption, even if this is achieved only n-bit by n-bit (see Figure 4). The effective encryption rate therefore equals the encryption rate V of the key stream generator. A simplified description has been chosen once again, because the feeding back of smaller units, as well as adjustments of word formats are considered in more detail in ISO 10116.

The OFB mode offers the benefit of a high data throughput but not a self-synchronization. Therefore this type is also called a synchronous stream cipher. The fact that the transmitted cipher text is not used for the generation of the key stream means that the cryptographic synchronization is completely lost, and also cannot be recovered after the occurrence of synchronization errors. On the other hand, no error propagation happens if bits have been modified during transmission.

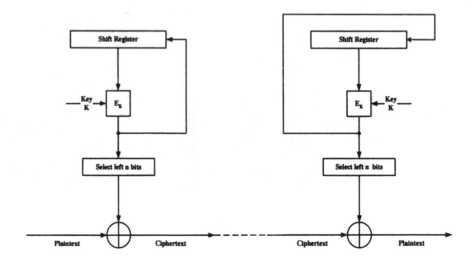

Fig. 4. OFB-Mode

4 The Statistical Self-Synchronization

The two described stream cipher modes of operation of block ciphers show big differences in their properties. The CFB is self-synchronizing, but only offers a low data throughput and error propagation. The OFB in contrast is not self-synchronizing, but rather shows no error propagation as well as a higher encryption rate.

The optimal solution would therefore be to combine the properties of both modes of operation. This is succinctly done by a new mode of operation, which we call statistical self-synchronization.

The statistical self-synchronization switches from one mode of operation to the other, and back whereby synchronization is reached between encryption and decryption by using the CFB-mode. OFB-mode is used between the synchronization phases. Loss of synchronization occurs in case of bit- or byte slipping. In order to re-synchronize, both sides need to be switched to CFB mode. The encryption and decryption are kept in CFB mode unless the input shift registers are filled with a complete block of ciphertext. This has to be identical on both sides. The content is used as a new starting value whereby OFB-mode is re-used afterwards (see Figure 5).

The decryption side, however, can not recognize when the synchronization has been lost. Both sides search for a fixed bit pattern in the ciphertext, as there is no additional communication capacity between the encryption and decryption entities to signal a switch in modes. This bit pattern occurs in a statistically distributed way in the ciphertext. Once the pattern is found, both sides switch to CFB-mode. The length of the bit pattern defines the probability of the synchronization and needs to be chosen in relation to the probability of bit slipping. The content of the bit pattern can be selected randomly, as all bit patterns of

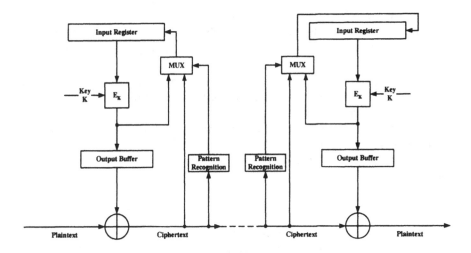

Fig. 5. Statistical Self-Synchronization

a fixed length are equally probable in the ciphertext. A bit slip causes a loss of synchronization, because the OFB mode is used between the synchronization phases. Encryption and decryption are out of synchronization till the bit pattern occurs in the cipher text. On the other hand, a switch into CFB-mode is achieved even in the case that no synchronization loss has occurred. This is the reason, why we call this mode of operation "statistical self-synchronization".

It should be emphasized again, that the bit pattern is generated by the encryption process itself as result of the encryption of the plaintext. No additional bandwidth is necessary to signal the synchronization start, or re-synchronization start, respectively.

A switching to the slower CFB mode implies for the encryption that during the operation in OFB-mode, as many key stream blocks need to be stored in the output buffer as are necessary to encrypt the plaintext during the next synchronization phase. Therefore, the encryption rate in OFB-mode must be higher than the transmission rate.

During a synchronization phase, another synchronization is not to be initiated. Therefore the bit pattern recognition is switched off during the synchronization process.

5 Implementation

Figure 6 shows the design of the encryption device. There are two essential components: the line interface, and the encryption and decryption module, respectively.

It is the task of the line interface to convert the 622 Mbit/s data stream into a format, which can be handled by the encryption/decryption module. During this process, plaintext and ciphertext are processed separately, i.e. different components are used (Red-Black-Separation). Otherwise it can happen that plain text could end up in the encrypted data stream due to malfunctioning of the interface components, or due to cross talking.

A direct connection between plaintext and ciphertext side exists only for the overhead, for different network alerts and for the high frequency reference clock, which are bypassed transparently.

An additional signal is passed together with the data stream from the line interface to the encryption/decryption module, in order to indicate which bytes do not have to be encrypted/decrypted as these positions contain overhead information.

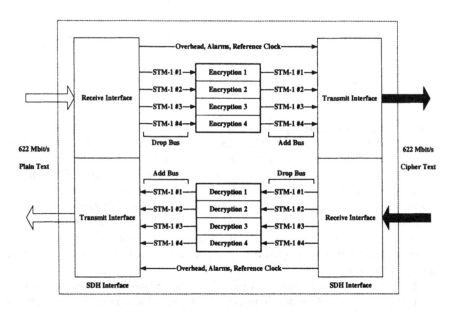

Fig. 6. Design of the Encryption Device

The SDH frame structure, which is supported by the encryption device, is shown in bold in Figure 1. It is a VC-4, which has been mapped into one STM-1 frame. One STM-4 frame consists of four STM-1 frames. This structure has been selected, as it is the most flexible one, as all tributary signals are transmitted in the VC-4. It is also used in ATM networks using SDH on the physical layer. The path via VC-3 and AU-3 is not very common in Europe and only serves as an adjustment to SONET.

5.1 Implementation of the Line Interface

The line interface has an optical input and output, as an electrical transmission is hardly possible at such high bit rates. The input signal is changed to an electrical signal and paralleled. This is followed by a demultiplexer which splits the STM-4 frame into four STM-1 frames and processes the STM-4 Section Overhead. The SOH bypasses the encryption and is reassembled at the transmitter side. There exist certain bits that have to be recalculated at the transmitter side. For each byte position of the SOH it is decided whether the received byte or the re-calculated byte is forwarded to the transmitted SOH (see Figure 7).

The four STM-1-frames are bytewise passed on to the encryption or decryption. The high frequency part is designed in ECL technology, whereas TTL technology is used for the other (approx. 20 MHz) one.

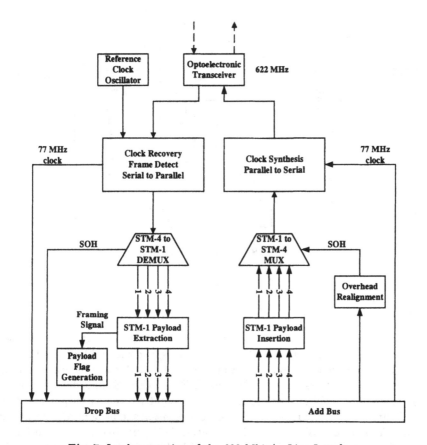

Fig. 7. Implementation of the 622 Mbit/s- Line Interface

No encrypted data is to be placed at byte positions that are used for overhead (SOH and POH). Encryption is therefore switched off during the presence of overhead data, i.e. overhead data passes the encryption in plaintext. The overhead bytes are recognized by analysis of the frame signal, evaluation of the pointer, and the counting of the byte positions.

Figure 8 shows the method for overhead bypassing. The SOH processing component has an interface that outputs the overhead in a serial form. The component supports receive clock, transmit clock, and a frame signal that indicates the overhead positions. It is required to synchronize the overhead on both sides. Two FIFOs are used for synchronization, which store the overhead, alternating frame by frame. Both FIFOs are used in an alternating way as synchronization is to be guaranteed even in the case of errors, i.e. interrupted overheads due to line drops. Each FIFO is reset after the overhead of a frame has been read out.

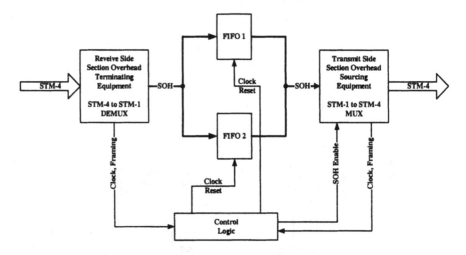

Fig. 8. Overhead-Bypassing

5.2 Implementation of the Encryption

A DES-encryption chip that offers an encryption rate of 160 Mbit/s is used for the key stream generation. DES is a block cipher with a block length of 64 bit [3]. The alternating mode that works with the double key length of 128 bit or 112 relevant bit respectively [2] is used. Each of the STM-1-data streams is encrypted separately and FIFOs are used as input as well as output registers. FIFOs are used for different reasons. Encipherment has to be done synchronously to the data transmission rate of 155 Mbit/s. The generated key stream has to be buffered as encryption does not work with this rate. In addition to this, the key stream generation works slowly (in CFB-mode) during the synchronization phase. For this reason, enough key stream bits have to be accumulated (in OFB-mode) during the encryption process so that there are sufficient key stream bits available for the then occurring synchronization phase. Therefore one encryption chip is not fast enough. DES-components are necessary which read alternate the input blocks from the input-FIFO. Thus, the input-FIFO has to be larger than one input block (see Figure 9).

In a standard case, i.e. not during synchronization phases, the encryption works in OFB-mode. A synchronization phase works like follows. The pre-defined bit pattern that triggers self-synchronization is constantly searched in the ciphertext. If this pattern is found, the multiplexer is switched over. The encryption works now in CFB-mode. Consequently, ciphertext is written into the Input FIFO unless a complete block has accumulated (64 bits). In the meantime, the plaintext continues to be XORed with the key stream bits which have been generated upfront in OFB-mode and which have been buffered in the active FIFO3 or FIFO4. This is continued until the generated and encrypted block in CFB-mode is available in the one FIFO that was not the last active one (this can be FIFO4 or FIFO3). Then the FIFOs are switched. The plaintext is now XORed with the key stream that has been generated in the CFB-mode. This process provides self-synchronization. The FIFO, active so far, is now reset and is prepared for the next synchronization phase. Please note, that the whole process is by far more complex if the layout is designed for a synchronous processing. Switching from OFB-mode to CFB-mode has to be done at the same bit position on the transmitter side and receiver side. Signal propagation and delay times have also to be considered. The fed back ciphertext bits can, for example, only be taken from the transmitted ciphertext after a certain period of delay. Furthermore, it is necessary to generate an additional key stream block in OFB-mode and store it in the new FIFO (directly behind the one in CFB-mode) before the output FIFOs can be switched. The bit pattern recognition that is required for the next upcoming bit pattern check, can only be released, once the FIFO has sufficiently been filled.

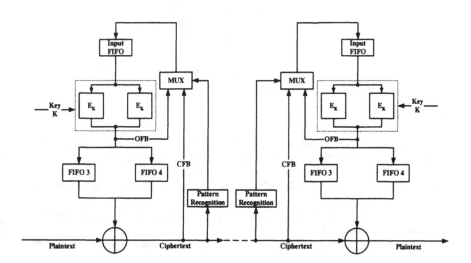

Fig. 9. Realization of the Self-Synchronization

There is an alternative concept for the realization of the statistical self-synchronization, which uses two separate encryption modules. Each of them contains two DES chips. The usage of two complete encryption modules has the benefit that the switching between the two different modes of operation is less complicated. One of two encryption modules works in OFB-mode the other one is idle as long as the bit pattern does not appear. If the bit pattern has been found, the second encryption module is initialized using the ciphertext in the input shift register. After the generation of one output block, in CFB-mode, this encryption module is switched to the OFB-mode. Now, this encryption module becomes the working unit. The benefit of this concept lies in the simpler relationships between the FIFOs and the encryption components. On the other hand, two complete encryption modules are required which results in higher costs per device.

6 Summary and Outlook

It has been demonstrated that it is possible to use encryption technology in high-speed networks. Additional channel capacity for synchronization purposes is consequently not necessary. The line interface has been developed in cooperation with the Worcester Polytechnic Institute and is available. At present the realization of the encryption module is in progress.

In addition to the used STM-4 frame which has been described in this paper, there exists another chained STM-4-frame, the so-called STM-4c. The STM-4c does not have a container with multiplexed four STM-1-frames but consists of one VC-4c-container which offers a transmission capacity of 600 Mbit/s. Currently we are working on an interface for the STM-4c. The challenges with the realization of the line interface for the STM-4c lies in the premise that there are no standard chips available which support the STM-4c interface. This means that standard components need to be complemented by programmable logic components (FPGA) in order to realize a STM-4c interface, e.g. to calculate the parity byte over the VC-4c. The encryption can be done through the use of the same modules that will be used in the device with the STM-4 interface.

References

1. ATM Forum: ATM Security Specification, Version 1.0, Final Ballot. January 1999
2. CE Infosys: CE99C003B Technical Reference (Priliminary). 1997
3. ANSI X3.92: Data Encryption Algorithm Standard, 1981
4. ISO/ICE: ISO 10116, Modes of Operation for an n-bit block cipher algorithm, Revision 1997
5. ITU-T Recommendation G.707: Network node interface for the synchronous digital hierarchy (SDH), 3/96
6. Schneier, Bruce: Applied Cryptography, 2nd Edition, John Wiley & Sons, 1996

Author Index

Lecture Notes in Computer Science

For information about Vols. 1–1640
please contact your bookseller or Springer-Verlag